OIL MAN

Works by Michael Wallis

Route 66
The Mother Road

Pretty Boy
The Life and Times of Charles Arthur Floyd

Way Down Yonder in the Indian Nation
Writings From America's Heartland

Mankiller
A Chief and Her People
(with Wilma Mankiller)

En Divina Luz
The Penitente Moradas of New Mexico

OIL MAN

The Story of Frank Phillips and
the Birth of Phillips Petroleum

MICHAEL WALLIS

Foreword by
John Gibson Phillips, Jr.

St. Martin's Griffin
New York

OIL MAN. Copyright © 1988 by John Gibson Phillips, Jr., and Michael
Wallis. All rights reserved. Printed in the United States of America.
No part of this book may be used or reproduced in any manner
whatsoever without written permission except in the case of brief
quotations embodied in critical articles or reviews. For information,
address St. Martin's Press, 175 Fifth Avenue, New York, N.Y. 10010.

DESIGNED BY VIRGINIA M. SOULÉ

Library of Congress Cataloging-in-Publication Data

Wallis, Michael
    Oil man : the story of Frank Phillips and the birth of Phillips
Petroleum / Michael Wallis.
        p.   cm.
    Includes index.
    ISBN 0-312-13135-6 (pbk.)
    1. Phillips, Frank, 1873–   .   2. Phillips Petroleum Company—
History.   3. Industrialists—United States—Biography.
4. Petroleum industry and trade—United States—History.   I. Title.
[HD9570.P52W34   1995]
338.7'6223382'092—dc20
    [B]   95-2131
                                                            CIP

First published in the United States by Doubleday

First St. Martin's Griffin Edition: May 1995
10 9 8 7 6 5 4 3 2 1

Dedicated to

Suzanne Fitzgerald Wallis

For her love, patience, encouragement, and inspiration but mostly for her magic

CONTENTS

FOREWORD

By John Gibson Phillips, Jr.

Growing up with a legend is not easy. I should know. I was raised in the shadow of Frank Phillips, the king of the oil-field wildcatters, the founder of Phillips Petroleum Company, and the classic American entrepreneur who was never afraid to take a risk.

Frank Phillips was my grandfather. As a boy and as a man I called him Father and until the day he died I respected him as though he was a god.

His moods were often exaggerated. He could be demanding, precise, intimidating. Sometimes he was even harsh and terrifying. At other times he could be tender and sentimental. He was always relentless in his passion for excellence and honesty. Just being in his presence was a tall order. But it was worth every second. I wouldn't trade my storehouse of rich memories for anything. I was proud to be a Phillips. I always will be. That pride comes from Father.

For Frank Phillips was a man who, despite the many demands he placed upon everyone, including himself, managed to keep a sane perspective of his relationship to people as a whole and serve as a permanent inspiration to legions of young men and women who believed that they too could rise to greater heights.

Father's life was extraordinary and exemplified the indomitable pioneer spirit. He was a country boy who had virtually nothing, including a formal education. But he took full advantage of the opportunities under the free-enterprise system and the guidance and nurturing he received from my maternal great-grandfather, John Gibson, to become a successful and enormously wealthy oil titan and industrialist of international fame.

He did it all with his own good judgment, boundless energy, and above all, faith in himself.

The story of Frank Phillips stands out as a shining example of rugged individualism—the willingness to gamble and risk everything. He took little advice and asked for no quarter. He entered the oil business at a time when the romance was in full bloom; when oil gushers were allowed to flow unrestrained; when fortunes were made and lost between a single sunrise and sunset.

One of the last of a breed, which included Getty, Sinclair, and Marland, he was on the verge of becoming a myth. For Father was not some faceless corporation executive or a self-centered tyrant. He was a flamboyant individual cast in the mold of the nineteenth-century entrepreneurs but with a twentieth-century style and an eye toward the future.

To this day, Frank Phillips stands for someone who took full advantage of each and every opportunity that came his way. His life is a blueprint for anyone interested in becoming a success. He was a man who knew how to use his courage and initiative and great administrative ability to create industry and wealth where there had been nothing. And by bridling imagination, he fostered an array of research projects and left a legacy that will always be a monument to his memory.

He could be hard and unrelenting. I learned that time and again. There was the summer when I was a boy working on his ranch and I tried to break up a fight between two of his prized swans with a rock and accidentally killed one. Father fired me on the spot and ordered me off the ranch as soon as I had my lunch. I was crushed. But then during the meal, after he had time to weigh both sides, he told me he had reconsidered, and because I came straight to him and told him the truth and didn't make up some sort of story or excuse, I still had my job. He even gave me a raise.

I was exposed to Father's quick temper on more than one occasion, and I also saw him take the starch out of the toughest executives if he felt it was necessary.

There's little doubt that the so-called corporate raiders—the ones who believe takeovers solve more problems than they create—wouldn't have stood a chance if they had to go toe to toe with Frank Phillips. And that most certainly includes Mr. T. Boone Pickens, the oil company executive who worked for Phillips Petroleum as a fledgling geologist for a few years after my grandfather died and whose own father had been a Phillips employee. Boone Pickens might strike fear in some hearts, but Frank Phillips would have written him off as just another oil-patch upstart.

Anyone who could stand up to Depression-era bankers, skittish directors, inflexible union leaders—and even "Pretty Boy" Floyd—the way Father did, wouldn't even flinch when it came to today's dealmakers.

It's also a safe bet to say that, although this biography of his life is superb in every detail and nuance, Frank Phillips would not have liked it. He would have read it from cover to cover, and secretly relished a great many of the chapters, but the bottom line is quite simple—as long as he drew breath no biography about his life could ever have seen the light of day. Too many secrets are revealed, and far too much information about himself and his personal life is exposed.

I myself had a very difficult time coming to grips with much of the material the author, Michael Wallis, has so skillfully woven into the book.

Reading *Oil Man* was a bittersweet experience. I was alternately sad and happy; exhilarated and shocked. The stories Wallis has included bring up both painful and happy memories. In the course of a single page I can both laugh and weep. I expect many readers will do the same.

But everyone who reads this book will be left with haunting memories of a time that's long gone; memories which provide all of us with valuable lessons that apply today and always. Memories of a man who built a lifetime into a petroleum kingdom.

ACKNOWLEDGMENTS

It's early morning and from my desk I'm listening to a steady rain drip on the big leathery magnolia leaves. There's no lightning to muddle my word processor, just enough rain to take the edge off the Oklahoma summer and keep the azaleas and ferns looking dapper.

Inside my office at treetop level, Beatrice, the wise mama cat, naps with her green eyes at half mast on a pile of papers. Her daughter, Molly, sits like a Sphinx on the floor just beyond my reach.

A second pot of decaf is brewed, I had my morning run before the rain arrived, and now there are no more excuses. I'm ready to spend another day with Frank Phillips, one of the last of the great oil tycoons and a genuine wildcatter.

A stack of yellowed obituaries and death notices tell me Frank Phillips died on August 23, 1950, in Atlantic City, New Jersey. I've stood inside the mausoleum where his body was entombed. But none of that matters. For me, Frank Phillips is as alive as the cats, or the trees, or the rain outside my windows.

For the past year and a half I found him wherever I went—from the steel and concrete canyons of Manhattan to the wild and free prairies of Oklahoma. I saw him strolling with his cane up Park Avenue, and I caught glimpses of him beneath the old elm at Pawhuska just before they cut it down. Frank kibitzed with me from the ghost rigs scattered across the oil patch, and I felt him at the town house on Cherokee Avenue and on the porch at Woolaroc and along the banks of the Caney River and in the lobby of the Empire State and in the garden behind Philbrook and in a thousand other places.

He came to life in the secret journals and diaries and letters I was privileged to find and read. He came to life in the dusty boxes I searched in the basement at Woolaroc and in the attic at his mansion. He came to life in the eyes of the hundreds of people I interviewed who knew him or had some information to share which helped me know him better.

But most of all I found Frank Phillips alive and well here in my office, this comfortable room filled with the notebooks, diaries, photographs, tape recordings, books, letters, and memories of the man. Every day when I climbed the stairs with Molly in the lead and Beatrice close behind, we found Frank Phillips waiting for us. He became alive and real. I felt him at my shoulder. I could smell the bay rum and cigars.

The seeds for the biography of Frank Phillips had been planted by my editor at Doubleday, Jim Fitzgerald, in December 1985, at Johnstone Park in Bartlesville over a winter picnic of turkey sandwiches and a couple bottles of a rather ordinary Chardonnay. By the following spring, those seeds had taken root.

During the long months of research required to prepare this biography I consulted hundreds of books, magazines, newspapers, private and public collections of documents and records, videotapes, and other pertinent material. I interviewed hundreds of Phillips relatives, surviving friends, former employees, and others.

Without the help and encouragement of countless individuals and entities I would not have been able to produce this book.

Certainly the early input I received from Frank Phillips' grandson, John Gibson Phillips, Jr., was essential. John and his lovely wife, Gloria, who live quietly with their sons, Christopher and Kevin, in the deep Southwest, joined me in Oklahoma at the start of the primary research phase of my work and were instrumental in opening doors and in locating priceless material, including several private diaries written by Jane Phillips. John, who has dedicated his life to the study of spiritual philosophy, maintained an intense interest in the book's development. Although it was often painful, he was able to unlock doors which shielded personal memories and candidly share them with me. For his enviable courage and invaluable assistance my heartfelt thanks.

Other Phillips family members who were instrumental in helping me include: Mary Low, Jane Begrisch, Mary Kate Phillips, and Elliott Phillips. Two of Frank's Bartlesville kinfolk—Ruby Cranor and her mother, Pearl Phillips Hill, a niece of Frank's father—were especially helpful by supplying critical information about the family background, particularly about the early years of Frank and his siblings.

This book was neither authorized nor solicited by Phillips Petroleum Company, but without the assistance of several key individuals in that remarkable organization my work would have been greatly hindered. Several years ago the company had the wisdom and forethought to videotape a number of key figures involved with Bartlesville's colorful past as well as many of the early Phillips employees, ranging from former top executives to men and women in lower-echelon positions. I was allowed to review and incorporate comments and observations from those invaluable oral histories in the book, including Anna Anderson, Phil Phillips, Bill Keeler, Paul Endacott, Clarence Burton, Billy Parker, Clarence Clark, Bill and Sue Angel, Kenneth Beall, and many others.

At Phillips Petroleum, my thanks go to J. Thomas Boyd, Vice President, Public Affairs; William C. Adams, Director of Public Relations; Dan L. Harrison, Director of Public Information; and Philip Caudill, Manager, Advertising & Business Promotion. A special thank-you to William C. Wertz, Director of Executive Communications, for allowing me to probe unfettered the company's historical files and providing me with his help in locating photographs and tracking down unanswered questions. Bill Wertz served as the capable editor of *Phillips—The First 66 Years,* an excellent company history published by Phillips Petroleum Company in 1983 and one of the guiding sources I relied upon as I prepared this book.

Robert Finney, a former longtime Phillips Petroleum employee, and in his retirement both the company's historian and an active member of the Washington County Historical Society, was a diligent and resourceful member of my research network. Bob was always there when I called for help and he never let me down.

Special thanks go to many individuals and organizations: Cindy Steinhoff Drake, of the Nebraska State Historical Society; Virginia Haberman of the Aspen Historical Society; Utah Historical Society; Anne Steinfeldt of the Chicago Historical Society; Illinois State Historical Library; Landmarks Preservation Council of Illinois; Chicago Landmarks Commission; Iowa State Historical Department; Iowa Genealogical Society; Joan B. Turner of the Westport (Conn.) Public Library; Dr. Brooke Collison of Wichita State University. In Creston, Iowa: James B. Harsh; Jane Beecher; Loudia Reed; Irma M. Miller; Charlene Hudson, of the Creston Public Library; Dwight and Harlan Mickey, owners and operators of one of Frank's barbershops. In Bartlesville: Mildred Moore; Lucille Endacott; Russell W. Davis; Jack Leonard; Norma J. Hettick, the capable curator of the Frank Phillips Mansion; C. Tom Sears; Homer Baker; Bill Jones; Bill

Angel and Sue Angel; Joe Billam and Grace Billam; Mary Crow; Robert Cotton; Edgar Weston; Elisabeth Seaton Edgerton; Lois Lynd; Arthur J. Boose; Susan C. Box and the staff of the Bartlesville History Museum & Archives at the Bartlesville Public Library. At Woolaroc: Robert R. Lansdown, Director of the Woolaroc Museum; Marjorie Sydebotham, Woolaroc Lodge Hostess; W. R. Blakemore, General Manager and Secretary-Treasurer of the Frank Phillips Foundation, Inc.; and the Board of Trustees of the Frank Phillips Foundation, Inc.; Richard Kane, Chairman; Leo Johnstone, Vice Chairman; Bob Phillips; Bill Creel; Dr. Vernon Lockard; Don Doty; and C. J. "Pete" Silas, who also serves as Phillips Petroleum's chairman and chief executive officer. In Tulsa: John Graham, the most intrepid of researchers from Info II, the excellent research service offered by the Tulsa City-County Library; and Marcia Manhart, Michael Sudbury, and Helen Sanderson of the Philbrook Museum of Art.

Nancy Marshall, of Muskogee, Fern Butler's niece, was an important and gracious source of information concerning the woman who had such a tremendous impact on Frank Phillips and the oil company he founded. Pat Patterson, the first director of the Woolaroc Museum, was also a critical source, as were Lewis and Gerda Fisher, of El Dorado, Kansas.

At Doubleday my sincere appreciation to Jim Fitzgerald, my patient editor, as well as thanks to Bill Barry, Laura J. Nixon, and Casey Fuetsch. Thanks for guiding me through the various stages of this book's development.

My gratitude to the partners and associates of Wallis Gideon Wallis, Inc., Public Relations/Editorial Marketing, especially my partner, Joyce Gideon, and our capable and faithful administrative assistant, Frances Merryweather. Thanks, Joyce, for believing in me and allowing me to take time from our partnership to produce such a book. Thanks to you, Frances, for all the typing, telephone calls, photocopying, and general scut work which tends to get lost in the shuffle when kudos are passed out.

There are two people who deserve my everlasting respect and appreciation because of their input, assistance, and encouragement to me as I developed this book. They are Paul Endacott, the former president and chief operating officer of Phillips Petroleum Company, and my wife, Suzanne Fitzgerald Wallis.

I am particularly indebted to Mr. Endacott for his wisdom and guidance, which kept me going when the going got rough. During the months I spent researching and gathering material, Mr. Endacott unselfishly gave me hours and hours of his valuable time. His candor, sense of history, and devotion to accuracy will not be forgotten. He is a true gentleman in every sense of the word and has my undying respect and admiration.

Finally, and most importantly, I salute my wife. Suzanne was the magician who kept me together throughout the entire process. Without her inspiration this book would not have been written. Suzanne went far above and beyond what a spouse should be expected to endure in order to help me produce this biography. My book became her book as she assisted me with tedious research, escorted me on more than one wild-goose chase, helped me celebrate major victories along the way, and wiped away the tears of agony I cried when the dead ends and scattered defeats every book has to endure took place. I also thank her for being conscious of the words of Aldous Huxley, who said, "If you want to write, keep cats." It was Suzanne who encouraged her own sweet familiars, Beatrice and Molly, to serve as my muses and climb the stairs with me every day to deal with Uncle Frank.

My lasting thanks to all of you.

Michael Wallis
Tulsa, Oklahoma

PROLOGUE

May 1950
Late Sunday afternoon at Woolaroc lodge
The Frank Phillips Ranch, fourteen miles southwest of Bartlesville, Oklahoma

Frank Phillips sits in his favorite chair on the front porch and listens to the sweet song of a mockingbird. It's the same bird he's heard all day—singing from the chimney before sunup and ever since from the oaks in front of the lodge.

The oak leaves are new and tender and the geranium blooms in the flower boxes on the porch railing are fiery red. All around the big log lodge, sunlight—the color of honey—flows through the leaves and melts on the ground. A faint slice of moon that's hung around for days peeks through the tree branches like a ghost waiting for nightfall.

Frank's cane is hooked on the chair back, and on a small table are kitchen matches, a glass of water that's sat untouched for hours, and an ashtray. Dressed in a dark business suit and bow tie, Frank watches summer close in on his ranch in the Osage. The rosy flush of spring is about to vanish.

Potatoes and gravy and a slice of beef he poked at during Sunday lunch soon take effect and Frank's eyelids grow heavy. After a few minutes of dozing he jerks awake, listens to the mockingbird some more, and manages to smoke most of the cigar before he nods off again. Soon he's back asleep. His Japanese valet, Dan Mitani, comes out on the porch and takes the cigar from between Frank's fingers and puts it in the ashtray. The old man stays deep in his dreams. The sleep has made him young again and he's on the prowl for crude oil in the everlasting hills of Oklahoma.

He's back in the oil patch with his brothers and they're building an immense empire by risking big money on their big notions. He's watching gusher after gusher come to life—a series of wildcat eruptions that shower the prairie with rich Oklahoma crude. He's hitting the first real wild boom in a land where Indians are rich, outlaws are respected, and oil is in abundance.

In his dream, Frank sees oil derricks covering the land, and then just as the oil rumbles from the earth once again and explodes in a great black plume into the sky, he cries out and wakes himself. At first, he's not sure where he is, but then he sees the shimmering lake and, beyond, the rolling hills where his herd of buffalo feed on tender spring grass.

Dan returns and with him the newest nurse who's been hired to care for Frank. The nurse is a pleasant young man, and he and Dan hover about to make sure Frank is comfortable. They brush the cigar ashes from the skirt of his suit coat and Dan offers to bring a fresh glass of water, but Frank waves him off. ·

Behind the lodge, Frank's museum, filled with Wild West art and treasures, has closed for the afternoon, and Sabbath visitors return to their cars. A few straggle down the hill to look at Clyde Lake and catch a glimpse of Uncle Frank Phillips, hoping he'll happen to be out on his porch for some afternoon air.

One family stops just short of where Frank is sitting. The young wife holds the hand of her three-year-old daughter and the child peers around her mother's legs at the old man sitting on the porch. The husband is a Phillips Petroleum employee, as was his father, and he nods out of respect to Frank. When Frank sees that the man is holding a baby, he reaches out his hand and the young man gives the baby to Dan, who carefully places the child in Frank's arms.

Frank holds the little girl for a few moments. He laughs and gently tickles the baby and then she is passed back to her father. Not a word is spoken. The family and Frank just smile at each other. After a few seconds the family leaves and Frank returns to his dreams.

His mind slips back in time and it's years before at Woolaroc lodge and Frank sees Indian and cowboy fires scattered over the hills as guests arrive the night before one of his "cowthieves reunions." He sees the proud Osage chiefs, wrapped in their colorful blankets, and the Indian dancers walking beside Pawnee Bill and Zack Miller and the other guests. He spies

Grif Graham in his best red shirt and a new ten-gallon hat and over at the edge of the crowd is Henry Wells and a few of the old-time desperadoes passing around a bottle of bootleg whiskey.

There are tables piled with barbecued buffalo and everyone is talking and dancing and fiddles are playing. Frank believes the music never sounded better.

It's one of his best dreams and he sees every guest that ever came through the ranch gates. It's as though all the parties and picnics he hosted through the years have combined into one fantastic blowout. The parade of guests goes on and on. There's Tom Mix, Herbert Hoover, Harry Truman, Perle Mesta, Rudy Vallee, Edna Ferber, Aimee Semple McPherson, Ernie Pyle, Will Rogers, Cardinal Spellman, and all the rest. He hears a humming sound and he looks up and sees Wiley Post circling in his airplane, and above him is Art Goebel in his plane and he's writing out Frank's name in huge white letters made of smoke.

Then the scene shifts from the ranch and Frank is back in Bartlesville at his fine town house on Cherokee Avenue, and he sees his family—his brothers and sisters and his son and foster daughters. And he sees Jane—his own Lady Jane—and she's moving through the rooms chatting with friends and family and making sure everyone's glass is filled. She's holding a cigarette in her own jaunty way just like always and Frank hears her deep throaty laugh. In his sleep the old man smiles and laughs out loud.

Inside the lodge, beneath the huge chandeliers and the lifeless animal heads staring from the walls, Dan hears the laughter and comes back out on the porch just as Frank awakens once more. Dan sits on the porch step and the two men talk about the disappearing spring and of the summer holiday Frank is planning to take in Atlantic City.

Dan tells him that Fern Butler telephoned during the afternoon to see how Frank is feeling. At the mention of her name Frank can picture his secret lover in his mind. He sees her bustling through the New York office and in her soft lounging pajamas at her apartment. Finally, he sees Fern astride Woolaroc, the spirited horse Frank sent her from the Osage. It's a pleasing image.

By now the afternoon is almost gone. Shadows stretch across the porch and Frank looks beyond the dark lake to the hills, but he can no longer see the buffalo grazing. Soon the oaks become silhouettes and the discussions of thousands of insects fill the night air.

Frank and Dan stay in their places for a few more moments, comfortable in their own silence, content with their own thoughts, at ease with themselves and the evening. Neither of them notices that at last the mockingbird has stopped its song.

OIL MAN

PART ONE

BEGINNINGS
1873–1903

"To be ignorant of what happened before you were born is to be forever a child. For what is a man's lifetime unless the memory of past events is woven with those of earlier times?"

—Cicero

CHAPTER ONE

Dawn slipped up quiet as a mercy killer. As first light broke over the land, Lewis Phillips saw in an instant that his world was gone. The crops and grass were chewed to the ground. Only broken roots and stubble remained. The cottonwood trees along the Loup River were bare as if somehow it had jumped from summer to the dead of winter.

The devastation and destruction he found that scorching July morning in 1874 moved Phillips to genuine tears of pain, something he hadn't felt since serving as a young artillery gunner in the Civil War.

Phillips kicked a mound of dusty earth and watched it blow away. North of his homestead, several forlorn cattle and a lean buckskin horse stood at the river's edge. They snorted, turned from the river and left thirsty. The fouled water was the color of varnish—stained from the excrement of the pestilence that descended on the land before returning to a roost in hell.

Grasshoppers. Billions of grasshoppers had darkened the sky. They arrived the day before—a clear, hot afternoon and suddenly a haze came over the sun and deepened into a gray cloud as the grasshoppers swept down upon the earth with a roaring sound like a rushing prairie storm. As far as the eye could see in every direction, the air was filled with them. The worst grasshopper plague in history, the swarm of insects reached from Texas to the Canadian border. They came in great clouds, almost a mile high, 100 miles wide and 300 miles long.

They crossed the Nebraska border near Scottsbluff and winged eastward. At Kearney, thick carpets of grasshoppers buried the Union Pacific rails. The tracks became so greasy from the mashed insects that the locomotives' wheels spun and couldn't pull the trains. It was difficult to drive a team across a field because the carpet of grasshoppers flew up in the faces of the horses and made them wild. Chickens and dogs, hunting for food, learned to eat the insects. Grasshoppers blanketed the hog lots, where Duroc and Hampshire sows gorged themselves. For a long time hams in this part of the country would have a peculiar taste.

Hot tears streaked down Phillips' cheeks. The crop of corn had been well along and the garden filled with vegetables for the winter. But now it was all gone. Every leaf, every stalk, was chewed up. There was nothing but bare earth. Even turnips and potatoes underground were eaten. The spade handle was gnawed where Phillips' palm grease had soaked the wood. Harnesses were cut in shreds.

Everywhere the earth was a gray mass of struggling, biting grasshoppers. Phillips slipped as he walked around his land, his boots slick with slime from their crushed bodies. He listened to the mechanical song of the dying hoppers as they laid their eggs.

Lewis Phillips would never forget the taste and smell of the grasshoppers. Most of all, he would never forget the feel of them moving over his body. He'd feel their spiny feet grasping his flesh as they crawled up his pant legs and down his shirt and over his neck and face. He'd feel them in his beard, inching toward his mouth and eyes and nostrils. He'd feel them fly into his chest and legs as he moved from the barn to the log cabin. Years from now, safe in a rocking chair on a cool porch in Iowa, he'd still be feeling the grasshoppers chewing up his life.

Inside his cabin, Phillips' wife, Lucinda, sat rocking her baby boy in his cradle. Her hands and arms were covered with stinging bites. The day before, as the cloud of grasshoppers approached their farm, Lucinda rushed to the garden and covered the plants with her aprons. She watched helplessly as the grasshoppers methodically chewed through the cloth and then ate the cabbages underneath.

Today seemed a bit better. At least some of the noise had stopped. More warm winds came with the sun and caressed the scarred homestead. Lucinda soothed her two daughters, who were still frightened from the day before when the resonant drumming of wings beat outside the door. The family had sat huddled in the cabin staring at each other as the grasshoppers came in endless waves. The insects rode the chinooks—the warm, dry winds that sweep down from the Rockies and fan the Great Plains. It seemed an endless siege. For a time, the Phillipses believed the grasshoppers would never stop coming.

"They've eaten all our crops and chewed the grass right down to the roots," Phillips told his wife, "but I think they'll soon be gone. There's nothing left for them."

Only two years before, Phillips had brought his wife and two daughters, Etta and Mary Jennie, from Iowa. They were homesteaders, part of the thousands spreading out across the prairies encouraged by the Homestead Act to settle western lands. Little encouragement was needed. Most of them, like Lewis Phillips, were young veterans mustered out of the Union Army, full of ginger and itching for adventure. They had a taste to leave Ohio or Illinois or Iowa and try new lands in the West.

Phillips picked Nebraska. In 1872 he and Lucinda, his wife of five years, put their little girls and the family's few belongings in a canvas-topped wagon hitched to a team of oxen and said goodbye to friends and family in Iowa.

They followed a nameless trail into Nebraska and the wide-open spaces of the Great Plains. The land had once been covered with buffalo—millions of shaggy-coated bison in herds so large that when they stampeded, the earth thundered and the horizon was black with lumbering bodies. Hide-hungry hunters and civilization were taking their toll. Buffalo herds still roamed the grasslands, but within a dozen more years they'd be gone.

Already the land knew change was at hand as hundreds of homesteaders followed their hearts, joining the dusty column of prairie schooners pushing westward. When the plows and spades broke the tough virgin sod, the soil yielded a sort of sigh, as if it knew the old days were gone forever.

The land Phillips homesteaded was pure frontier, in the North Loup River valley near the geographic center of Nebraska. Sioux Indians roamed the countryside, often raiding farms or stealing cattle. Sometimes they took scalps. Wolves and coyotes hunted in the valley; rattlesnakes bathed in the strong prairie sun.

Shortly after the Phillips family arrived, they were joined by more homesteaders—war veterans, miners, and laborers from back East, immigrants from Ireland, Germany, and Denmark. Sod houses and cabins began to dot the landscape.

Phillips stayed busy clearing his land and building a sturdy log home of cottonwood cut from a river island and fragrant red cedar carefully hewed and hauled from nearby canyons. Money was scarce, so in addition to his crops, Phillips earned a living cutting scrub cedar posts. Lucinda (or Josie, as she was known) picked wild onions to season rabbit stews and roasted prairie chicken suppers. She learned that the roots of the yucca growing on the sandy hills, when rubbed in water, made a soapy lather, and she cut handfuls of Queen Anne's lace, wild asters and sunflowers to brighten the tidy cabin. Her bone-white muslin curtains in the windows were the envy of every white woman around.

Phillips was a popular man on the frontier, especially among his fellow Civil War veterans. He was proud of his fancy certificate signed by the governor making him a second lieutenant in the Greeley County Guards. On hunting trips for large game or during the eighty-mile trek from the canyons of red cedar to Grand Island, no man was better liked or trusted than Lew Phillips. A family history records that "his genial personality and generous brave spirit made him a favorite companion of the scouts and plainsmen found in the vanguard of civilization."

In 1872, Phillips helped organize Greeley County, Nebraska, and the town of Scotia was selected as the county seat. Lew was appointed tax assessor, and when the first general election was held in November 1873, he was elected county judge. He helped build the one-room school which also served as the county courthouse. The judicial duties were minimal. In the year or so he served, Phillips issued only one marriage license. Marriageable couples in those parts were rare as college professors.

Pioneer life was not easy. Constant fear of hostile Indians, long trips to frontier towns and plain old loneliness took a toll. Needlework and reading the Bible and three-week-old newspapers from Des Moines helped a little, and when Lew's aunt and uncle and some cousins moved to the Loup valley, there were more friendly faces to see. Their houses were five miles apart, and with telescopes they could keep watch over each other and detect any snooping Indians. That brought some comfort. But on harsh winter nights, with the door heavily barred and icy winds howling down the chimney, time passed very slowly.

Blizzards, prairie fires, drought, hailstorms, and rattlers could be endured. No Indians raided the Phillips cabin or stock pens and the farm escaped harm's way until now, when what would be known as the Grasshopper Scourge of '74 struck. It was the last calamity he cared to face. Lew Phillips knew, as sure as the grasshopper eggs would hatch, it was time to go home to Iowa.

Looking for more comfort, Phillips reached for the Bible, opened it to Exodus, and read the story of the locust plague. "They covered the face of the whole earth, so that the land was darkened; and they did eat every herb of the land, and all the fruit of the trees." It was just like ancient Egypt, he thought. Phillips went to Ecclesiastes and found another passage he remembered: "The grasshopper shall be a burden." He read the verse aloud and then he turned to the front of the Bible and, as if for reassurance, read the family names recorded there.

Phillips ran his finger down the entries until he came to the name that had been entered less than a year before. A smile came back to his troubled face when he read: "Scotia, Nebraska. November 28, 1873. Born to Lucinda and Lewis Phillips, a baby boy, their first son . . ."

Phillips still knew every detail of that November evening. He recalled scrawling the entry by lamplight and fanning the ink dry. He remembered closing his eyes and his silent prayer of thanks before turning to the doctor to proclaim, "My first son."

The doctor, bone-weary and mindful of the ride before him, shook Phillips' hand, closed his black bag, and buckled on his pistol belt. He rode off in the early-morning darkness and went straight to the hitching post in front of Scotia's general store to share the good news with the sturdy pioneers who gathered there. Soon the word spread up and down the Loup River valley. "There's a new baby boy at the Phillips place. They've named him Frank."

CHAPTER TWO

Frank Phillips came from good stock. Both sides of his family provided a classic American lineage including a *Mayflower* celebrity; French Huguenot and Welsh immigrants; New England colonists; veterans of the Revolution, the War of 1812, and the Civil War; sturdy midwestern pioneers; and a long line of teachers, soldiers, clergy, merchants, blacksmiths, and farmers.

Frank, instilled by his parents with a keen sense of history and family pride, was ever mindful that he was a direct descendant of Captain Miles Standish, the most adventuresome of the Pilgrim fathers. Like other children, he learned the tales of the *Mayflower* and Plymouth Rock, and he pictured the noble Standish as presented by Longfellow. What Frank didn't know was the story of the expedient Captain Standish who invited a party of hostile Indian chiefs to a conference, promptly killed them, and then defeated their confused warriors in a brief skirmish. Nor did Frank learn that Standish was excitable and passionate, known for promptness in making decisions, and unperturbed by sudden danger—all traits that would pass through nine generations to himself.

The Phillips and Standish families combined with the marriage of Frank's paternal grandparents—nineteen-year-old Marilla Standish, a New York State native and the daughter of Ester Curtis and Matthew Kettle Standish, and twenty-six-year-old Daniel Phillips, son of Spencer and Susanna Phillips, hardworking Welsh people and early settlers in colonial America.

In his early years, Daniel, remembered by his family as "a tireless worker" who "possessed unlimited energy," labored as a lumberman and raftsman along the Ohio and Mississippi rivers. He helped take great rafts of logs as far south as Natchez and, on occasion, New Orleans.

Marilla and Daniel married on June 2, 1839, in Potter County, Pennsylvania, and settled near the town of Pomeroy in Meigs County, Ohio. Here the first three of their eight children were born. Lew Phillips—born Lewis Franklin Phillips on January 4, 1844—was the third child and first son. According to family records, Lewis was "a promising child and a great favorite with his [maternal] grandfather Standish." In 1848 the Phillips family moved to Jackson County, Iowa, where Daniel took up farming, while Marilla turned a room in their home into a school to teach her children and others from neighboring farms. When Lewis was eleven, the family moved to a farm north of Des Moines, where the boy finished his schooling and learned the carpenter's trade.

Lewis was a seventeen-year-old apprentice carpenter when news of the Union defeat at the first battle of Bull Run sent him to Des Moines to enlist, on August 1, 1861, as a buck private in the 4th Iowa Infantry. Soon his company was transferred to the artillery, and became the 2nd Iowa Light Artillery. Phillips was assigned to Battery E and during the next three years rose to the rank of corporal.

Phillips fought throughout the war, mainly in the western theater under Major General A. J. Smith. The young soldier participated in the siege of Nashville and the Vicksburg campaign, and even managed a brief love affair with a local girl when his outfit wintered in La Grange, Tennessee. In a privately published memoir, Lewis Phillips recalled a hot July day in 1864 when he was slightly wounded while serving as a gunner at the battle of Tupelo, Mississippi: "In that blazing sun this was work for giants to do, but the 18- and 20-year-old boys from Iowa were giants in those days. As long as there was an enemy in sight our boy was as full of fight as a young bulldog. We hardly need tell you that every time this boy went into action it took the tightest grip he could get on his nerves to hold him up to the work, but after he had heard the swish of a few of their death-dealing missiles and fairly gotten the scent of battle, his feelings underwent an entire change and all he thought was to 'lick 'em.' "

Shades of Captain Standish.

Phillips was released from the Army on August 7, 1865, at Davenport, Iowa, and resumed his life as a farmer and carpenter.

Frank's mother was a Faucett, a family that could trace their roots back to William Faucett, a French Huguenot émigré who fled France as a youngster and grew to manhood in Ireland before settling in 1783 in North

Carolina, where he married a local girl and raised a family. In 1815, William's son, James Faucett, born April 19, 1789, married Elizabeth Jeffers and proceeded to sire a dozen children. The eighth-born—arriving August 31, 1826—was Thomas Linch Faucett.

When Thomas was six, the family moved to Indiana and roosted near French Lick. His father maintained a 200-acre farm and operated a tavern with sleeping rooms for highway travelers passing on the stage route between Louisville and Vincennes.

Thomas was twenty and his boyhood sweetheart, Mary Jane Tate, sixteen when they married October 24, 1846. His young bride hailed from a family of Scottish descent who settled in North Carolina before the Revolutionary War and later moved to Indiana, where Mary was born September 27, 1830.

Thomas, with a strong back and a meager education, took up the blacksmith's trade, farmed, and even peddled grindstones for a time in order to support his wife and dozen children.

The family moved from Indiana to an 80-acre farm in Illinois and finally in 1864 to Iowa. Faucett obtained the contract for steel welding for the first railroad line to enter Des Moines and worked a 160-acre farm. A deeply religious man and an evangelist at heart, Faucett enjoyed working as a blacksmith six days a week and climbing to the pulpit on the Sabbath to deliver sermons and tell his faithful flock how "I hammer iron all week and hammer the Gospel into people's hearts on Sunday."

The Faucetts' second-oldest child, Lucinda Josephine, was born in Orange County, Indiana, on August 13, 1849. She followed her family to Illinois and Iowa, where she met Lewis Phillips shortly after he returned from his service in the Civil War.

He was dazzled by Miss Josie, who was known to be "the most beautiful girl in this section of Iowa." She was likewise impressed with Lew Phillips, who was a handsome war veteran. After a brief courtship, they married in Des Moines on July 3, 1867. Lew was twenty-three, Josie a month shy of her eighteenth birthday.

They set up housekeeping near Mitchellville. Josie's good cooking and artistic touches around their modest home were appreciated, but Lew Phillips wanted something more. The prospect of farming and working as a carpenter in Iowa all his life wasn't enough. Even after his first child, Etta, was born in 1869, followed by a second daughter, Mary Jennie, in 1871, Lew Phillips was still eager to explore new lands. He wanted to move his family westward. In 1872 they left for Nebraska.

And now they were homeward bound. After only two years, they had their fill of the frontier. But Phillips felt satisfied. At least they had given it a try. He didn't feel like a failure. After all, the life they were returning to would be far gentler for his daughters and baby son. He pointed the wagon eastward and eagerly watched for the green fields of Iowa.

CHAPTER THREE

The sky bending over the prairie was clear, but when clouds gathered, they cast their shadows in fantastic shapes. Barely a year old, Frank Phillips bounced along in the wagon watching the clouds chase one another. He shook the string of dried gourds clenched in his fist at the great clouds, making Josie and the girls laugh and Lew turn his head and smile.

The wagon creaked and groaned its way down the trail snaking through tall prairie grass toward Omaha and the ferry landing on the broad Missouri River. Council Bluffs and the deep and fertile soil of Iowa waited on the other side.

Once across the river, Lew Phillips coaxed his team to turn southeast. The family went to southernmost Iowa, to Taylor County, named in honor of Zachary Taylor, the old Army general who had served as President shortly before the county was organized in 1851. Situated sixty miles east of the Missouri River and perched just north of the Missouri line, Taylor County possessed rich soil, a fact that caught Lew Phillips' attention. He stopped the wagon when he reached the 40 acres of farmland he bought a few miles east of the town of Conway. He said a quiet prayer and told his wife and children they were home. Then he hacked away the scrub oak and hazel brush and turned under the wild sod to plant rows of corn that would march toward the horizon.

The Phillipses made new friends and tried to put their past behind them. "The grasshoppers ate the crops and chewed the grass down to the roots. When they started on the spokes of the wagon wheels, I decided it was time to try our luck elsewhere," was how Lew described their time in Nebraska to his neighbors.

The Taylor County land was productive and in prehistoric times had been swept by tongues of ice, flood, and fire. A deposit of rich soil, often five feet deep, covered the entire county, especially along the rivers and creeks in the bottomlands.

In nearby Conway, a new town laid out by the railroad in 1872 and not incorporated until 1878, were banks, grocery and general stores, druggists and doctors, a livery barn, a creamery, and even a Masonic Lodge. Soon there was also a carpentry shop, owned by Tom Daniels and his partner, Lew Phillips, an ambitious young farmer anxious to earn more money than raising corn and hogs could bring.

When the Phillipses' oldest child, Etta, fell ill with diphtheria and died one crisp October day in 1875, just three days shy of her seventh birthday, Lew and Josie grieved and wondered if the bad times in Nebraska had hitched a ride in the wagon and followed their family to Iowa. But time and prayer healed their wounds. Comfort soon followed when Josie's mother and father, now an ordained Methodist minister, moved to a 200-acre farm in the county.

Josie planted beds of violets outside the white frame farmhouse and spent hours over a scorching cookstove. Lew rolled up his sleeves and worked the land. They turned to the Scriptures in the evening and visited with family and neighbors.

On August 8, 1876, another son arrived at the Phillips farm. They named him Lee Eldas and paid the doctor from Conway a tidy seven dollars for his services—five dollars for delivering the baby, a dollar and a half for coming out to the farm, and fifty cents for a bottle of colic medicine. Frank now had a little brother to play with and someone to help with the chores.

In 1879 a third son, Ed, was born, and on January 19, 1883, Dr. H. B. Liggett rode out from Conway to tend Josie and deliver identical twin boys named Waite and Wiate.

In 1886, Lew sold the farm, borrowed $800, and bought an 80-acre farm north of Conway. It was here that the last three Phillips children were born: Nellie, 1886; Fred, 1889; and Lura, 1893.

Not blessed with great wealth, Lew and Josie taught all ten of their children to be proud, never to turn their backs on family or friends, and, most important of all, to work hard. With six sons to help him, and burdened with the hardship of the new farm debt to repay, Lew extended his work as a master carpenter, amateur architect, and building contractor and built homes and farm buildings across the county.

Big brother Frank and his siblings grew up smelling horse manure and fresh-cut clover. The boys had callused hands, and Josie's daughters could sew as well as any farm girl in Iowa. The new Phillips home was a two-story frame house with a small front porch, fireplaces, and a privy out back. Besides his growing fields, gardens, and orchards, Lew kept bees, dairy cows, hogs, and poultry.

As the oldest son, Frank assumed a leadership role from the start. There was plenty of work and plenty of bodies to help. Frank churned butter until his muscles burned, but he made sure that the others all took their turn—a hundred cranks apiece. After the coal and kindling were hauled to the house, the corn was husked, the cockleburs cut out of the pasture, the cows milked, and the other chores completed, Frank led his brothers to the best swimming holes the One Hundred and Two River offered. They found time for hunting and trapping and for adventures in the thick woods of black oak, walnut, and poplar, where, according to legend, great Indian council meetings had been held when painted warriors chose these forests as their favorite hunting ground.

There were crayfish under every rock in the creeks, blackberries to pick and devour on the spot or carry home to be smothered in cream, and even time to experiment with handmade cigarettes, although when Frank accidentally started a grass fire and got a stout strapping, the boys lost some of their urge for manly pursuits.

A resourceful lad, from the start Frank never seemed to let anything get in his way. Not even a hole in his britches. The family liked to laugh about the time a pal from a nearby farm called at the Phillips place and asked if the boys could go to Conway with him. "Frank can't go because he hasn't any clean clothes," said Josie. But Frank wanted to go, so he dashed upstairs and found a pair of clean overalls in his mother's work basket. He didn't care if the seat had been trimmed out and Josie was planning to patch them. They were clean pants. He slipped them on and, standing with his back to the kitchen wall, said, "Look, Mother. How are these?" Josie gave him permission to leave, and with his friend right on his heels, Frank went to Conway. By the time he returned home, Josie had discovered his ruse and she applied a slipper to the area on Frank's pants to be patched. He didn't care; he escaped the farm for the afternoon. It had been worth a good country licking.

During the spring and summer, the Phillips youngsters worked side by side with Lew on the farm. But when the fall and winter months arrived, Frank and the older children tramped across the fields of corn stubble to the one-room Schwemley School, taught by Mary Ann

Schwemley, the young daughter of a local farmer. It would be the only formal education Frank would receive.

Frank wasn't much interested in Miss Schwemley or her school. He learned his letters and numbers and read the prescribed poetry and the stories about Ivanhoe and Christopher Columbus and even the old Indian fighter himself, Miles Standish. But most of the time Frank's mind managed to slip out the window of the stuffy little schoolhouse. His thoughts would escape and go straight to those same great clouds still chasing over the countryside.

Frank was caught up in the American dream. This was the era of Horatio Alger, the prolific and enormously popular novelist who wrote stories for and about young men and produced biographies of Abraham Lincoln and James Garfield and other self-made American heroes. These books illustrating how bright young Americans had conquered poverty to become great figures enthralled Frank. He wanted to be an "Alger hero." He believed it was his destiny.

Poring over the latest Alger book by lamplight, Frank could actually feel himself strutting down a sidewalk in New York, or cruising the South Seas, or working in a circus—all locales especially appealing to a country boy in late-nineteenth-century Iowa.

And there were others who fueled the boy's American dream.

William Frederick Cody no doubt had an impact on young Frank Phillips. Born in Iowa in 1846, Cody became a Civil War cavalryman, gold miner, trapper, stagecoach driver, Pony Express rider, Indian fighter, and scout. He was a crack shot and an expert horseman and could speak fluent Sioux. About the time Frank was squalling into the world on the Nebraska frontier, Cody had already killed enough buffalo to provide meat for hungry railroad workers to earn the nickname "Buffalo Bill."

A flamboyant impresario with goatee and fringed buckskin suit, Cody formed his "Buffalo Bill's Wild West Show" in Nebraska in 1883, featuring such box-office attractions as Annie ("Little Sure Shot") Oakley, Sitting Bull, and a cast of trick-riding cowboys and war-bonneted Indians who re-created the battle of the Little Big Horn before Queen Victoria and in hometowns across America. Cody became the subject for countless dime novels. Here was a man who left Iowa and actually rubbed shoulders with Jim Bridger, Kit Carson, and Wild Bill Hickok. A farm boy could sink his teeth into a man like Buffalo Bill Cody, the quintessential Westerner.

The romance and glamour of Buffalo Bill and his colorful outdoor extravaganza and the other circuses and carnivals of the day crisscrossing the Midwest were what really captured Frank's imagination.

When a traveling circus set up its billowing big top near Conway one year, there wasn't enough money to send the entire tribe of Phillips children. Frank was delegated to attend and bring back a full account of everything he saw. He was ecstatic. The wild animals, gaudy clowns, fancy ladies riding bareback, jugglers, tightrope walkers, and sideshow barkers left him breathless. He would have gladly hopped into one of the brightly painted wagons and never looked back. Instead, Frank trudged home to the farm. He held court at the kitchen table and kept the other children, and even Josie and Lew, spellbound long into the night with stories of elephants and snarling cats and of the smell of straw and canvas and roasted peanuts.

Frank managed to stay content with his dreams, but he wanted to go to the places he read about in the Ned Buntline and Horatio Alger books. Lew's stories of Captain Standish and the other family ancestors, of his own experiences fighting rebel soldiers in Mississippi and the hard years in Nebraska, stirred Frank's imagination. He realized he was a true son of the West. Born just four days after a patent was obtained for barbed wire—the surefire method to fence the open plains—and the same year Oliver F. Winchester developed the Winchester .44-40, the powerful weapon popularly known as "the gun that won the West," Frank felt the history of the land in his bones. As impatient and eager to see new sights as his kinfolk had been, he also knew he did not want to spend his life watching corn grow. When economic hardships on the family ended his school days at age twelve, Frank didn't mind paying his own way. He hired out to area farmers to dig potatoes for ten cents a day. But this work wouldn't last forever. Frank had glimpses of the world outside the farm. He'd make his dreams come true. He knew that as sure as he knew himself.

CHAPTER FOUR

There had to be a way to get off the farm. Frank dug potatoes and bided his time. He didn't have long to wait. During a Sunday family excursion to Creston, the neighboring Union County seat, the youngster found the opportunity he needed when he spied one of the town's barbers wearing the flashy striped pants that were in vogue. Resplendent in trousers and spats, the dashing barber tipped his hat to every lady he passed during his leisurely Sabbath stroll. Frank was impressed. "I made up my mind that I wanted to earn enough money so that I could afford to wear striped pants even on weekdays," he later said.

He called upon a stylish barber, an acquaintance of the Phillips family, and asked him just how a fourteen-year-old farm boy could become a successful tonsorial artist. "Why, I know you're a hard worker; I'll give you a job myself," said the man. Frank's farm days were over.

A quick learner, with solid work habits and an engaging personality, Frank's apprenticeship went as smoothly as a gentleman's shave. He swept up locks of hair, ran errands and watched the barbers as they deftly handled shears and straight razor and still managed running conversations with their customers. He learned about oils and pomades and the array of grooming devices. He discovered the value of a good quinine wash and that the best remedy for preventing dandruff was to beat together an egg yolk, a pint of rainwater, and an ounce of rosemary spirit and rub it into the scalp. He was taught how to strop a razor, work up a fine lather in a china shaving mug and apply it with only a few strokes of a soft badger-hair brush.

Frank was comfortable in the masculine retreat of leather reclining chairs, amid the reek of cigar fumes and bay rum, where men congregated to browse the spicy pages of the *Police Gazette,* and talk politics, business, and weather, while they waited for their fifteen-cent shaves. For the next few years Frank made the rounds, picking up experience and learning more wherever he went. He worked at barbershops in Conway, in the town of Clearfield, just east of his father's farm, and in the larger city of Council Bluffs on the wide Missouri.

As the Gay Nineties blossomed, seventeen-year-old Frank, tired of only looking across the broad river to the open spaces beyond, decided to leave Iowa. In 1890 there were more than 85,000 barbers and hairdressers in America, and Frank, always confident, felt he was one of the best and needed to share his talent with the rest of the country. He went back to the family farm to say his goodbyes and struck out for the West.

Frank boarded a Union Pacific train and followed the Platte River beyond Nebraska into Colorado, where the thousand miles of the American prairie ends. The train chugged into Denver on the high plains, just east of the Front Range of the Rocky Mountains. The thin mountain air made Frank light-headed as he took in the sights of "the Queen City of the Plains."

Colorado was in the midst of its mining boom, and although other mining areas got an earlier start, Colorado had emerged as the greatest mining district in the West. Mineral riches pulled from the depths of the mountains made Denver throb with excitement. The busy streets were filled with prosperous businessmen and their ladies, all beneficiaries of the gold and silver bonanzas.

There was much to see and do. A sprawling entertainment complex called Elitch's Garden featured a zoo filled with wild animals, variety shows, fireworks displays, and balloon ascensions. The Tabor Grand Opera House, built by silver baron Horace Tabor, drew overflowing audiences, and the Windsor Hotel, its taproom floor studded with 3,000 silver dollars, featured high-stakes poker games that never stopped and an elegant dining room offering frog legs, mountain trout, venison, and bear meat.

Frank listened to the stories about the gold and silver kings. He heard that Tabor, now one of the richest men in Denver, was only a furniture store owner who netted millions from a $17 grubstake. Frank enjoyed the stories of the miners—eternal optimists who believed a great store of riches was only a stroke of the pick away. He was told that many of the early prospectors were so fresh from the farm when they arrived that, according to a common saying of the time, they "mined with pitchforks." Dreams of unimaginable wealth lured all sorts of men and women to the mining towns and camps. All of them wanted their share of the great treasure lodes. Many would also be needing haircuts and shaves. Denver was exciting, but Frank wanted to be closer to the action.

The young barber headed southwest across the Western Slope to the mining districts. He considered Creede, Telluride, Cripple Creek, and Ouray, but he settled on Aspen, the silver boom town only thirty miles from Leadville, where Tabor had struck it rich with his Matchless Mine. Aspen was one of the main towns on the "Silver Circuit."

Frank located a comfortable boardinghouse and had little trouble securing a job. He took over a chair at the Silver Dollar Barber Shop, a name that boded well with young Mr. Phillips. On February 22, 1891, the Aspen *Times* noted that "Joe Schwendinger and Woody Fisher, the proprietors of the barber shop at the corner of Mill Street and Cooper Avenue, are now prepared to meet the wants of the many who desire a clean shave or haircut by men who thoroughly understand their business. The services of Messrs. H. C. Shoemaker and F. F. Phillips have been secured, and their many friends will be pleased to call on them at their new quarters. With such a quartette of tonsorial artists, this new barber shop will not be long in taking the lead as the favorite resort of Aspen gentlemen."

The newspaper was right. The barbers of the Silver Dollar stayed very busy. The shop became an oasis for miners and prospectors, surveyors and merchants, railroaders and teamsters. Facial hair had become stylish in Victorian America. Mustaches—especially the virile handlebar variety of the West—were popular as ever. But by the 1890s beards were starting to go out of fashion, except in rip-roaring mining towns where thick whiskers kept a prospector's face warm when he faced the Rocky Mountain winds.

Frank trimmed beards, waxed mustaches, and cut hair all day long, six days a week. The stream of clientele at the Silver Dollar seemed endless. Each day the railroads brought more people. Aspen was one of the fastest-growing towns in the area. Located in an irregular valley in the rugged mountain country 8,000 feet above sea level, Aspen was also a far cry from the prairies of Iowa.

From its earliest days as a mining town, Aspen was famous. Founded in 1879 as Ute City, the mining settlement became rich from the reef of silver hidden beneath the land. Lost in the wilderness, the town was renamed Aspen in 1880 and, by the time Frank arrived, the Smuggler Mine made Aspen the world's largest silver camp, with a population of more than 8,000 that would swell to almost 12,000 at Aspen's peak in 1892. The town boasted a courthouse, streetcar lines, the first electric lights in the state, a hospital, several churches, a half dozen newspapers, banks, and the Hotel Jerome, considered by some to be "the best west of Kansas City."

Frank didn't miss the Iowa corn fields or the swimming holes or even the thick forest where he and his brothers played amidst Indian spirits. Here there were trout streams, solid rivers of ice, and steep blue mountains covered with shimmering aspens and ponderosa pines.

There were also hearty souls in Aspen. Many were Frank's best customers. Miners and prospectors, some who had survived Indian attacks and were used to wading in snow up to their necks, climbed up on Frank's chair. They were rugged men, weary of living off beans, flapjacks, and sourdough bread. A trip to Aspen for more provisions—flour, bacon, vinegar, sugar, beans, and bullet lead—also meant they could chew on juicy beefsteak, sip whiskey, and get a refreshing shave and haircut.

Luxury and entertainment abounded. A favorite recreation was the Bathing Train, which made runs every evening on a $2 round trip to Glenwood Springs, forty miles away. The Wheeler Opera House was the setting for light operas and plays. A young man out on the town could listen to actors recite Shakespeare or take in a popular melodrama. There were countless prizefights, medicine shows, fortune tellers, troupes of blackface minstrels, snake charmers, magicians, and brass bands. Almost every dance hall had its share of female partners ("for dancing only") and drinks for the princely sum of twenty-five cents. Brothels, saloons, and gambling halls stayed busy with miners to please. Besides sporting girls with fancy derringers strapped on their wrists, there were "respectable ladies" arriving in Aspen. They formed a literary society, established Sunday schools, and enjoyed sleigh rides with "respectable gentlemen." Aspen was thriving, and everybody, including Frank Phillips, thought the future was bright.

Then in 1893, as sudden as a mountain storm, disaster struck. Only three years before, in the continuing battle over silver and gold coinage, Congress had passed the Sherman Silver Purchase Act, which called for the government purchase of 4,500,000 ounces of silver each month. The steady decline in the price of silver bullion, coupled with the economic recession, strengthened the political weight of silver. Aspen grew fat and sassy. But beginning in June 1893, four years of deep depression swept the nation, starting with the crash of the New York stock market. The Silver Purchase Act had badly drained gold reserves, and with silver hitting an all-time low of 77 cents per ounce, the Colorado producers agreed to shut down all their mines and smelters until the government could take measures to support the silver miners. But with the Panic of 1893—one of the worst depressions to ever strike the country—that help never arrived. By November, the Silver Purchase Act was repealed after a bitter congressional battle. The fortunes of the mining states collapsed.

The silver kings lost their crowns. Horace Tabor would wind up with worthless mines and die penniless in a Denver hotel before the end of the century. Mining camps across the Rockies turned into ghost towns almost overnight. Idle mine shafts filled with water and shattered dreams.

As the nation changed from silver to the gold standard, Aspen's mines ceased operations. The town began its long downward slide. Banks closed, business dried up, and nearly 2,000 miners were out of work. Money once spent for poker chips, new clothes, and a haircut and shave went toward buying one's ticket out of town.

One of the first Aspen refugees was Frank Phillips. He headed due west, to Utah Territory, where he lived in Ogden and spent time in a railroad construction camp called Terrace.

When he arrived, Ogden was a thriving community with drugstores, hotels, and a dry-goods emporium. Travelers could get a decent meal at the Keeney House for about fifty cents. Ogden was also where the Continental Oil Company's first business was conducted, and the city served as a major distribution and marketing center for kerosene, axle grease, and other petroleum products shipped from the East.

Located on the Weber River between the Wasatch Mountains and the Great Salt Lake, Ogden was in the land where a quarter century before as many as 1,000 Shoshones had camped; where mountain men and trappers rendezvoused with Indian wives "healthy as bears"; and where Mormon settlers established the second-largest city in the territory. But the most significant event in Ogden's history was the coming of the railroad. In March 1869, the first train steamed into the city. It was part of the first transcontinental line and met with the Central Pacific and Union Pacific rails at Promontory on May 10 of that year.

Ogden became the principal railway center of the intermountain region and, perched on the dividing line between the Mountain and Pacific time zones, became "the place to reset watches."

Frank took up barbering and, deciding to extend his range of experiences, also worked as a brakeman for the Central Pacific Railroad. This line of work took Phillips to Terrace, a railway maintenance center and the largest of the last few construction camps built by the Central Pacific as the tracks neared Promontory.

Situated in the northern edge of the Great Desert 150 miles from Ogden and, as one observer of the day put it, "10,000 miles from the rest of mankind," Terrace was home to hundreds of railroad workers, including a good many Chinese who stayed on after helping to build the original tracks. The Chinese population lived in one section of town, in dugouts and shanties. On weekends the sticky, sweet aroma from their opium pipes drifted down the streets, and when the Chinese New Year arrived, the men of Chinatown made strips of white coconut candy for the children of Terrace. It was desolate country, but Phillips enjoyed the new friends he made in the drafty wooden bunkhouses. He was popular with his fellow workers, and in the evenings after supper, Frank sharpened his barbering skills trimming hair and beards.

Back in Ogden, Phillips and his pals, many of them also young Midwesterners out seeking adventure, took part in some of the city's social events. Social clubs hosted balls where a well-groomed gentleman could dance with one of the city's young ladies to orchestra music and enjoy a supper for a mere fifty cents. It is clear from correspondence Phillips received from old friends that he found his short stint in Utah pleasant. Yet in later years, Phillips never discussed the time he spent in Ogden and Terrace, or even Aspen, and hardly anyone knew he ever worked on the railroad and cut hair in the Rocky Mountains.

In March 1895, Dr. C. B. Moffett, a young physician known to his friends as "Doc," wrote Phillips in Ogden. In the letter, Moffett, with a newly established medical practice in Monroe City, Missouri, reminisces about working with Phillips in Utah, where "we certainly had happy times." He also tries to convince Frank to move to Missouri and consider a career in medicine.

> Now Frank, I wish you could come down on a visit. If you will look at the map, whenever you feel like you wanted to take a vacation, come down and spend a couple of months with me, or with us, you know there would be noone more welcome, and I wish you could open up a shop here, there are only two, and are far from being first class. I have not had a decent hair cut since I left Ogden. I still wear my whiskers for effect and have got quite use to them. But Frank, send me a good haircut, or else come down and open

a shop and it won't be long until you have plenty to do, and in the intervals you can study medicine. I can fix it so your tuition at college will only be $25 a course. I am going to operate on a harelip case now in a day or two, I wish you could make up your mind to study medicine. I could arrange for you, your room and board at St. Louis for the six months would only be $100, so that you would be getting off cheap . . . and I'll take you in as full partner, and by that time I will have a good business established and my word on it, you will be making more money than cutting hair. I have made, since I've been here, as much as $15 in one day. You are capable of studying medicine as nine-tenths of all the medical students I have had the pleasure of seeing. A profession, as you are aware, is better than a trade if the individual will stick to it. You have everything your way, you are just at the proper age to begin, are exceedingly blessed with good looks and all else pertaining to it, and a good talker. I wish I could get you to give it serious thought, and in after years you would not regret it. But anyway I hope you are doing well where you are, and hope you will curb that propensity to roam and settle down as I have been compelled to do.

The letter never reached Phillips' Washington Avenue residence in Ogden. Doc, aware of Phillips' "propensity to roam," was smart enough to jot "please forward" on the envelope. Just how Phillips responded to Doc's offer and the suggestion of studying medicine is not known. What is known is that he didn't move to Missouri. Nor did he become a doctor. Through with adventuring for a while and feeling poorly after an illness, Phillips wired his family for money to come home.

By the winter of 1895, Phillips, a strapping twenty-two-year-old, was back in Iowa, reading Doc's letter, sipping Josie's vegetable soup, and contemplating his future.

CHAPTER FIVE

Snow, silent as sleep, covered the land. It stayed, like an uninvited guest, all winter and into spring. That snow then became rain. Frank, bundled in flannel quilts in the warm farmhouse, listened to the rain tap at the window. In the morning every limb in the orchard and the dead weeds guarding the fences were wrapped in ice, sparkling like crystal in the sunlight, looking as though they shouldn't be cold but hot to the touch.

The aroma of gingerbread and fresh coffee mixing with the smell of burning oak and ash, the sounds of his sisters' and mother's laughter and his little brothers' deepening voices, and the sight of his father hoeing the cold black soil back to life, made Frank realize that he really had missed his family. It was good to be home.

Tired from his years on the road and the journey back to Iowa, and still weak from his illness, Frank convalesced while his eight siblings, especially the feisty twins—Waite and Wiate—pestered him with questions about the West. They wanted to know about the mining towns and about working on the railroad, and if he had seen any train robbers or Indian war parties.

Frank, grown tall and slender, ran his fingers through thinning brown hair and his blue eyes flashed when he told them about the wild animals at Elitch's, minstrel shows, bear-steak dinners, the crusty characters he shaved at the Silver Dollar Barber Shop, and how the Great Salt Lake had looked shimmering in the desert. They were good yarns, even better than the ones from years before when Frank went to the circus for the whole family.

Lee Eldas, or L.E. as he was called, hadn't strayed from Iowa yet. He was anxious to tell his brother about the money he earned picking and selling gooseberries and hiring out as a farmhand. Only three years younger than Frank, L.E. was also proud because, at nineteen, he was well on his way to receiving a pretty fair education by country standards. He had completed grammar school and had gone to Shenandoah, where he not only saw his first electric light bulb and indoor toilet but received training in bookkeeping and penmanship at Western Normal College. L.E. paid the tuition by working as a janitor and waiter and was able to stretch $5.80 over one ten-week period. Like his big brother, L.E. was convinced that he was not cut out to be a dirt farmer. Years before, he and some of his chums founded the Never Sweat Club. The entire membership pledged that they would do everything in their power to ensure a sweatless life off the farm. L.E. told Frank that they had just renamed themselves the Anchor Club and most of the members, including L.E., were off to become bona fide schoolteachers. L.E. was bound for the Taylor County Teacher's Institute to earn his certification.

Jennie, the oldest of the Phillips children, had blossomed into a young woman while Frank had been gone. She contributed to the family by working as a seamstress. Ed, a lanky teenager, helped Lew on the farm and took a part-time job as a section hand on the railroad. The younger children were busy with school. Except for the restless twins, who envied Frank's wandering, they all appeared content with their lives.

Before spring was in full bloom and Lew could recruit him for chores, Frank moved back to the town of Creston in Union County. He had no trouble finding employment at A.B. Tucker's Climax Barber Shop, and soon Frank became known as one of the town's up-and-coming young businessmen. He squirreled away his money and attracted an impressive number of customers. His decision to return to Iowa was wise. Coming back to familiar territory was a tonic Frank needed. He could hone the skills and experience he picked up in Colorado and Utah and put them to good use in Creston.

It was more than twenty-five years before when Creston sprang into life as a railroad town, built on the summit, or "crest," of the prairie—the highest point on the Burlington railroad line between the Missouri and Mississippi rivers. The railroad established a huge complex there, including a roundhouse, a station, and machine shops. Farmers came from far and near to trade and buy supplies and goods, and sell produce and stock.

Agriculture was important to Creston, and an organization called the Blue Grass League of Southwestern Iowa was formed to publicize the "unexcelled advantages of the blue grass region." The League erected a towering Blue Grass Palace, built entirely of native grasses, and filled the interior with exhibits of grains, grasses, fruits, bricks, lumber, and manufactured articles from the area. A second palace was built and the League held expositions for the public, with railroads running special trains at reduced rates.

In the spring of 1895, Creston, with a population of about 6,000, had an opera house, a thriving chapter of the Independent Order of Oddfellows, and a gas and light company. Local grocers delivered even a single spool of thread to their best customers, and when the bills were paid, folks expected to receive a sack of candy. There was a chautauqua, a literary circle, several city parks, and a zoo with a collection of bears, elk, antelope, and wolves.

After living in a dusty railroad camp Frank enjoyed these comforts of Creston, and his reputation as a skilled barber spread. In no time, he purchased the Climax Barber Shop, later called the Climax Shaving Parlor, located in the basement of the Creston National Bank Building, an imposing brick structure at the intersection of Pine and Montgomery. Frank was his own boss. He wore striped trousers every day of the week and on Sundays hired a horse and rig to take him on a round of social calls.

The Phillips establishment was quickly labeled as the fanciest barbershop in town. The largest mug rack in Iowa, filled with personalized hand-painted shaving mugs, many emblazoned with the owner's name and occupation, commanded an entire wall. Potted palms and dark leather chairs added a touch of class. He hired Joe Dill, Fred Warren, and Earl Russell— the best barbers he could find—and transformed them into salesmen, paying them on a commission basis. Clad in their immaculate white coats and black bow ties, the Phillips barbers carried business cards and spent time outside the shop drumming up clientele. Frank catered to the customers' needs, and at the same time used the basic principles of merchandising, offering his customers hot towels, cigars, chewing tobacco, and other extras.

By the time he celebrated his twenty-fourth birthday, Frank had built up enough capital to hire more barbers and buy a second barbershop. This one, located in the new Summit House, a three-story hotel popular with travelers, was built shortly after the worst fire in Creston's history destroyed the original structure. Frank divided his time between the shops, acting as general manager and greeting customers.

Not satisfied with controlling a sizable share of the town's barbering trade, the young entrepreneur decided to broaden his product line. Patent medicines and cure-alls promising solutions to every known aliment were the rage. Victorian America could choose from Kickapoo Indian Salve for corns, pimples, and itching piles; Ayer's Sarsaparilla for chronic fatigue; and Egyptian Regulator Tea, a brew that would bring "graceful plumpness" to flat-chested girls. There was Parker's Ginger Tonic, Mrs. Winslow's Soothing Syrup, and Dr. Morse's Indian Root Pills. But Frank had a remedy all his own in mind.

The inspiration came straight from the hog lot. Ever since he was a farm boy, Frank noticed that while many men went bald, no pig was hairless. Most even had a thick stand of bristly hair running down their backs. "Nothing except rainwater gets on a hog," Frank explained to his barbers. "So it must be that rainwater that keeps the hair growing."

Frank claimed that rainwater must have some mysterious quality to help prevent baldness. The notion of taking a basic element of nature, slightly altering it, and then marketing it to the general public was appealing. After hours, when his shops were closed and the day's receipts were counted, Frank hung up his suit coat, rolled up his sleeves, and mixed perfumes and lotions with water from the rain barrel. In no time he developed a concoction, composed mainly of scented rainwater, that he believed would appeal to the customers. Although Frank didn't know for sure if the stuff would cure baldness, he was certain it sure couldn't hurt.

Frank's own barbers snickered at the notion of splashing sweet-smelling rainwater on scalps and they warned him that his reputation was on the line. But Frank was a gambler. He remembered the high-stakes poker players and silver prospectors he met in the Rockies. He knew only the risk takers ended up winners. Frank dubbed his discovery Mountain Sage. Prominently displayed on the barbershop shelves, the tonic sold like hotcakes.

Decades later, Frank, himself bald as a light bulb, had tongue in cheek when he addressed the Des Moines Chamber of Commerce. "Gentlemen, I see out there many bald heads that I treated with my hair tonic." There are no further testimonials to the effectiveness of Mountain Sage, but Frank's business flourished as town merchants, railroaders, and prosperous farmers flocked to his barbershops.

One of Frank's best customers at the shop in the bank building at Pine and Montgomery was a banker himself. John Gibson, one of Creston's movers and shakers, frequently left his rolltop desk at his busy bank across the street to get a shave and hot towel or have his snow-white hair and mustache trimmed. Gibson enjoyed the flamboyant young shop owner with a flair for salesmanship. He knew that anyone with a rapidly retreating hairline who could manage to sell bottles of rainwater to stop baldness had to be one smart fellow. Gibson listened to Frank's easy banter. He liked what he heard and he liked the way the man carried himself—tall and erect, shaking hands and offering fat cigars and Sen-Sen to the customers. Gibson decided to keep his eye on the young barber.

CHAPTER SIX

John Gibson was right people. All of Creston knew that, including the young man who cut his hair. Gibson was the epitome of a successful small-town American. People used words such as "important," "prosperous," and "respectable" when they spoke of him. Everyone—from the prim tellers at his bank to the porter at the Bon Ton Saloon—looked up to John Gibson. He was as trusted as a railroader's watch.

His father was Josiah Gibson, a Methodist minister born in 1817 in Virginia and educated in Scotland. His mother was Elvira Ann Ebbert Gibson, born in 1821 to John and Mary Ebbert, a prominent Terryopolis, Pennsylvania, couple. Josiah and Elvira married on New Year's Eve 1845. They lived in Ohio, Pennsylvania, and Illinois. John Ebbert Gibson, born October 2, 1849, in Wellsville, Ohio, was the oldest of their seven children.

In Illinois, the Gibson family first settled in Joliet and later moved to Dixon, where, during the Civil War, young John attended the local high school and on every fourth Friday afternoon entertained his classmates by reciting "Spartacus to the Gladiators," "Spartacus to the Roman Envoys," "Regulus to the Roman Senate," and other passages of prose and poetry he memorized. Gibson made it a habit from boyhood on to rise at five o'clock every morning and spend at least two hours before breakfast memorizing selections that he loved.

Blessed with good looks and a sharp business mind, Gibson met and courted Matilda Jane Martin, born November 12, 1851, to a father from Virginia and a mother from West Virginia, who also settled in Illinois. John and Matilda married in the tiny town of Alma on June 18, 1870. Gibson became a lawyer in 1872, was admitted to the bar, and launched a thriving law practice in Pueblo, Colorado, where he also devoted some of his time to learning the mining business.

In 1877, the Gibsons decided to return to the Midwest and moved to Creston. John opened a law office with John A. Patterson, a partnership that lasted four years, until Gibson withdrew to focus his attention on his growing number of financial interests. By 1883, Gibson was selected president of the Iron Mountain Company of New York, owner of the celebrated Iron Mountain at Durango, Mexico, at the time the largest iron mine in the world. During his presidency, Gibson saw to it that more than $400,000 was invested in a new plant at the mine site. But weary of being gone from home much of the time, Gibson left the mining business, abandoned his law practice, and quickly became one of Creston's premier bankers.

The Iowa State Savings Bank had been organized in Creston in 1883. Three years later, Gibson took over as president, and in the next few years the bank rapidly increased in size. Owing to differences in policy, Gibson, always aboveboard in all business dealings, disposed of his stock in the bank in the summer of 1890. By the following February, Gibson's successor suddenly died and at the same time it was revealed that the cashier had involved the bank in unauthorized and questionable real estate investments. The state auditor moved in, made a thorough examination of the bank's affairs, found it solvent, and reopened it for business. In January 1892, satisfied that the institution's good name had been restored, Gibson again became bank president.

When Frank opened his first barbershop in Creston, the Gibsons were already firmly ensconced on the top rung of the town's social ladder. John, with distinguished white hair and full mustache, not only headed one of the town's leading banks but also served as president of the Nebraska Central Building and Loan in Lincoln, vice president of the Iowa Central Building and Loan Association in Des Moines, and treasurer of the Anchor Mutual Fire Insurance Company in Creston. He established the Gibson Investment Company and acted as one of the town's primary boosters and civic leaders.

Gibson also built a showcase residence on Sycamore Street on Creston's North Hill where his wife, called Tillie by her friends, presided. The Gibsons had lost both a daughter and a son in infancy during the first two years of their marriage, but two sons—Josiah, born 1873, and John Martin, born 1875—filled the void.

On August 8, 1877, shortly after they moved to Creston, a daughter was born who quickly became the darling of the Gibson household. When the German midwife attending Tillie handed Gibson the squalling baby girl, he was ecstatic. His family was complete. They named her Jane but to her friends she was Jennie, or Jenn, and a few called her Jerg. To Gibson she was always Betsie—his sweet, precious Betsie—and he remained devoted to her for the rest of his life.

A bright and energetic child, Jane had her mother's dark eyes and hair and gracious manners. She also inherited her father's intelligence and curious mind. Raised in a home of gentry, the young woman observed the rigid rules of etiquette which dominated life in Creston. But despite a pair of watchful big brothers and her doting parents, she managed to fit in a fair share of fun and frolic. Never without young gentlemen, attracted by her fresh looks and wit, to accompany her to socials when she reached the proper age, Jane counted among her array of acquaintances and friends a neighbor girl named Minnie Hall as her closest confidante. Jane and Minnie spent hour after hour in Jane's bedroom playing with china dolls and dreaming about the world outside Creston. They went through grammar school and high school together and made a pact that their firstborn daughters would be named after each other.

Life was pleasant in the 1890s on Sycamore Street in Creston, Iowa. Especially if you were a wealthy banker's daughter with no worries other than discreetly keeping track of the town's most eligible bachelors. The Gibson women stayed in style in wardrobes provided by local seamstresses and supplemented by mail-order catalogues offering the latest fashions straight from Chicago. A hired girl kept the house spotless, dusting the knickknacks and piano in the parlor and cutting handfuls of lilacs and roses to fill the vases.

All seasons were treasured. Arbor Day and Independence Day were observed; elaborate Valentines were exchanged. Bicycle races and bowler hats were popular and calling cards were de rigueur. Sousa marches, arias, and devotional hymns were popular in the Gibson household. Home manuals tendered advice covering the amenities. These books provided suggestions such as having servants wear thin-soled shoes (". . . the noise of their footsteps being unpleasant"); advice about placement of finger bowls, oyster forks, and goblets; codes governing etiquette at the dinner table ("Never sup soup noisily; never use a steel knife for fruit; never put potato skins on the table cloth") as well as etiquette for mourning, letter writing, courting, shopping, and family life. There were even rules for "society small talk," and important questions about dress ("What kind of scarf is it proper to wear with a tweed suit?") were answered ("A blue polka-dot or solid color four-in-hand").

These were the days of Indian summer picnics; and at Thanksgiving and Christmas, John Gibson, in his finest banker's suit, solemnly sliced roasted turkeys and hams baked in cider. Springtime brought masquerades and dances, lawn parties and church socials. During the drowsy after-noons of summer after a round of croquet, Jane and Tillie, smelling of talcum powder and lavender sachet, lounged on porches smothered in ivy and morning glories and festooned with potted ferns. They munched mac-aroons from the Model Bakery, sipped lemonade and iced tea spiked with stalks of mint from sweating glasses wrapped in napkins, and, their eyes always on their sewing, whispered gossip and secrets.

Gibson doted on his daughter. So he became very concerned when he found that Ben Williams, the young man who sold Frank Phillips the furniture for his shaving parlors, had introduced the barber to his sweet Betsie. Though impressed by the young man's business ability, the banker had no room in his comfortable life for a son-in-law who barbered.

Jane had other ideas. In her eighteen years she had never met anyone quite like the flamboyant Mr. Phillips. A dandy, dressed in fancy suits and splashed with scented rainwater, Frank, with his rimless spectacles and receding hairline, looked scholarly and much older than his twenty-two years. As one of his cronies put it, "Frank looked forty all his life." Jane was smitten. She found him very appealing. Frank, taken with Jane's thoughtful eyes and bashful smile, knew he had found a true lady, unlike any woman he had encountered as a farm boy or during his journey through the West. Despite her parents' disapproving words, Jane encour-aged Frank to call on her. He responded.

There were occasional social dalliances with other young men and women from the community, but when the summer of 1895 was in full bloom both Frank and Jane were fast approaching total commitment. The talk on Creston verandas turned to the romance between the energetic blue-eyed barber and John Gibson's daughter. Frank squired the pert Miss Gibson to Pratt's Opera House and to numerous balls and dances, and he made sure the Gibsons spotted him bellowing hymns every Sunday morn-ing at the Methodist church. The young couple kept in touch via hand-written notes and letters carried by messenger between Frank's boarding-house on Maple Street and the Gibson residence. When Frank sent a gift to mark her nineteenth birthday on August 8, 1896, Jane replied immediately with a polite note, telling her suitor to "please accept my thanks on paper and I shall be glad to have you call at your earliest convenience and I shall then endeavor to thank you personally."

Jane's beguiling letters, written on onionskin or on her father's office stationery (with the banker's eternal question printed below his name: "Are you saving something for old age?"), were usually filled with invitations to visit the Gibson home or thank-you's for gifts and for escorting her to a social function. Frank answered with scrawled notes which Jane carefully stashed in a pouch made of pillow ticking embroidered with red thread and ribbons. By 1896 the "Dear Mr. Phillips" became "My dear Frank," and "Sincerely yours, Jane Gibson" turned to "I am as ever, with love, Jane." And although he didn't stop using "Respectfully yours" as a sign-off, Frank did manage to drop "Dear Miss Gibson" in favor of "My dear Betsie Jane" as a salutation.

The correspondence continued even when her parents sent their love-struck girl on trips to Chicago or extended visits to relatives in Nebraska, where Jane wrote Frank she "was almost crazy to see someone from home." Frank faithfully replied and worked eighteen-hour days while Jane was away from home. Work proved to be good therapy. The barbershops, like saloons and cigar stores, were bastions of male dominance, and Frank's shops were the most popular in town. A steady flow of teamsters, clerks, lawyers, cigar makers, ministers, and, of course, bankers helped occupy Frank's mind during his sweetheart's absence. When he learned that Jane was "somewhat under the weather," he wrote: "My advice to your kind auntie would be to bundle you up in a feather bed, or any old thing, and send you home. I can very candidly inform you it would be somewhat of a satisfaction to at least one person to even know you were in the city of Creston."

Frank had no intention of letting Jane Gibson slip away. She was attractive and he liked her spunky attitude. But also of importance, Jane's father, a successful entrepreneur, inspired Frank to dream of a business career beyond the barbershop. Gibson's vast resources and contacts could be of enormous help.

As Gibson himself was still not fully convinced of the soundness of the relationship between his daughter and his clever barber, he didn't favor the announcement of their formal engagement. He did, however, remain impressed with the young man's potential and natural business savvy.

The Gibsons also knew they could not keep their daughter away from Creston forever. In the face of the inevitable, a solution was devised to appease everyone. If Frank would give up barbering and join in the banking business, Gibson would not stand in the way of the young couple's plans to marry. Jane was delirious with joy. Frank was only too happy to comply.

The marriage—a double wedding ceremony uniting Frank and Jane and her brother Josiah Gibson with Miss Marcelene Miller, "one of the best young ladies of Council Bluffs"—took place February 18, 1897. It was a major social affair for Creston and attracted much attention. As Frank later declared: "The two leading events in 1897 were first, my wedding in Creston, and second, the discovery of gold in the Klondike!"

The wedding took place on a Thursday afternoon before 1,500 friends and neighbors at the Methodist church. The Creston *Gazette* reported that more than 2,000 men, women, and children turned out to witness the ceremony. The newspaper described Frank as "bright, intelligent, of good character, temperate habits and good standing in the community, he has a bright and promising future." The newspaper's remarks about Jane were also glowing. "His bride is a charming lady who has the love and respect of a large circle of acquaintances. Reared in one of Creston's best homes she has had the best of care and training by a devoted mother who is regarded as one of the noblest and best ladies of this city. Jennie is an exceptional girl and in every way worthy the heart and hand of a bright, promising young man."

All the wedding arrangements were kept simple. No formal invitations were sent out, but a special invite was issued by John Gibson to church and Sunday-school members the Sabbath before the nuptials. The Gibsons wanted their daughter and son to have a plain wedding that would not offend anyone and would make all the guests feel comfortable.

Just before 5 P.M. the bridal parties appeared. Seats had been reserved in the front of the church for close relatives. Hugh Fry, Jim Donovan, Harry Spencer, and the Thompson boys—friends of Frank and Josiah— served as ushers. Just after the Phillips family and the Gibsons were seated, the brides and grooms entered the church. Frank came first with Jane at his side, followed by Josiah Gibson and Miss Miller. Frank Betts, a young Crestonian acquaintance of both couples, played Mendelssohn's wedding march on the organ as the party walked down the main aisle. Two pretty little flower girls, Myrtle Grubb, a daughter of Frank's sister Jennie, and Ethel Burket, Jane and Josiah's cousin, preceded the two couples as they marched toward the altar. Jane and Frank stood on the right and left sides of the altar, as Josiah, in a black suit, and his bride, dressed in silk and lace with carnations in her hair, moved side by side to the center. The flower girls, carrying bouquets of roses, stepped behind and held a white bridal ribbon across the aisle.

When the final notes of the wedding march ceased, the Reverend A. E. Griffith, pastor of the church, and the Reverend T. L. Faucett, Frank's blacksmith/preacher grandfather, took their places before the bridal party. The Reverend Faucett, retired from the ministry but remembered from his younger years as one of the pillars of western Methodism, was pleased to come from Conway for the wedding. Except for a few muffled coughs, the congregation fell silent.

The Reverend Griffith pronounced the words joining Josiah and Marcelene as husband and wife. They then stepped aside, and Frank and Jane stood before his grandfather. Jane was radiant, gowned in a light sage-green brocade with a bodice overlaid with heavy cream lace and an elaborate trim of silk and velvet. She wore white gloves and cream roses in her hair. Frank wore a black suit and white bow tie. The old man smiled at the young couple and led them through the simple vows. They promised to love, cherish, and protect each other until death part them. They were pronounced husband and wife. Then the Reverend Faucett, using his best preacher voice, implored the benediction of Almighty God and asked for blessings from on high to fall on the newly married couples.

The ceremonies were brief, according to the best traditions of the Methodist Episcopal Church. With the flower girls in front, the couples left the church and, accompanied by close relatives, took shiny carriages to the Gibson residence for a quiet wedding supper. The evening passed around a roaring fire, with the families getting to know each other.

By eight o'clock Josiah and his bride were off to the train depot to take the evening express for Chicago. After studying at Cornell College and, as the newspaper pointed out, being a young man who "inherits from his father a capacity for business and a love of all that is good and pure," Josiah had landed a position as a draftsman in a civil engineer's office in Chicago.

"These four young people," said the *Gazette*'s account of the wedding, "start out in their married life under the most favorable circumstances and with bright prospects for the future. And they all deserve prosperity and happiness for they are worthy of all that is good and pure in life. They are model young people, all members of the Methodist church and all possessed of beautiful characters such as will win them the confidence and respect of people with whom they associate. May they live to comfort and cheer each other when time shall have silvered their locks and their feet tread near the banks of the silent river over which all must cross."

The *Gazette* also reported that Frank and Jane Phillips would reside at the Gibson home on North Hill. Frank's boardinghouse days had come to an end. He packed his few possessions in a large trunk and had it delivered to his new abode. Next came a visit to the best tailor in town. It was time to put aside the fancy striped trousers. A conservative banker's suit was in order. Mountain Sage and two-bit haircuts would soon be memories. Frank's dreams of becoming a Horatio Alger hero were coming true.

CHAPTER SEVEN

If Frank learned anything by the time he was twenty-four it was the importance of timing. Although he was as anxious as a caged wolf to become involved in Gibson's banking and investment enterprises, he also knew to bide his time. Frank wanted to be absolutely certain that none of his circle of family and friends, especially Jane and her father, took him for an opportunist only interested in climbing to the top on John Gibson's coattails.

Frank politely thanked Gibson for the $20,000 he bestowed on the couple as a wedding gift and he accepted the hospitality of residing in the Gibson residence, but he took his time when it came to selling off the barbershops. He wanted to remain his own man. Frank put the banker's suit in a cedar closet and held on to his striped trousers a little longer.

Gibson went along with Frank's plan. He respected his new son-in-law's business acumen and admired his ability to handle investments. He decided Betsie Jane was wise to choose Frank as her husband. Perhaps, Gibson thought, his first hunches about the resourceful barber were correct after all. He appeared to have the makings of a fine banker. Gibson knew it wouldn't be long before he would be able to give Frank just the opportunity he needed. As Frank continued to oversee his thriving barbershops, Gibson launched his informal banking education. Frank was introduced to other bankers and lawyers and prominent businessmen from across the Midwest who had dealings with one of Gibson's financial institutions. He spent hours listening to his father-in-law explain the strategies he used as a successful entrepreneur.

At the dinner table Gibson freely divulged his many daily victories and the occasional defeats. The talk was lively and the stories vivid. Frank

especially enjoyed the conversations about the rags-to-riches business tycoons dominating the American scene. He listened to Gibson speak of
Andrew Carnegie, the Scottish immigrant who started as a bobbin boy in a
cotton factory and proceeded to climb the economic ladder until he mo-
nopolized the steel industry and became a multimillionaire. He heard the
story of John D. Rockefeller, who saved his pennies, invested wisely, and
not only became the richest man in America but realized his life's obsession—the absolute control of oil. Frank was fascinated. He devoured the
accounts chronicling the lives of Hearst, Pulitzer, Guggenheim,
Woolworth, Wrigley, and the others who learned the fundamental princi-
ple of capitalism: make money work for you, not you for money.

Frank did find other ways to occupy his time when away from work
besides listening to his mentor. During the scorching summer of 1898, as
the Spanish-American War raged, Jane, uncomfortably pregnant, required
her share of tender loving care. In the evenings after dinner, Frank kept his
young wife company on the front porch while he read the latest war dispatches detailing the exploits of Teddy Roosevelt and his Rough Riders.
At the local band concerts the national anthem and "Rally Round the
Flag, Boys" encouraged the home folks to keep the banners waving. But
the tune most heard in the saloons, as the war approached a conclusion,
was the lusty "There'll Be a Hot Time in The Old Town Tonight." Frank
whistled it for months as Jane waved her funeral parlor fan and counted
the days.

On December 9, 1898, the day before the Treaty of Paris was signed
by the United States and Spain, formally ending the "splendid little war,"
a plump baby boy arrived at the Gibson residence on North Hill. Jane and
Frank proudly named their firstborn John Gibson Phillips. Now there was
real cause to celebrate. Family and friends flocked to the Gibson house.
Lew and Josie brought the Phillips clan from Conway for a visit. For more
than a week no customer could buy a cigar in Frank's barbershops. With
the name of his dynamic maternal grandfather along with the proud Phillips surname, Frank and Jane were convinced their son was a child of
destiny.

A few months later a rare Phillips family portrait photographed at the
farm near Conway shows Jane holding son John dressed in his best white
gown. All nine Phillips children are in the photograph as well as Lew and
Josie. Frank and his brothers, looking like somber judges, are sitting in a
row. There's not even a hint of a smile on any of the women's faces. The
family dog, Spot, is perched on a table behind the twins. Lew Phillips, his

beard graying, is staring at something out of sight, as if he's remembered a moment from the past. It would be one of the last times the entire family would be together. They'd soon be scattered across the country.

L.E. Phillips, Frank's brother, had become a country schoolteacher and even taught his twin brothers for a spell. Poor health kept L.E. from serving in the Spanish-American War and he was getting weary of living off meager teacher's wages. After putting up with rough-and-tumble farm kids at five schools in less than two years, L.E. was ready for a new occupation.

Although Lew and Josie advised him to remain in the teaching profession, by 1899 L.E. had moved to Creston, where he sold insurance. After several months he met George B. Rex, and before long L.E. went to work as a traveling salesman for $50 a month, plus expenses, at Rex's coal company. During the next three years he covered a wide territory, including parts of Wisconsin, Ohio, Missouri, Kansas, and Nebraska. By 1902 L.E. left Rex and became an original incorporator of the Hawkeye Mining Company, a coal-mining operation based in Knoxville, Iowa. That same year he married Lenora Carr. Called Node (pronounced Nodie), she was an accomplished music teacher from Taylor County, whom L.E. had escorted to hayrides, picnics, and charade parties for several years. Four members of the Anchor Club and Frank and Jane Phillips were among the wedding guests. L.E. was on top of the world.

By the late 1890s, Jennie, Frank's sister and the oldest Phillips child, was also married and raising a family of her own. The younger Phillips children were content on the farm. Except for the twins. As impatient as their big brother Frank, they were gypsy-footed and ready to roam.

In the autumn of 1899, Waite and Wiate began a two-year odyssey through the West. The sixteen-year-olds left Conway for a pause in St. Joseph, Missouri, before working as farmhands outside Leavenworth, Kansas. Quickly tiring of doing the same work, the boys went on to Kansas City and secured railroad passes to work for one of Frank's former employers, the Union Pacific Railroad at Rock Springs, Wyoming. From there they worked their way to Ogden and Salt Lake City, where they became messenger boys for Western Union Telegraph. In Salt Lake the boys lived with an older man who encouraged them to accompany him to the Alaska goldfields near Nome. They departed on a cold winter afternoon, pausing only for short working stints at a smelter in Montana and a tool shop in Idaho to earn grubstakes. By spring they reached Seattle to rendezvous with their Salt Lake friend. When he failed to show up, the twins abandoned their Alaska trip.

Still anxious to see as much of the West as possible, the brothers moved through a series of jobs. They worked as laborers building a railroad tunnel near Everett, Washington; took a ship from Seattle to San Francisco to work on a highway construction gang and as kitchen helpers at the Golden Gate Hotel; returned to Washington for jobs in a shingle mill; visited Vancouver, British Columbia, and Spokane; and finally decided to split up for a short time. Wiate went to the Pendleton, Oregon, Roundup and then on to Montana, while Waite worked in a lumber mill and then took up with a Dutchman named Louis Vrolick, a Missouri lad named Frank Hawkins, and a grizzled old-timer called Bitter Root Bill. The unlikely foursome outfitted for the mountains and took packhorses and provisions into the high country to trap pine martens. In the rugged mountains of Washington, they blazed trails, built a cabin, and set their traps.

The food supply soon dwindled and Waite was chosen to hike thirty miles to the nearest settlement to get more supplies. When he reached the town, he found Wiate waiting with frostbitten feet. The brothers returned to the mountains with fresh packhorses and more supplies, only to find the camp empty. Bitter Root Bill had refused to pay his share of the expenses and deserted the party. The boy from Missouri accidentally shot himself in the arm while deer hunting. The Dutchman rushed him to a doctor in town. But Vrolick returned and he and the twins stayed on in the cabin until the venison and beans ran out. They left the mountains, visited their wounded friend, now recovering, minus his arm, in a Missoula hospital, and sold their pelts.

The Phillips brothers went to Spokane and then back to British Columbia to work in the Black Bear Mine before heading to North Dakota for good-paying jobs in the wheat fields at harvesttime. They then traveled to Minneapolis and Chicago and to Indiana to work in a glass plant. Next on their itinerary was a visit to St. Louis before hiring on at a lumber camp in the backwoods of Arkansas, where their foreman used fictitious commissary charges to cheat the boys out of most of their wages.

As 1902 began, the twins went to Fort Smith and took railroad construction jobs in Oklahoma Territory and in Kansas, where they again became separated. They were reunited by chance in a Denver restaurant and immediately left for Cheyenne. A kindly railway freight engineer befriended the boys and insisted they accompany him on his run to a small

Wyoming town named Rawlins. To pass the time, the twins went to a saloon and broke the gambler's bank playing blackjack. They figured they had won enough money to return to Ogden and open a cigar store. But after going back to Utah to rent hotel rooms and buy new duds and pocket watches, they found their funds depleted. Always a gambler, Wiate took the pocket change from their last dinner bill, put it in a slot machine and hit the jackpot. He tried to increase their new winnings during an all-night gambling binge, but by dawn their small fortune was lost.

With one silver dollar each in their pockets, the twins went to Nevada to build culverts for the Central Pacific Railroad. Still hungry for the open road, they took their blanket rolls and headed over the Sierra divide to work at a brick plant in California and then to Portland for jobs on a riverboat. From Portland they went to Yakima and worked on an irrigation project and then to Sprague, Washington, for positions at a local hotel.

In Sprague, the abdominal pains Wiate felt months before in Portland returned. When the pain became severe, a doctor visited their hotel room and ordered the young man placed on a cot in a baggage car and shipped back to Spokane and the Sacred Heart Hospital. There, Waite learned his twin's problem was appendicitis. An immediate operation was ordered but, as usual, the boys had no funds. Waite wired his parents for help. Frank supplied the money and the emergency operation was performed. It came too late. On July 16, 1902, Wiate died of peritonitis. The twins' adventures in the West had come to a halt.

Waite, devastated by his brother's death, took the body back to Iowa. L.E. met the train in Council Bluffs and accompanied the forlorn Waite and their dead brother to Taylor County and the rest of the Phillips family. Wiate was laid to rest in the Gravity Cemetery and his twin returned to the farm.

While his younger brothers were off on their two-year nonstop tear across the country, changes were also occurring in Frank's life. Significant changes. Frank had finally decided the time was right to make a move. He sold off his barbershops, pocketed a profit, and took the banker's suit out of the closet. At last the flashy striped trousers went to the rag bin. By the turn of the century Frank was rightfully employed by his father-in-law, John Gibson.

Aware of his son-in-law's innate ability to sell, and with his own range of financial and investment activities extending well beyond Creston, Gibson believed Frank would make a crackerjack securities and bond salesman. After a brief initiation into the world of high finance at the venerable

Iowa State Savings Bank, where Gibson also maintained his own investment offices, Frank was handed his first major assignment, and it was not
an easy task.

A syndicate of Iowa and Chicago investors wanted to build a new
coliseum in Chicago. The group included Charles F. Gunther, a prominent
candy manufacturer; A. G. Spalding, the star baseball player who went on
to make a fortune selling sporting goods; and John Gibson. The proposed
building was ballyhooed as the city's finest events and sports palace. It was
to be a monument of civic pride. They planned to erect the coliseum on
South Wabash Avenue, on the site of the old Libby Prison building, an
infamous Confederate prison in Richmond, Virginia, that had been dis-
mantled brick by brick, shipped to Chicago, rebuilt, and opened in 1889 as
a museum filled with Civil War relics. The Libby Prison Museum was a
popular Chicago attraction, especially during the World's Fair of 1893, but
toward the end of the century people lost interest in the Civil War and
attendance at the prison museum had dwindled. The businessmen who
financed the museum wanted to turn their former tourist attraction into a
more profitable venture. Their plans took on new meaning on Christmas
Eve 1897, when the original Chicago Coliseum, located at Sixty-third
Street and Stony Island Avenue, was destroyed by fire.

The city needed a new convention and exhibit hall and the Libby
Prison Museum investors realized their building's site was ideal. By 1899
those original investors, joined now by John Gibson and others, formed
the Chicago Coliseum Company to build the new stadium. Gunther was
appointed president, Spalding secretary and Gibson treasurer. Estimated
construction cost was a half million dollars. Architects were commissioned
to develop a design for the coliseum, and the old prison walls were disman-
tled one more time. The huge wooden doors and some of the bricks were
given to the Chicago Historical Society, but some of the ornate wall on
Wabash Avenue would be incorporated into the new building. Everyone
was ready to begin construction. All that was needed was the money. And
that was Frank Phillips' assignment.

Gibson gave Frank a satchel packed with Chicago Coliseum Company prospectuses and a list of potential investors. He wished the eager
young bond salesman good luck and sent him on his way.

Frank traveled through the upper Midwest, checking names off the
list and perfecting his sales pitch as he moved eastward. The Panic of '93
ended two years before and belts were being loosened. This, combined with
the recent U.S. clobbering of Spain and the popular William McKinley

sure to win a second term as President, helped set the stage for anyone out trying to sell projects that reeked of patriotism and national pride. A spirit of optimism moved like an invisible fog across the country. Frank first felt that spirit in Ohio and Pennsylvania. It made it easier to put up with skimpy meals and furnished rooms.

Next he scoured New England—the nation's principal reserve of private investment. Wherever Frank went in his rented buggy, he was greeted with open doors. The New Englanders were not at all what he had expected. They were shrewd, and demanded to know all the facts, but their straightforward, no-nonsense approach to business appealed to Frank. Likewise, the Easterners found the zealous Iowa bond peddler refreshing. The stack of coliseum prospectuses decreased. In their place, Frank crammed into his satchel thick bundles of promissory notes and signed investment contracts.

He returned to Creston, Jane, and infant son John like a victorious warrior with his booty. Everyone was amazed. In less than six months Frank had managed to sell the majority of the bonds needed to finance construction of the coliseum. He was bursting with pride. Gibson and the other investors were delighted. They were only too happy to hand Frank a tidy $75,000 in commissions for his effort. The Phillips bank account was growing.

Construction of the coliseum began immediately. There were few problems, except in August 1899, when twelve immense iron arches supporting the roof tumbled like a row of dominoes and crashed to the ground, killing eleven workmen and critically injuring fourteen others. But work continued and within a year the grand building was complete.

On August 26, 1900, amid a monster military concert, the Chicago Coliseum—a rambling Gothic building covering two acres—was dedicated. President McKinley, little more than a year away from dying at the hands of an assassin, delivered an eloquent address to the cheering crowd. Cannons boomed in the distance and Frank's face flushed with pride. He was convinced this event was occurring because of his stellar salesmanship. It was the most satisfying day of his life to date.

There was little time to gloat. The famed Ringling Brothers Circus of Baraboo, Wisconsin, was booked to open the coliseum and inaugurate the first season. Frank was placed in charge of ticket sales, and his primary mission was to make certain that the owners, including his father-in-law, received their fair share of the take.

He was in his glory. After standing in the main ticket cage before each matinee and evening performance, peering over the ticket taker's shoulders, counting receipts, and taking the management's percentage, Frank was free to watch the Ringlings' three-ring extravaganza. It was like locking a drunkard in a saloon. Frank was intoxicated. The circus was one of his passions. Getting paid not only to watch the show but also to rub shoulders with some of the Ringling brothers was too good to be true.

The Chicago Coliseum was off to a fine start. It would soon become a favorite spot for sporting events and political gatherings. From 1904 until 1920 Republicans nominated five of their presidential candidates at the coliseum and sent three to the White House. Buffalo Bill Cody brought his Wild West riders to perform; and boxing matches, bicycle races, and marathon social bashes quickly followed. From the huge building's opening, society horse shows drew big crowds, attracting such luminaries as J. Pierpont Morgan; Robert Todd Lincoln, son of the fabled President; and Teddy Roosevelt's daughter, Alice. The project had paid off in big dividends.

Still basking in the glow of his part in the coliseum's instant success, Frank was soon summoned back to Creston. Gibson had new assignments and he was more convinced than ever that his son-in-law was the man to get the work done.

He gave Frank a book developed by George P. Ahern, an Army veteran in charge of the Forestry Bureau in Manila, called *Important Philippine Woods*. Ahern's book contained descriptions of the characteristics of the leading commercial timber species and spelled out the procedures for securing licenses to harvest the timber. Gibson smelled money. He recognized the potential worth of the millions of acres of virgin forest covering the islands.

Already successful in law, mining, banking, and real estate, Gibson knew that the importation of fine lumbers, especially mahogany, into the United States could be yet another bonanza. After reading Ahern's book and listening to Gibson's plan for establishing such a timber importation business, Frank had to agree. With the Philippines an official U.S. possession, and because of strong business and political connections—enhanced by Gibson's term of service as an Iowa state representative in the late 1890s—the establishment of the Philippine Lumber and Commercial Company went without a hitch. The new firm was headquartered in Chicago with Gibson as president and Stewart Spalding, one of the coliseum investors, acting as secretary. Frank remained in Creston. He worked out of the bank, taking trips through the Midwest and East to sell securities and promote the new lumber operation.

While Frank labored on the home front, occasionally shuttling between Creston and Chicago, Gibson, accompanied by Tillie, went to the Philippines to look after the firm's lumber interests. It was to be a year's journey. After concluding business in the Philippines, the Gibsons planned to visit Japan and China and then make a tour of Palestine, the Holy Lands, and Egypt, returning to the United States by way of Europe. Gibson also announced that once the trip was completed, he intended to resign from most of his vast business interests and spend more time in Creston.

The Gibsons bid friends and family farewell in March 1902. They traveled to San Francisco and in early April sailed on the steamship *Gaelic* into the warm Pacific winds. After a stop in Honolulu, they proceeded to Japan and then cruised through smooth seas to their next destination, Manila. Throughout their voyage, John and Tillie wrote of their homesickness in letters to Jane and their grandson, John. In one letter to four-year-old John, his grandfather described a visit to the Temple of Buddha in Tokyo, where the Gibsons wrapped their feet in cloth so their shoe soles would not touch the sacred floor. To sweeten the epistle, Gibson enclosed a check for five dollars, along with instructions so John could endorse it and deposit the money in his account. In another letter to John, sent after their arrival in the Philippines, Gibson sent more checks for both the boy and his mother, and again told his grandson how much he missed him. "I am thinking of you so much every day. I could not keep account of the number of times I think of you and your mamma every day. When I wake in the morning and before I get out of bed, I wonder how you are and whether you are well and asleep. You know when I wake about seven o'clock it is about 10 o'clock the night before with you. Time good boys are in bed. When I go to breakfast I wonder if John is taking oatmeal or shredded wheat biscuits or malta vita for his breakfast—and so it goes all day."

On May 15, in a letter Tillie wrote to the ladies of the Bancroft History Assembly in Creston, she described their arrival in the Philippines: "Considerable anxiety was felt on board the steamer in regard to the cholera which we knew was prevailing in Manila. We found on arriving that conditions were not so bad, no alarm seemingly among Americans. From 15 to 30 new cases per day with a fatality of 81 per cent. We are forbidden to drink any but distilled water and have but little fruit served to us, which is an affliction where fruit is so good. If we get home safe and all goes well

I suppose it will be a source of pleasure to know I have made this trip. The question has come to me over and over, why does anyone who has a good home with enough of this world's goods to be comfortable in a christian land willingly leave loved ones and subject themselves to sea sicknesses, Chinese cooks and what not. It at least adds new meaning to the words of the songs 'Home, Sweet Home' and 'America.' "

Tillie Gibson never returned to Iowa. Shortly after she posted the letter to her friends in Creston, she collapsed with violent diarrhea and vomiting and severe muscular cramps. The diagnosis was cholera. On July 17, Stewart Spalding, at the Philippine Lumber and Commercial Company offices in Chicago, received a cablegram from Manila. "Mistress Gibson ill. Condition grave. Advise children." Spalding contacted the Iowa State Savings Bank and before nightfall all of Creston was praying for Tillie. The following afternoon more word came. Another brief message from Spalding was delivered to Frank Ball, cashier at Gibson's bank. "Mrs. Gibson died of cholera. Body embalmed. John is quarantined." Ball, who was at his residence having lunch when the message arrived, left at once for North Hill to break the news to Jane Phillips. She could not be consoled. Frank was in Ohio on a business trip, so family and friends did their best to soothe Jane until he returned.

The newspaper account of Tillie's death reported:

> The shock which this news has caused to Creston can hardly be realized as yet, as the news is so recent. But the community seems paralysed to the point of being unable to do more than ask each other if the message can possibly be true, and if there is not a possibility of a mistake. The death of Mrs. Gibson and the blow which has fallen on this household has touched every heart in Creston and over a large part of the state of Iowa. Mrs. Gibson was one of the noblest women who has ever breathed the air of heaven and her charity and goodness was as universal as it was unostentatious. The hearts that have known her consoling sympathy in the hour of distress will bleed for her loss and the mourning which Creston will undergo in this hour is too deep for mere words to portray.

Within several days, Jane learned her father had escaped the infectious disease and was en route home. This news gave Jane something to look forward to and anticipate. When Gibson finally returned to Creston, Jane was greatly relieved. He told her that her mother had been placed in a tomb in the government cemetery in Manila. Gibson said it was a lovely resting place—a tropical garden of Oriental plants where only the songs of birds could be heard. Jane pictured in her mind the palms and orchids and climbing vines. She was comforted.

Life and business continued in Creston, and before long, Gibson was back in full form—rising early, as always, to read and memorize poetry and prose, and plan new ventures. Except for occasional business jaunts, Frank was at his side at the bank. But Frank was also getting restless again. Confident from his series of financial successes, he was itching to move into new areas and even take big risks with some of the money he had quickly accumulated. The goldfields in the Klondike caught his eye and for a time he considered heading north to try his luck there.

Then, in 1903, as he was returning to Creston from a business meeting in Chicago, Frank decided to take some time to stop in St. Louis and visit the World's Fair. Officially called the Louisiana Purchase Exposition, the fair marked the centennial of the purchase of the Louisiana Territory. Frank had been hearing about the fabulous fair for two years, as St. Louis business and civic leaders cleared acres of forest and invited nations from around the globe to participate. Although the World's Fair was not due to officially open until 1904, throngs of visitors, including several business-men from Creston, were already visiting St. Louis to watch the construc-tion of the palaces and exhibition halls. Frank was eager to get a sneak preview of his own.

As he roamed the fairgrounds, intent on finding the Iowa Building in the Plaza of States, Frank stumbled into Rev. C. B. Larrabee, a Methodist preacher from Creston whom Frank knew from frequent Sunday dinners at the Gibson residence. Brother Larrabee was recently returned from ministering to the faithful and trying to convert the unfaithful in Indian Territory, that sprawling hunk of wild country, just east of Oklahoma Territory, inhabited by cowboys and Indians and hearty pioneer stock but also infested with thugs, rustlers, tinhorn gamblers, bootleggers, whiskey peddlers, whores, and thieves. In Indian Territory every imaginable des-perado could find safe refuge from the law. For Larrabee, a man of God eager to serve as a missionary to the unwashed heathen, it was the ultimate challenge and he gleefully accepted the call.

Frank gobbled up the minister's accounts of Indian Territory. Larra-bee explained that the Methodists' main purpose was to share in the moral regeneration and advancement of the small towns where houses of worship could be built and souls saved. He spoke of helping another pastor, Rev. Jesse Morgan, gather tithes to build a church in a town called Nowata, the Delaware Indian word meaning "welcome," and he told Frank about Pawhuska, capital of the Osage Indian Nation. The stories of the colorful Indians and tough little towns were interesting, but when Larrabee described the town of Bartlesville and spoke of the fledgling oil businesses there, Frank's ears perked up. Bartlesville was where the first oil strike in the territories took place. Men were moving into the area, drilling holes in the prairies, and bringing up rich strikes of crude oil.

"The oil is flowing out of the ground like water," said Larrabee. "It's liquid gold and men are getting rich. Very rich."

By the end of the afternoon Frank had his fill of the World's Fair preview. He had seen and heard enough. He headed straight to Union Station and caught the first train bound for Iowa. All the way home, he scribbled notes to himself, trying to remember every word Larrabee used to describe Bartlesville and the oil operations there.

By the time he reached Creston, he had roughed out a plan of action. He ran from the train depot to the Iowa State Savings Bank Building and burst into John Gibson's office. The banker looked up from his pile of papers, but before Frank could utter a single word, Gibson saw in a flash a look in his son-in-law's eyes that told the whole story. Gibson had seen that same look in the eyes of miners who had sniffed out the mother lode and were about to strike it rich. He knew at once Frank had found a golden opportunity. Gibson shook Frank's trembling hand and took his hat. He motioned for him to sit down and told someone to send word to Jane that he and Frank would be late for dinner. Then Gibson closed the door and listened.

PART TWO

THE OSAGE
1903–1917

"The treasure which you think not worth taking trouble and pains to find, this one alone is the real treasure you are longing for all your life. The glittering treasure you are hunting for day and night buried on the other side of that hill yonder . . ."

—From *The Treasure of the Sierra Madre,* by B. Traven

CHAPTER EIGHT

It all boiled down to a single question—land. The Indians had it and the whites wanted it.

The issue of land in America was a point of contention from the start. The struggle over ownership of the land—a notion foreign to Indians—grew in the early colonial days, when Miles Standish was dispatching Indian warriors, continued through the American Revolution, and reached a full head of steam during the nineteenth century as land-hungry whites pushed westward.

Little by little the white pioneers encroached into what had been forever the homelands for countless tribes of Native Americans. In the early days of tribal relocation, the fledgling U.S. government set aside for colonizing tribes a vaguely defined area on the western margins of the Louisiana Purchase lands. It became known as "Indian Country," and by 1808 tribes from both south and north of the Ohio River began emigrating to the West. But before long Indian Country was greatly reduced in size by frontiersmen also intent on settling on those same lands.

Congress, fueled by cries of Manifest Destiny, approved the establishment of a frontier line to mark off an official area for the Indians in the so-called Great American Desert. In truth it wasn't a desert at all. This strip of land ran from the Platte to the Red River, west of Missouri and Arkansas in what would become the states of Kansas and Oklahoma.

A vast land of prairies and mountains crisscrossed with streams and broad rivers, the region was one of the most appealing portions of the Louisiana Purchase because of the rich commerce in furs, wolf pelts, and buffalo robes developed by French traders among the Osage, Quapaw, and Wichita tribes already living there.

It was wild land. Wild and free. For many it represented a last chance; for others it was a new beginning. For everyone it served as the junction of expansion to the West. It was the last American frontier and would become the catchall for the many tribes being driven from their ancestral hunting grounds east of the Mississippi River. The U.S. government named it Indian Territory.

For the most part, tribes from north of the Ohio settled the northern portion of Indian Territory, and tribes from south of the Ohio settled the southern portion. The Indian newcomers were moved into a land frequented by Plains Indian nomads. They were also joined by half-savage frontiersmen who married into tribes and built cabins and camps along the streams. Trading companies, especially the great Chouteau family of St. Louis, established remote outposts to carry on their business with the Indians.

The population of the area grew considerably starting in 1829, when President Andrew Jackson, the fiery Indian fighter, asked Congress for legislation to remove the Five Civilized Tribes of the South—Cherokees, Choctaws, Chickasaws, Creeks, Seminoles—to Indian Territory, where they would be guaranteed life, liberty, and the pursuit of happiness "as long as grass grows and the waters run." They had become known as the Civilized Tribes because many lived like whites in log cabins, wearing homespun clothes, tending livestock, and plowing their fields with teams of oxen. Some of these Indians, like their white brothers, even owned black slaves. But Jackson, a resolute patriarch, felt they had to go. Indian Territory would be their new home. In 1830 the Indian Removal Act was passed by Congress and the mass exodus from the Southeast to the West began, supervised by armed dragoons.

During the years it took to remove the Five Civilized Tribes, the Indians underwent great hardships. By 1838 the bulk of the proud Cherokee Nation had been rounded up, often at gunpoint, from their farms in Georgia and Tennessee and herded west. More than 14,000 Cherokees were relocated. Most were forced to make the 800-mile trek on foot. During the six-month journey one-fourth of the Cherokees died. Forever, this infamous forced march would be known as "the Trail of Tears."

Once settled in Indian Territory, each of the transplanted tribes established its own capital and carried on tribal business. All were recognized by the federal government as legitimate political entities, or "nations," capable of maintaining a separate existence as U.S. protectorates.

The boundaries within the Indian Territory would change frequently during the years, as land was divided and assigned to new tribes. And in 1854 the territory was greatly reduced by the creation of Nebraska and Kansas territories. More than sixty tribes would settle in the Indian colonization zone. All of the tribes were different. Some were peaceful farmers; others migratory hunters. Contrary to white beliefs, not all Indians were alike. Intertribal wars occurred, such as Osage attacks on Cherokee villages, and U.S. Cavalry troops stationed at isolated frontier posts made their presence felt. But as the years passed, and more of the good western lands were opened and occupied, white settlers, denied entrance into Indian Territory, crept into the area. Soldiers changed from being Indian guardians to Indian fighters.

When the Civil War erupted, the Five Civilized Tribes, still southern in spirit, sided with the Confederacy. The South went down in defeat, and the paternalistic federal government decided the tribes had to be punished. The Civilized Tribes forfeited many treaty guarantees. They were also forced to free all their slaves and some tribes had to grant the slaves freedom as Indian citizens with full property rights. The western part of Indian Territory was taken from the southern sympathizers and assigned to Indians transferred from other states.

New tribes moved in—more Osage, Arapahoes, Cheyennes, Wichitas, Kiowas, and Comanches. The newcomers took over the western half of the territory and the government built a string of forts to house more cavalry troops. In 1867, most Delaware Indians were moved into the Cherokee Nation from Kansas and a few years later more than two thousand Osages entered the eastern portion of the Cherokee Outlet, or Strip, an extensive ranching region along the Kansas border. Indian Territory quickly became known as the official Indian dumping ground.

But the land still appealed to outsiders. A good many whites coveted the lush Indian ranges where buffalo once fed. The post-Civil War cattle booms attracted giant trail herds from Texas ranches south of the Red River headed for railheads in Kansas. Between five and six million Longhorns alone trampled up the Chisholm, the East Shawnee, and other great cattle trails that ran the length of Indian Territory. At first, the Indians' tall grass only helped fatten the great herds. But before long Texas cattle barons managed to secure grazing rights, either by leasing land from the tribes in honest dealings or through devious means. During the early 1870s, railroads were built across the Indian lands and more settlers, many squatters defiant of tribal laws, opened up farms. The white immigration could not be stopped. With the rapid settlement of the neighboring states, the isolation of Indian Territory vanished.

Before the territory could be opened to white homesteaders, tribal title to the land had to be removed since the ownership of all the Indian land was vested in each tribe. To get around this obstacle and at the same time completely alter Indian culture once and for all, the government determined that the reservation system was a failure. It was ruled that all tribal governments would be abolished and the lands broken up. In 1887, Congress passed legislation providing for the division of the Indian reservations into small tracts. Government agents assigned each Indian—except members of the Civilized Tribes—a 160-acre homestead or allotment. The remaining land was declared surplus to be opened for settlement by whites.

The wave of the future was washing over the territory. Would-be settlers drew a bead on the two-million-acre region of "unassigned lands," taken from the Civilized Tribes, in the western part of Indian Territory. Many jumped the gun and had to be driven out. But the desire to move into these lands became irresistible and eventually white pressure grew too intense for Congress. President Benjamin Harrison proclaimed a "run" on the lands to start at noon on April 22, 1889. The lawless still didn't wait. Many of these "Sooners," as they became known, crossed the border before the soldiers could fire their pistols and blow the bugles signaling the start of the run. Towns sprang up in a matter of hours as thousands of homesteaders entered the territory on foot, on horseback, by train, in covered wagons, and even on bicycles and raced to stake their claims.

More land runs followed and in 1890 Congress acted yet again. The Organic Act was passed creating Oklahoma Territory. Coined from the Choctaw phrase for "red people," Oklahoma Territory was all the land in the western portion of the original Indian Territory.

A few years later Senator Henry L. Dawes of Massachusetts was selected to head a commission organized to persuade the Civilized Tribes to go along with the other tribes and exchange their common holdings in Indian Territory for individual allotments. The Dawes Commission traveled to Indian Territory and spent several years attempting to change the land system. But the tribal leaders had respect for tradition and clung to their common land ownership. Finally, after Congress increased its clout, the Dawes Commission was empowered to assign allotments to Indians without the approval of the tribal leaders. The Indians saw the struggle was futile. By 1902 all of the Civilized Tribes agreed to sign allotment agreements and every acre of land in Indian Territory was allotted to tribal citizens.

Statehood was just around the corner. But it was because of the creation of Indian Territory in the first place that Oklahoma's natural development from a frontier wilderness into an emerging state was stymied. It became known as an area of arrested growth; a limbo buffering the Midwest and West.

Still, it was a land with more than its share of excitement. Leaving the safety of the outside world for a jaunt through "Injun Territory" was not recommended. It was always intense; usually dangerous. This was especially true during the decades between the Civil War's end and the push toward statehood in 1907. It was one of Oklahoma's most colorful periods —the heyday of the Indians, cattle trails and cowboys, soldiers and outlaws. Lots of outlaws.

The lawless nature of the territory was provided by the swarms of desperadoes who rode into "the Nations" to find safe haven from the strong arm of the law. The list of renegades who made the territory their stomping ground read like a Who's Who of frontier crime. The James gang, the Youngers, the Daltons, the Doolins, Belle Starr—"the Bandit Queen"—and her lover Blue Duck, Ned Christie, and countless more found sanctuary in Indian Territory. Some had colorful names to match their tainted reputations. There was the "Verdigris Kid," the "Nighthawk Rider"—a sharpshooting lady bandit who rode with the worst of them and used a bullwhip with deadly effectiveness—and the notorious "Cherokee Bill," a mixed-blood outlaw whose actual name was Crawford Goldsby.

In the rugged hills and on the grassy plains where these outcasts, killers, and thieves found refuge, the main preoccupation was staying alive. But there was also time to nurse wounds and divide up loot, and opportunities for a wolf shoot when a hunter's moon appeared. Plenty of bootleg stills did a booming trade, turning out fresh batches of whiskey to help kill the poison in the night air. Murder, rape, robbery, and looting were common. Ruthless men, who would cut a baby's throat just to see if their blade was keen, traveled in packs. Gentlemen and ladies from St. Louis and Chicago did not venture into the territory, unless they were skirting the edges in a fast-moving train.

An immense wilderness, Indian Territory was bounded by Arkansas on the east, Kansas on the north, Oklahoma Territory and "No Man's Land" on the west and northwest, and Texas on the south. It was a criminal's paradise and a legal and jurisdictional nightmare. One newspaper explained that Indian Territory was "the rendezvous of the vile and wicked from everywhere."

In these parts absolutely nothing worked in favor of the law. The Indian tribes abided by their own rules. The tribal courts took no notice of the steady stream of horse thieves, prostitutes, whiskey peddlers, and bushwhackers pouring into lands where there were no courts for whites. The few peace officers that dared show their faces were often reformed outlaws, and some of the outlaws had been lawmen. A very thin line separated good and evil.

The common saying of those turbulent years in the territory was: "There's no God west of St. Louis and no law west of Fort Smith." No God, perhaps, but finally there came a judge.

For twenty-one years, from 1875 to 1896, Issac C. Parker, a stern Methodist from St. Joseph, Missouri, served as judge of the U.S. Court of the Western District of Arkansas. He represented the law west of Fort Smith and held exclusive jurisdiction over almost 75,000 square miles of Indian Territory and 60,000 people. During his tenure on the bench at Fort Smith, Parker did everything in his power to bring justice and order to the territory.

For openers, he rounded up more than two hundred marshals, armed only with tin badges, Colts, and courage. They were sent to Indian Territory with arrest warrants and orders to bring back as many outlaws as possible—dead or alive. Sixty-five, or one-third of "the Men Who Rode for Parker," as those intrepid lawmen were called, died themselves in the line of duty. The survivors fought on. Tough lawmen—Jim Rhodes, Charlie Colcord, Wiley Haines, and the three "Oklahoma Guardsmen"—Chris Madsen, Heck Thomas, Bill Tilghman—and many others, did their best against overwhelming odds.

If an outlaw made it back to Parker's courtroom, he found little mercy. During his reign of justice, the judge's gavel slammed down on more than 9,000 convicts. The lucky ones received long prison terms, but 160 of them were sentenced to hang. Of that number, seventy-six actually swung from a noose fashioned of well-oiled Kentucky hemp rope by dour George Maledon, a Bavarian immigrant called "the Prince of Hangmen" and the good judge's faithful executioner. Journalists from around the world came to Fort Smith to cover the trials and hangings. They dubbed the well-used scaffold "the Gates of Hell." Parker was good copy. The relentless jurist, who kept his court open from early morning until nightfall, six days a week, for more than two decades, earned a moniker all his own—"the Hanging Judge."

But despite the diligence of Parker and his spirited marshals, law enforcement remained an uphill battle. For every outlaw jailed or hanged, two more appeared. During the final decade of the nineteenth century, crime in the twin territories reached its peak before finally beginning a slow decline.

The last stronghold for the outlaws was concentrated in the northeast reaches of Indian Territory, present-day Oklahoma. Here, among the blackjack oaks studding the hills and canyons, desperate men could relax. Some local people, by their nature wary of authority and hostile to any outsiders, even gave shelter to outlaws or provided them with ammunition, fresh horses, and hot suppers. In the upper reaches of the territory, a deputy was lucky to get a kind word, much less find a friendly witness willing to go all the way to Fort Smith for a trial.

This area had long been the land of the Indian. In some ways it was a predictable place. Thick carpets of bluestem grass sprouted across the plains every April and grew tall from August to October. Prairie chickens danced in the strong summer sun and migrant bald eagles came each year to nest. The prairie teemed with quail, turkey, and deer and was on the fringes of the feeding grounds of the buffalo—the great monarchs of the plains. For many years Indian horsemen rode across the emerald prairie of bluestem grass, singing to their horses, on their way to buffalo hunts. They worshipped the buffalo as a sacred animal and praised its spirit around evening fires. The bison supplied everything the Indian needed to stay alive. Hide, horns, hooves, and bones were made into tools, weapons, clothing, and ceremonial objects. The buffalo's hump and tongue were choice cuts to eat. But the Indians didn't eat the heart. They believed the buffalo's heart contained mystical powers that would help regenerate the depleted herd. Hearts were always left behind. Unfortunately, the hearts didn't prove to be strong enough medicine. The huge buffalo herds eventually vanished.

But the tall bluestem grass remained. Even after most of the tallgrass prairie of the southern plains was plowed under or destroyed, the sweet grasses of the prairie in northeastern Indian Territory, just below the Kansas line, flourished and fed wild game, livestock, and spirited ponies. As far as the eye could see, stands of bluestem spread undisturbed. It was all that remained of a prairie that once ranged over a quarter billion acres.

For many years—long before the invasion of white settlers and outlaws—this distinctly American landscape served as home to an Indian tribe whose ancestral lands were along the wide Missouri. Early explorers found these Indians arrogant and imperious. They named them the Osage, the white version of Wah-Sha-She, the Water People, the Children of the Middle Waters.

The Osage lived in what became western Missouri and they roamed the Mississippi Valley and the Great Plains hunting buffalo and fighting other tribes. They were statuesque people and wore elaborate earrings of bone and shell, and adorned their bodies with tattoos to commemorate successful hunts and victories in battle. Skilled warriors, the Osage were feared by other tribes because of their willingness to fight, their prowess in combat, and, most of all, their fierce courage. When the buffalo herds played out, the Osage moved into new lands, including northeastern Indian Territory, which they eventually were forced to give up to make room for the Cherokees. The Osage were moved to reservation land along the Neosho River in east-central Kansas. But shortly after the Civil War, when Indian lands in Kansas were beginning to open to white settlement, the Osage reservation was sold and the proceeds used to buy 1.5 million acres of rolling bluestem grass from the Cherokees situated near the Arkansas River in the eastern end of the Cherokee Outlet. The Osage had come home to Indian Territory.

By 1872, more than 1,500 Osages moved back to the territory and were settled on their new reservation—an area where their fathers once chased game and rode to battle. Although spacious, their new land was mostly rough meadow and hill country, apparently fit only—as it had always been—for grazing. Most whites considered the Osage lands poor and not very desirable.

Osage soothsayers and men of magic knew the land was special and the hills were filled with mystery. They also believed that below the coal-black soil, locked in the deep recesses of the earth, was a treasure that could help the tribe if they held on to their land. One visionary said he saw, clear as the summer sky, the death of the old ways. He saw visions of more white men coming and he could even picture their strange machines, snorting and bellowing as if they were iron buffaloes. He said the white men would come like a wave of water. "You cannot stop that wave," he told the old Osage chiefs. They nodded and sat silent beneath the night sky.

Before long the Osage recovered from their forced move and molded their tribal government to fit the white man's system. They adjusted to the government boarding school and the Quaker schools and the Catholic schools and the missions. They occupied themselves with raising stock and crops and found that what were thought to be the poorest lands of any in Indian Territory were actually the best because of the rich bluestem grass. If the grass could tempt buffalo, it would also please hungry cows. As the cattle industry grew, creating a demand for more rangeland, the Osage became landlords of thousands of acres of pasture. All the while, interest was accumulating in the tribal trust fund established with the remainder of the $9 million they received from the sale of their lands in Kansas. The Osage were becoming prosperous—an unusual situation for an Indian tribe in America.

They settled in separate communities scattered throughout the Osage Nation, worked the land, and banked the proceeds from the pasturing leases divided on a per capita basis among the members of the tribe.

Word of the Osage success spread like prairie fire. Cowboys tending herds and outlaws headed for their favorite hideouts were likely to say they were "going up to the Osage." The land there had a certain charisma. It was different. Maybe even special. The only problem was that smelly dark ooze that seeped from some of the springs and covered the creeks with a thin scum.

It was oil. Pure crude oil percolating from the earth. Centuries before, the Plains Indians noticed the oil and scooped the greenish-black liquid from the springs or used feathers to skim it from tiny pools formed by buffalo hooves in the mud. They found it a good cure for rheumatism and rubbed the oil into cuts and sores on their hunting ponies.

When the Civilized Tribes arrived in the territory they also spied the oil trickling from beneath rocks and traces in the springs. Like the Plains warriors, they used the oil as a treatment for chronic diseases. When a large pool of oil was found, they'd spread an old blanket over the surface, soak up as much as possible, and then wring the oil into a pot. Some enterprising Cherokees even sold the stuff as a natural elixir. But besides its miraculous curative powers and using cattails soaked in oil for torches, the Indians had no other real use for oil. It fouled the streams for the stock and game and had little value.

In 1875, Jasper Exendine, a Delaware Indian cowboy, and George B. Keeler, a white trader married to a Cherokee, were rounding up stray cattle in the Osage Hills. It was the dry month of August and the riders turned toward Sand Creek to quench their thirsts and water their mounts. When they reached the stream the horses refused to drink and the men saw that the creek's surface was covered with something that smelled of coal oil. Neither man had ever seen crude oil before, but they knew what it was. Exendine frowned and pushed back the scum with a stick so the horses could drink. "Oil no good. Make water bad," he said. Keeler wasn't so sure. He had heard the stories about oil and understood it had more uses besides swigging for aches and pains and slapping on animal cuts.

Late that afternoon, after the cattle roundup was over, the two men headed home. All the way back, Keeler thought about the oil they found. Word of that discovery would one day be the main topic of conversation in the tiny settlement where he lived. Not really yet a town, it was little more than a trading post and gristmill on the banks of the Caney River. An old soldier and Indian trader owned most of the place. His name was Jake Bartles and they called the collection of buildings his village.

CHAPTER NINE

Frank felt the excitement brewing in Indian Territory even before he arrived. From everything he heard, it was the place for him to be. The sooner, the better.

Rich crude oil—black gold—was oozing out of the earth. With a little start-up capital and some luck, ambitious souls could get rich in a hurry.

The likelihood of making a fortune in the oil business was ripe. But other facets of the territory were also attractive. Frank believed all the right ingredients were present. The land possessed the potential for an honest living. There were also wide-open spaces and land stretching to the horizon, and enough cowboys and Indians to outfit a thousand Wild West shows, seasoned by a supporting cast of desperadoes, wildcatters, and gamblers. The land had an edge to it—always on the brink of danger, yet always alluring. It was definitely a place for risk takers.

The prospects left Frank eager to explore this new "promised land," and he brought this excitement into John Gibson's office in Creston. Gibson could feel it in the room as Frank recounted Brother Larrabee's description of Indian Territory. There was a passion in Frank's eyes and a wisdom in his voice that made Gibson's intuitive juices flow. He liked what Frank was telling him. Before they arrived home for their late supper with Jane, both men had booked rail tickets to Indian Territory. They would explore the possibilities for themselves. The Horatio Alger hero lurking inside Frank Phillips was off and running. He'd never look back.

Bartlesville in 1903 was located in the land of the Cherokees, less than a pistol shot from the border of the Osage Nation. The town was just beginning to realize the full potential of the petroleum industry. Although the land would soon become the oil man's dream, Indian Territory still loved its Wild West image. Cowboys—and often outlaws—were heroes. But the word was that folks willing to roll up their sleeves and do some work could make a nice living from crops, cattle, and, now it appeared, that dirty greenish scum as well.

The kerosene age was about to come to a close. New factories, trains, ships, and those machines Henry Ford was starting to turn out would usher in the long reign of a different fuel. Oil was due to become king. Bartlesville was destined to become one of the favorite realms in the kingdom. Frank Phillips would become a crown prince.

His first scouting trip to Bartlesville confirmed what Frank had hoped to find. Larrabee's stories and the reports of the oil boom in Indian Territory circulating through Iowa's southern counties were true. There had been little exaggeration. Oil derricks were everywhere. The town was filling with roustabouts and speculators and the feeling of prosperity was in the air.

Bartlesville was a young town—not much older than the oil industry itself. The founder, and town namesake, was an enterprising trader named Jacob Bartles. Son of Joseph Bartles, the man who established the first telegraph wire in New York City, the young adventurer worked on a Missouri River steamboat before settling in Kansas. After service with the 6th Kansas Volunteer Cavalry in the Civil War, Bartles married Nannie Journeycake Pratt, widowed daughter of Charles Journeycake, a consecrated Baptist preacher and chief of a remnant of the Delaware Indian tribe that had been granted equal rights in the Cherokee lands.

Persuaded by his Indian princess wife to move closer to her family, in 1873 Bartles gave up his prosperous Kansas farm and became the third white man to move into what was known as the Coo-wee-scoo-wee District of the Cherokee Nation. They first settled at Silver Lake, where Bartles worked for the Chouteau Trading Company, and then moved to a new homesite on Turkey Creek.

In 1875, Bartles bought a corn gristmill built several years before on the banks of the Caney River by Nelson Carr, the first white man to settle in the area through his marriage to a quarter-blood Cherokee. The mill was on the north side of a sharp horseshoe bend in the Caney near a natural limestone dam and close to trails used by the Indians. Bartles may have bought the buildings and the huge millstones that had been shipped all the way from New York, but the land itself belonged to the Cherokee Nation. All individual Indians, as well as the whites who bought their citizenship or married into the tribes, could erect all the buildings they wanted, but they couldn't buy the real estate. Land was owned only by the tribe as a whole.

Bartles quickly enlarged the mill to make flour, and hauled in a dynamo to produce the first electric light to glow in Indian Territory. Later he added a boardinghouse, a livery stable, and a blacksmith shop. A water system was installed and Bartles' home even had an indoor privy. Bartles cut ice from the frozen Caney River and stored it in sawdust for use during the sweltering summer months. A Swiss cabinetmaker was lured to the settlement to turn walnut logs, cut from the river bottoms, into fine furniture. Bartles' home became a social hub, and his cluster of buildings on the banks of the Caney, a center of commerce. Throughout the northern portion of Indian Territory, Bartles' village served as a mecca for early settlers who came to trade with the Osage, Delawares, and Cherokees.

One of Bartles' clerks, George Keeler—the young trader whose horse didn't like the taste of oil—watched his employer become a prosperous businessman. Keeler longed to be in that same condition. Born in Illinois and fluent in the Osage tongue and Indian sign language from his days spent buying buffalo robes for the Chouteau fur company, Keeler had visions of grandeur all his own.

He befriended William Johnstone, a young transplanted Canadian, who also worked for Bartles. Johnstone happened to be married to Bartles' niece, the granddaughter of Chief Journeycake. Aware that more people were settling on the south bank of the Caney, and much to Bartles' chagrin, Johnstone and Keeler quit their jobs in 1884, and the pair opened a competing store of their own directly across the river from Bartles. Each store tried to outdo the other, and a vigorous rivalry developed.

By 1896, the community south of the river was growing. Keeler and Johnstone's strategy paid off. More businesses and homes sprang up as more settlers moved to the area. Frank Overlees, another former Bartles clerk, already a newspaper publisher and owner of a lumberyard, built his own general merchandise store not far from the Keeler-Johnstone store. Doctors hung out their shingles, a drugstore opened, and children started attending schools. A railroad company was organized, tracks were laid, and bridges were built across the Caney. In 1897 the population of the settlement reached 200 and it was time to incorporate. Despite continuous haggling over the exact location, they named the town Bartlesville, for the early pioneer who had given so many their start.

Ironically, even though the village he helped create was named for

him, Bartles decided to move. He was disappointed that the competition
on the south banks of the Caney had attracted the lion's share of the
population, the post office, and the railroad depot. The determined Bartles
took a year—using block and tackle, winches, and a crew of stout men—to
haul his intact store and home on log rollers to the middle of a wheat field
he owned four miles north of Bartlesville. There, he founded the town of
Dewey, named for the resolute admiral and national hero who had recently won a glorious victory over Spain at Manila Bay. A few years later,
Bartlesville's first newspaper, *The Magnet,* still smarting over Jake Bartles's departure, rallied an effort to change Bartlesville's name to Petrolia,
a name, they editorialized, "of character and distinction." Another group
favoring a name change suggested calling the town Boudinot, after a
prominent Cherokee citizen. Both efforts fizzled and died natural deaths.
But the oil business wasn't dying at all. Far from it. Bartlesville was about
to take off.

It had been years since George Keeler's horse refused to drink creek
water covered with oil in the Osage foothills. In that time much had
transpired, but Keeler hadn't forgotten the experience. He did his best to
convince others that if there was oil in the area it would be a huge boost to
the local economy.

Still, the oil business had a hard time getting started in Indian Territory because all land leasing and development of oil wells in Indian lands
were snarled by ownership laws and government regulations. Some early
oil pioneers were quickly discouraged when they faced the snarl of tribal
and federal government red tape. The faint of heart didn't stay.

But not Keeler. He persisted and was able to convince his partners—
Johnstone, Frank Overlees, and a few others—to search for a commercial
vein of oil. They pooled their money and went to Tahlequah, tribal capital
of the Cherokee Nation, to see about prospecting for oil in the Indian
lands. It took several trips and many hours of negotiating, but they finally
obtained a lease for drilling on fifteen square miles of Cherokee territory.
The white men would be permitted to drill a well on each of the one-mile
sections in the sprawling lease. To seal the agreement, they posted a hefty
$100,000 bond and also agreed to pay a 3.5 percent royalty per barrel of oil
to the Cherokees.

John Overfield, a traveling salesman for a meat-packing concern, connected the novice oil men with a branch of his firm called the Cudahy Oil
Company. Overfield had been active in other Cudahy oil operations, and
he was picked to develop the lease and supervise the drilling sites at Bar-

tlesville. A contract was inked with McBride and Bloom, an Independence, Kansas, drilling outfit, and after two weeks of tugging and pulling, fourteen teams of horses moved the drilling rig seventy miles to the west bank of the Caney River near the horseshoe bend. Drilling began just fifteen days after the town of Bartlesville was incorporated. After several months there was no sign of oil, but the local boosters were not discouraged.

Finally, on April 15, 1897, after drilling to a depth of 1,303 feet and hitting what would become known as the "Bartlesville Sand," they struck pay dirt. When it was time to "shoot" the well, Jenny Cass, stepdaughter of George Keeler, was given the honor of dropping the "go-devil," the explosive torpedo used to set off the blast of nitroglycerin which would open the well and permit the flow of oil.

There was a muffled thud followed by an eruption of water, rock, and earth. Then, within a heartbeat, came a blast of oil. Pure crude, the column of oil flowed high over the top of the wooden derrick and rained down on the crowd of curiosity seekers gathered around the rig. Oklahoma's first commercial oil well was born.

Named the Nellie Johnstone No. 1, for William Johnstone's young stepdaughter on whose land allotment the drilling site was located, the gushing well was quickly capped since there were no refineries or storage tanks in the area. Also, the land was still under Cherokee control and the tribal government would not allow the sale or removal of oil to refineries across the border.

In effect, the oil business was shut down until legislation was passed which permitted each owner of a producing well to hold 640 acres for production.

The Cudahy Oil Company chose the 640 acres immediately surrounding the discovery well, which was estimated to be capable of producing 50 to 75 barrels of oil per day. Even after it was shut down, townspeople filled buckets with oil that leaked from the metal casing pipe. It wasn't until August 1903, after the laws governing the sale of oil were changed and the Santa Fe Railroad started shipping tank cars of oil to Kansas refineries, that the well was opened again and allowed to produce. In a half century of production more than 100,000 barrels of crude oil flowed from the Nellie Johnstone.

Before long Cudahy had more than 105 producing wells under lease on 4,800 acres, totaling an average daily production of 2,000 barrels a day. They shipped their oil to refineries in Kansas on the Santa Fe tracks, and the railroad erected facilities near the Bartlesville depot for loading tank cars. The Missouri, Kansas, and Texas Railroad, called the "Katy," built another rail line from Coffeyville, Kansas, to Oklahoma City, and on the southwest outskirts of town they constructed loading vats which moved the rich crude from pipelines to tank cars bound for the North.

By the time Frank Phillips and John Gibson first set foot in Bartlesville, the Nellie Johnstone No. 1 was practically old news. In the half dozen years since that first commercial well blew in, a lot of water had flowed under the Caney River bridge and thousands of barrels of crude had been sucked from the ground. Bartlesville was well on its way to becoming Indian Territory's first oil boom town. The sound of pumping wells echoed down the busy streets and out across the Osage Hills.

In May 1903, a local newspaper noted: "The city is again filling up with prospectors and oil men," and described one former resident who "abandoned his former haunts" and returned because "the odor of petroleum and the decidedly metropolitan air which are the distinguishing features of Bartlesville are irresistible to him."

The smell of crude oil was also proving irresistible to an assortment of promoters and schemers. That spring, a New York reporter toured the Bartlesville field and sent home dispatches claiming the oil in Indian Territory was superior to the Kansas product, and he wrote of the colorful Indians he encountered, including some Osage with shaved heads and scalp locks, wearing moccasins and blankets. "The best wells on the Osage are close upon the line of the Cherokee lands," he wrote. "We visited several wells and found one flowing fifteen barrels natural. Rattlesnakes are numerous and antidotes are forbidden fruit in this country."

But talk of rattlers and Indians with shaved heads didn't stop the flow of oil or halt the steady stream of newcomers. Each day, more and more oil men flocked to Bartlesville, many of them with visions of Titusville, Pennsylvania, swimming in their minds. The oil industry had its beginnings at Titusville on August 27, 1859, thirty-eight years before the Nellie Johnstone well blew in on the banks of the Caney. On that date, the world's first commercial oil well was successfully drilled by a former railroad conductor named Colonel Edwin Drake. The vision and determination of Drake launched the petroleum industry. Drake himself profited little from his accomplishment. But many others did.

Pennsylvania had a ready market, good transportation and storage facilities, and—best of all—no legal problems concerning ownership of the land to hinder oil men trying to obtain leases. More oil finds were located in neighboring states and the petroleum industry spread. When word of the early oil discoveries in Indian lands traveled east many of the same oil men who helped build and develop the industry in Pennsylvania, Ohio, and West Virginia headed west to Bartlesville.

The town was suffering from growing pains. No homes could be bought or rented and tents appeared on every spare hunk of real estate. The newspaper in nearby Vinita declared: "Bartlesville is a wonder. It is the dirtiest city in Indian Territory . . . the mecca for men who believe more in grease than grace. A boom town if there ever was one."

Frank felt a sense of urgency to become part of the boom. He didn't want to miss jumping on the oil bandwagon and allow an opportunity to pass by without at least having a chance to see if he could become an oil man. Satisfied with what they found in Bartlesville, he and Gibson returned to Creston, but Frank only stayed long enough to pick up a clean shirt and head for Des Moines. After he arrived, Frank checked into the Kirkwood Hotel and wired his brother L.E. and told him to join him at once. Frank had important news to share. He paced the hotel lobby and waited.

L.E., living with his wife and baby son, Philip Rex, in Knoxville, southeast of Des Moines, dropped everything and raced to meet his big brother. When he walked into the hotel he found Frank beaming from ear to ear. "I've got a proposition for you," said Frank. "Come here and look out this window. You see that contraption down there?" Frank pointed at an automobile sputtering down the street. "I think people are going to buy quite a few of these new buggies and they need gasoline to make 'em go. It may be the thing of the future. There might only be a few gas wagons now, but someday there will be millions of the things. Oil is only twenty-eight cents a barrel right now. But what will it be in ten years?"

L.E. had no answers. Not only that, he wasn't so sure he even knew what Frank was talking about. The idea of getting involved in yet another moneymaking scheme wasn't what L.E. had in mind. All the way to Des Moines he had been hoping Frank was going to come up with a solid business venture that L.E. could sink his teeth in and stand behind. He was dubious. Making the price of crude oil their guiding star didn't make much sense.

But L.E. listened on. He had no other choice. After all, he was flat broke. The Hawkeye Coal Company, where he served as treasurer and general manager, had proved to be an unsound venture. L.E.'s $1,500 investment had evaporated in less than two years and now he was out of a job. To make things worse, Node was expecting their second child. Pressure was mounting. Although the memory of his failure in the coal business was still painfully recent and the idea of exploring for oil sounded a lot like mining for coal, L.E.'s interest increased as his brother painted a picture of the possibilities that waited in Indian Territory.

Frank first told L.E. about his trip to the St. Louis World's Fair and his meeting with Brother Larrabee. In St. Louis, Frank had seen more motorcars than ever before in one place. Frank said the World's Fair had popularized the automobile and there was a single display of more than one hundred shiny gas buggies, including one remarkable auto that had traveled all the way from New York to be part of the Exposition.

Next he described his first trip to Bartlesville with John Gibson. Frank painted a picture of the forest of wooden oil derricks and the crowds of people rushing to the Indian lands to get their own share of the wealth. "We can be among them, brother," Frank said.

L.E. still wasn't sold on the idea. "I demurred strongly against going out into the Indian Territory, especially to engage in the oil business, for the very good reason that I didn't have any money," L.E. wrote years later in his personal memoirs. "Besides, I had the idea, which is still prevalent with some people, that the Standard Oil Company owned, by some sort of divine right, all the oil in the country and no one could be successful in venturing in that line of business."

Frank, not one to mince words, told L.E. that his way of thinking was plain foolish. He also reminded him of his options and asked if he'd prefer to go back to the farm to milk cows and cut cockleburs. Frank's points were well made and his eagerness was contagious. With really no other prospects to rely on, L.E. succumbed and decided to join in the grand plan. "Not having any choice . . . and seeing a possibility of adventure, as well as making money, I agreed to join Frank and go there to look into the possibilities," he wrote.

With his brother in tow, Frank returned to Creston and developed the next steps in their plan of action. "We were to furnish the ideas and the enthusiasm and someone else was to furnish the money," wrote L.E. Naturally, they turned to John Gibson.

He didn't let them down. Satisfied that his son-in-law's sales ability and force of will could make an oil operation succeed, Gibson agreed to help all he could. "Well, boys, we'll organize a company," he said. Gibson called on some of his associates from the Chicago Coliseum venture and explained the situation. Several wanted to get involved. "All right, we'll put in $15,000," Gibson told Frank, "and you boys sell the balance of the stock on a commission." It was a done deal.

The Anchor Oil and Gas Company was organized with authorized capital of $100,000 and the paid-in capital of $15,000. The Anchor offices were located on the eighth floor of the busy Merchants Loan and Trust Building on Adams Street in Chicago. Gibson was president; Homer K. Burket, president of the Valley State Bank in Omaha and Gibson's brother-in-law, vice president; Stewart Spalding, of the Coliseum Company, was made secretary; and Frank Phillips became the treasurer. Directors of the new company were Gibson, Burket, and Spalding, as well as L. E. Phillips; J. A. T. Hull, a congressman from Des Moines; John C. Martin, a wholesale merchant from Salem, Illinois; and H. D. Reeve, secretary of the Military Committee at the House of Representatives in Washington, D.C.

According to the company agreement, neither Frank nor L.E. was asked to put in a cent of investment. But they were required to sell the stock. At first, they had minimal success, but eventually they managed to sell off $50,000 between them, often taking their commission in Anchor stock.

Frank returned to Indian Territory and started buying leases. The April 16, 1904, edition of the Bartlesville *Examiner* noted: "Mr. Frank Phillips, of Creston, Iowa, son-in-law of an old friend of Reverend C. B. Larrabee, has been in Bartlesville the past week looking over the field and invested to quite an extent. He returns to Iowa firmly impressed with the business future of Bartlesville."

Less than two months later Frank was back, this time with Gibson and some of the other Anchor officers, for his third visit of the year. Each trip allowed Frank a deeper insight into the world of oil. He liked everything about it.

During one of his early solo visits to Indian Territory, Frank was met at the train by none other than Brother Larrabee himself, by this time pastor of the local Methodist church. Always the booster, Larrabee acted as a sort of unofficial chamber of commerce and made it a point to go to the train depot every day to welcome visitors and lease hounds. When Larrabee saw Frank stepping down from the train with his well-worn suitcase, the minister was ecstatic. His "missionary work" at the World's Fair had paid off—here was another good Christian businessman coming to Bartlesville. Larrabee grabbed Frank by the arm and dragged him across town to meet Albert Latta, publisher of the Bartlesville *Enterprise*. By the end of the afternoon, Frank not only signed up to have the newspaper mailed back home to Creston but also bought a year's subscription for Larrabee and transferred his membership to the preacher's church.

For the next several months, Frank stayed busy as a one-armed man in a corn-shucking contest. He sold stock and kept tabs on the leases earmarked for future development, and only took off a little time from his duties with Anchor when he was appointed a sergeant at arms at the 1904 Republican National Convention. It was held at a familiar site—the Chicago Coliseum. He liked hobnobbing with his Iowa and Chicago cronies, but as soon as Teddy Roosevelt won the bid for a second term of office, Frank went right back to peddling stocks and learning as much as possible about the oil.

More trips to Bartlesville followed. In early June, Gibson and several of the Anchor officers, including Frank and L.E., met in Creston and took off for Indian Territory to check on their investments. At Parsons, Kansas, they transferred to a stuffy fifteen-coach Katy train, crowded with Indians, salesmen, preachers, and gamblers for the last leg of the trip. Once in Bartlesville, the party inspected potential oil leases in the area and introduced themselves to local bankers and businessmen. Before leaving, they secured several more leases in what they hoped would turn out to be oil-rich lands north of town.

There was much talk in Bartlesville about George Keeler's latest success—the Keeler Number 5, a well developed by Cudahy Oil Company in the Cherokee side of the Bartlesville field, less than a half mile from the business district. Keeler's well required sixty quarts of nitro to shoot and was still flowing at a rate of 300 barrels a day, making it the best producer in Indian Territory, with the possible exception of some wells in the Osage Nation.

Whenever he was in Bartlesville sizing up opportunities, Frank took a room at the popular Right Way Hotel. Opened in 1898 by Frank Overlees and Ola Wilhite, for years owner of the fastest racehorse in town, the hotel was situated on a prominent downtown corner. The Right Way housed investors, drillers, tool dressers, and other members of the oil fraternity.

Frank appreciated the comaraderie he felt with the other guests at the Right Way. He enjoyed checking into the comfortable hotel, renting a horse and buggy, and riding out in the countryside to survey the untamed landscape. Frank found bustling Bartlesville—by now swelled to a population of more than 3,000—more appealing every visit.

Following a brief stop in December 1904, the Bartlesville *Enterprise* noted: "Frank Phillips, of Creston, Iowa, who has extensive interests in this field, and who, by the way, has been instrumental in turning a number of people towards Bartlesville . . . is as firmly imbued with the idea that investments in this field are as good a thing now as they were the first day he landed here, several months ago."

Back in Creston with Jane and son John for the Christmas holidays, Frank devoured the New Year's Eve edition of the Bartlesville *Examiner.* In the bullish front-page story, illustrated by a drawing of the city's newest hotel, the Almeda (which employed a real French chef), it was suggested that "the new year which begins tomorrow has much in store for the oil and gas metropolis of the twin territories, and the next anniversary of this publication will find here a town that has no leaders and few peers."

Frank sensed 1905 was going to be a good year. But, as always, timing was important, and he believed the time had come to stop fiddling with leases and glad-handing other oilmen. Frank was anxious to put a drill bit in the ground.

He was also ready to make a commitment to the town where he hoped to become a wealthy oil man. Frank moved Jane and their six-year-old son, John, to Bartlesville and rented rooms for his family at the Almeda. The day they arrived, Jane was greeted by the sight of a prone cowhand lying dead on the hotel floor. The man had been shot in an altercation. Jane braced herself. While the deceased was removed, she took John by his hand and went outside to take in deep breaths of air coated with the smell of oil. She seriously considered gathering up her child and taking the next train back to Iowa. But then Jane thought of her mother lying beneath the damp Filipino soil. If the beloved Miss Tillie could face danger, and even death, and go with her husband across the seas to a heathen land, then Jane could do little else for Frank. The time had come to leave tranquil Creston and be at her husband's side. Jane decided to gut it out. She told herself that Bartlesville was where she belonged.

But she didn't stay long. There was, after all, still young Master John to consider. While Frank planned his first oil well and supervised the construction of a proper residence for the family, Jane took her son on an extended visit to her aunt and uncle, the Homer Burkets, in Nebraska.

With his wife and son gone, Frank immersed himself in his work. There was no shortage of it. By February 1905, Frank and L.E. had rented a small office, hired a driller, and started the development of their first oil well—the Holland No. 1.

It was all wildcat—a well drilled in unproven territory. On June 23, the Phillipses struck oil. When Frank heard the news he screamed like a banshee and raced to the Western Union office to spread the word.

L.E. was back in Iowa, checking on his wife and their three-week-old son, Lee Eldas, Jr., when he received the telegram from Frank telling him that the well had been successfully completed. Still saddled with serious cash-flow problems, L.E. had been forced to borrow $100 from his father, close up his house in Knoxville, and move his wife and sons into his in-laws' residence in Bedford. When he learned of the oil discovery, L.E. was on cloud nine. But the celebration was short-lived. The Holland No. 1 didn't last long. Located on only a small "pocket" of gas and oil, the Phillips boys' first real oil well played out and dried up.

More bad luck followed. The next well they drilled also turned out to be a duster—a dreaded dry hole. They were disappointed but decided to keep trying. "We had had a taste and felt sure our fortune lay in that direction," said L.E. As if to show the gods—and especially brother Frank —that he hadn't lost any confidence, L.E. packed up his household and moved Node and his two little boys to Bartlesville. After all, the third time's a charm. Or so they say.

The Phillipses' third well came in dry as a desert wind. "It began to look," noted L.E., "as though there was no place in the oil business for anyone except the Standard Oil." Frank didn't agree. He wasn't about to let down John Gibson and the other Anchor investors. Instead of cutting their losses, taking what was left, and running back to Iowa, Frank decided to become like those other risk takers he so admired who made it to the top. If he could peddle Mountain Sage and convince people to invest in a building in Chicago and unproved oil land in Indian territory, then, by God, he could find oil.

There was just enough money left to drill a fourth well. This one would have to be a winner. Drilling commenced in late summer in the tallgrass about seven miles north of Bartlesville, just southwest of the trading center of Copan, a short distance from the fork of the Big and Little Caney rivers.

It was Delaware land, and the site was part of Anna Anderson's eighty-acre allotment. An eight-year-old Delaware girl, Anna lived with her grandparents about two miles from her property. Since Indian tribes were allowed to provide allotments of land to individual tribal members to lease for mineral mining, Anna's grandfather arranged for two allotments —one for himself and one for little Anna.

Frank held the rights to drill on the lease. He had sniffed out the site and, using the most reliable industry practice of the day, figured it was about as close as he could get to the nearest producing well. A rig was moved in and Frank and L.E. pitched pennies and prayed for luck while the bit pounded deep into the prairie soil.

On September 6, 1905, Lady Luck came through for the Phillips brothers. The Anna Anderson No. 1 roared to life. It was a gusher. By nightfall, the bluestem grass around the rig was black and greasy. The well came in making about 250 barrels of oil a day. Anchor Oil and Gas Company had struck it rich and so had the 100 or so shareholders who purchased more than $100,000 of additional Anchor stock. Without even knowing it, Anna Anderson had just become the richest Indian girl in the territory.

Frank's prayers were at last answered. His luck had changed. He'd taken the cards dealt him, played them well, and won the pot. Now he was in the dealer's chair and all the wild cards were for him to name. He vowed to never surrender the deck again.

CHAPTER TEN

The Anna Anderson No. 1 wasn't to be a flash in the pan. Frank sensed that immediately. Even while the oil, black as licorice candy, spurted into the sky, Frank was deciding where to locate his next well. And the one after that. He was ready to make up for lost time and wanted to drill as many producers as he could, as quickly as possible.

The Phillipses' drilling crews, drenched in oil and sweat, paused only long enough to allow a photographer to snap a portrait of the gushing rig. Jane happened to be back in town for a visit. She and Frank, joined by L.E. and Node holding their older boy, Phil, lined up, barely visible, in the far background of a dramatic portrait which shows the imposing plume of crude oil blowing from the top of the rig.

As it turned out, Frank was dead right about their turn of fortunes. The Anna Anderson truly was only the beginning. Everything seemed to be falling into place. In a phenomenal show of oil-field savvy, aided by a healthy dose of luck, Frank and L.E. managed to drill a string of eighty-one proven oil wells in a row without a single dry hole. All of them were winners. The Phillips boys were really on a roll. Years later, when he could afford to be cavalier, L.E. liked to say, with a touch of drama: "If we hadn't hit with the Anna Anderson, we'd have been back in Iowa."

Over his brother's dead body. Frank was confident that even if the Anna Anderson had turned out to be yet another dud, he and L.E. had an ace in the hole that would keep them afloat in Bartlesville.

Back during the scorching summer months of '05, as Frank and L.E. watched the good money of the Anchor investors disappear down the trio of thirsty dry holes, Frank began to take stock of the banking business in Bartlesville. Banks were familiar turf to Frank, after his short tenure as a director at the Iowa State Savings Bank in Creston and the several years he spent handling fiscal arrangements for the various Gibson enterprises. He was comfortable in the presence of fellow bankers and enjoyed talking to them about the world of finance—especially high finance.

But after meeting the men who managed the three banks operating in Bartlesville, Frank was less than impressed. He didn't care for their attitude, particularly when it came to financing oil ventures. They were willing to risk money on mules, crops, and hogs, but wouldn't lend on oil. Frank believed that conservative banking practices and procedures could sometimes be appropriate, but there were other occasions when a banker had to take a chance, especially if the emerging industry in the area was controlled, at least in part, by risk. And that was definitely the case when it came to oil.

Frank had grown accustomed to the dynamic John Gibson—a man willing to bet on a long shot after he had studied all the available facts and if he believed in the players. Gibson was an ideal role model. For years he was the one man from whom Frank took his cues; a teacher who nurtured the young entrepreneur's business skills and helped mold his future.

But by the time Frank was taking a serious look at the banking scene in Indian Territory, Gibson had again left the country. Anxious to personally oversee his growing Filipino lumber interests, Gibson married his widow's niece, Bertha, resigned from the bank in Creston, and returned to the Philippines. He and his new wife would remain there for the next several years. Although she wasn't particularly fond of her cousin, especially in the role of a stepmother, Jane reconciled herself to the fact that her father would have a companion to help him face life in the Philippines. She bid the newlyweds goodbye and prayed no tropical disease would take her father's bride.

When he realized his best source of counsel was gone, and after it became clear that there was no one with Gibson's foresight in Bartlesville, Frank determined that the role was for him to assume.

Following careful investigation, and several conferences with local businessmen, Frank and L.E. decided to open a bank of their own. This would be their ace in the hole—a backup, just in case the oil business didn't pan out. More important, Frank was convinced that a bank which catered to oil interests "would probably prosper with the new town." The Phillipses figured they could be both bankers and oil men. If the oil business in the territory was going to expand, it would take lots of capital. And if there was no capital, then someone had to be willing to extend credit to the oil operators. This was not the time, and definitely not the place, to be suspicious. The bankers who could roll the dice with the oil prospectors and hedge their bets, could come out smelling as sweet as a fresh vein of Osage crude.

In July 1905, Frank and L.E. organized the Citizens Bank & Trust Company with a capitalization of $50,000. They had to charter the bank under Arkansas banking laws, which were still being followed since Indian Territory had no laws of its own. It was said that the Arkansas banking statutes "countenanced practically every sort of business except selling whiskey to the Indians." That suited the Phillips brothers just fine.

Frank had no problem coming up with the money to buy his shares of stock. But L.E., still feeling the money pinch brought on by the ill-fated coal mining company collapse back in Iowa, was forced to borrow $1,000 from the Guttenberg State Bank in Guttenberg, Iowa, so he could purchase ten shares.

The two brothers matched pennies to see who would run the bank. Frank won. He became the bank president and L.E. was elected the treasurer and bank cashier. H. J. Holm, a likable businessman whom the Phillipses met on one of their first visits to Indian Territory (I.T.), was tapped to become the vice president. At the time, Holm was a city councilman, and also managed the Bartlesville Vitrified Brick Company—the outfit turning out hundreds of red paving bricks stamped with "Bartlesville, I.T." Holm also happened to be a veteran banker and for more than fourteen years had been the sole owner of the Higgins Bank in Higgins, Texas, a railroad town on the eastern edge of the Panhandle. D. L. Ownsley, with twelve years' experience at the Bank of Commerce in Louisville, Kentucky, was chosen secretary. Joining them to form the bank's board of directors was T. S. Ives of Guttenberg, Iowa.

In late August, the Phillipses bought a downtown lot at 107 East Third Street and started construction of the bank building. Only local workmen were employed, and except for some handsome mahogany veneer trim shipped from Gibson's lumber operation in the Philippines, the Phillipses saw to it that all the building materials came from their new hometown. Before they opened the bank doors, they wanted to be certain that they had the goodwill of the community, and that meant supporting potential depositors. It was a horse-sense strategy which would quickly pay off.

By autumn the finishing touches were being applied to the two-story bank, with its pressed-metal ceiling, three vaults, marble counters, shiny brass spittoons, and a fireplace in the rear offices. There was even a telephone booth installed beneath the stairway leading to second-floor office space, where Frank would oversee the oil business.

That October, just weeks after the Anna Anderson hit and a couple of months away from officially opening the bank, Frank scribbled a letter on the new Citizens Bank & Trust Company stationery to son John, still staying with relatives in Nebraska. In the rambling two-page epistle, he praised John for his legible handwriting, inquired about the boy's marble collection, and described life in Bartlesville at Uncle L.E.'s house, where Frank now lived and enjoyed playing with his two nephews. "I put men to work on our new house today and expect to have it all ready for you and your mother soon as she arrives from the West," wrote Frank. "You must not get lonesome. We will make up for our absence from each other when we are together again and it won't be long I hope until this can be realized. Write me again John, I love to read your letters. With a squeeze and kisses for the best lad I know, I am with love your devoted Pa."

As Frank had predicted—and hoped—Jane and John soon returned to Bartlesville. Before long they were able to move into the family's first residence, located on a lot on the corner of Tenth and South Johnstone Avenue. The property alone cost Frank $2,600, but it was worth every penny. It was a comfortable two-story frame house with a front porch large enough to accommodate a hammock. On pleasant evenings Frank stretched out in the hammock to do his after-dinner thinking, enjoy a cigar, and tell Jane about the day's events in the hustling, bustling oil patch.

John, who lugged along his bowl of goldfish and huge collection of agate shooting marbles from Nebraska, was enrolled in nearby Garfield School and appeared each Sabbath in his Little Lord Fauntleroy suit at the Methodist Sunday school. To make up for the time they were apart, Frank gave his eight-year-old son a pair of roller skates and bought him a pony, which he hitched to a fancy buggy for trots through the neighborhood. If he couldn't be with John as much as he wanted because of the demands of the oil field and the bank, then Frank was determined to lavish as many gifts as possible on the boy.

Jane was glad to be back with her husband. She stayed busy setting up the Phillips household and making new friends. After spending her life under John Gibson's roof, and on extended visits with various relatives, she was pleased to at last have her own home.

Shortly after Frank's family was settled in their residence, the Citizens

Bank & Trust was ready to open for business. The *Examiner* instantly labeled the new bank "the finest in the city" and described in detail the exquisite mahogany trim and fine furniture. An advertisement in the newspaper announced the bank's opening and boldly told the citizens of Bartlesville that Citizens Bank "Solicits a Share of Your Business."

But, on December 5, 1905, when opening day arrived, Frank and L.E. were nervous. They had good reason. In some of the business circles in Bartlesville, the establishment of a new bank meant competition for the old reliable bankers whom everyone knew and trusted. There was considerable talk about the newcomers who were muscling their way into the community—drilling oil wells and now opening a bank. A few locals referred to Frank and L.E. as "those Iowa upstarts."

The time of reckoning finally came due. Dressed in their best suits, stiff white collars, and dark banker's neckties, the Phillips brothers and their tellers anxiously watched the front door for signs of life. At nine o'clock sharp the first customer walked in, went straight to a teller's cage, and plopped down hard cash to open a new account. More depositors followed. Lots more. The bank's front door almost wore out by noon.

Chief Baconrind, the principal leader of the Osage, accompanied by a retinue of squaws, was one of the opening-day visitors. The Chief, his skin the color and texture of dried tobacco, stalked the dimensions of the bank. He managed only a few nods as he inspected the furniture and fixtures, smoothed his palm across the cool marble counter, and peeked inside the telephone booth. After a brief tour, Baconrind paused for just a moment and his eyes locked with Frank's. When the young banker didn't flinch, Baconrind gave a soft grunt of approval. Then, without uttering a single word, the Osage delegation paraded out. It was only the quiet beginning of a relationship which would last for decades and would make both the tribe and Frank Phillips enormously wealthy.

At the close of the first day of business, Frank scrambled to lock the door and pull the shades. L.E. hurriedly tallied the day's deposits—$80,000. A princely sum for a couple of upstarts. Frank and L.E. were as thrilled as the day the Anna Anderson roared to life. They felt like they had hit another gusher.

Despite some lingering predictions that "those young squirts from up north won't last a month," the Citizens Bank & Trust prospered. Within only sixty days, Citizens was so strong that the Phillipses decided to deposit their own money in the bank. As the Phillips bank attracted more attention, and more importantly, added more depositors, the town's other banks grew more resentful. It didn't take long for Frank and L.E. to figure out that the other three banks had joined forces, intending to drive Citizens Bank & Trust out of business.

The competition never stood a chance. The odds and the tide of public support were definitely with the Phillipses. In addition, their shrewd emphasis on building a solid base of oil customers instantly brought results. Oil men, many who knew Frank and L.E. from staying at the Right Way Hotel or else met them out in the oil patch, flocked to the new institution. Oil operations required a large capital investment which called for constant financing and refinancing. The Phillips brothers were only too happy to accommodate. Ironically, one of the competing banks would eventually fail and the other two would be bought out by the Phillipses. The oil industry was expanding, and so was the new bank. It wouldn't be long before the Phillipses would rule the banking market in Bartlesville.

Still, the bank demanded a great deal of energy. Because of the Department of Interior regulation stipulating that no single company could hold more than 4,800 acres of Indian land, it became necessary for the Phillipses to resort to creative camouflaging of the books. As a result, the Phillipses soon had fifteen different corporations established, all of which had their own oil-lease property requiring their own individual set of books. Frank concentrated more on the oil ventures, while L.E. devoted most of his time to keeping the bank's books by day and working on the oil company's books in the evening. This meant abiding by an incredibly hectic schedule, with the workday starting at the bank at 7 A.M. and lasting until at least 11 P.M. and often 2 A.M. Sleep became a luxury. Young John Phillips' gifts piled up.

In order to provide some relief, more employees were hired, including Fred J. Spies, an acquaintance from Creston who had worked for John Gibson at the Iowa State Savings Bank. Spies had impressed Frank, and the young man was appointed assistant cashier at Citizens.

Shortly after joining the bank, Spies received an Indian Territory baptism of fire when a lad named Ernest Lewis strolled up to a Citizens teller and pulled out a pocketful of oil royalty checks. Spies had heard some of the tales about the bandits in the area, and when he spotted the fourteen notches carved into the handle of Lewis' six-gun, he became concerned. Reassurances that Lewis also had about the same number of oil leases didn't comfort Spies. He went straight to Frank's office to warn him about the "undesirable" customer.

Frank heard out Spies before he examined the royalty checks that

Lewis wanted to deposit. He didn't see any problem. It appeared to be a straightforward piece of business. Notches on the butt of a pistol didn't particularly bother Frank. He looked for the color of greenbacks and that's exactly what he saw in this instance. "His money is as good as anybody's," Frank told Spies. "Remember, we want depositors. We're bankers."

There were more dealings with the desperate elements that called Bartlesville home. The first customer seeking a loan at Citizens was a hesitant young Cherokee half-breed who approached L.E.'s desk one morning. "I'd like to hire some money. Five hundred dollars."

L.E. glanced up. The young man before him, hat in hand, had clear eyes the color of crude oil and a strong jaw. He looked like a hard worker, but still L.E. and Frank had already decided it was going to be bank policy to at least ask a few questions of borrowers who wanted more than a hundred dollars. Never anything too embarrassing of course.

"What's your name?" asked L.E.

"Henry Starr."

"What do you do?"

"Oh, I just operate around here."

"Do you have any collateral?"

"What the hell's that?"

L.E. explained and the young man gave him the names of a pair of Bartlesville businessmen who Starr claimed would "go the note" with him. Satisfied with all the answers he received, L.E. wasted no time in drawing up the simple loan papers. The young man signed and L.E. handed him $500 cash.

It wasn't until the next day that L.E. found out that his first loan customer, Henry Starr, was one of the most notorious outlaws operating in Indian Territory. The nephew of the infamous Belle Starr, young Henry liked to boast that "I've robbed more banks than anyone in America." Twice he was sentenced to die on "Hanging Judge" Parker's terrible gallows and twice he escaped, thanks to a Supreme Court reversal and a presidential commutation. Starr, who could rob two banks in the same town in less than an hour, thrived on the bandit life. "It tingles in my veins," he said. "I love it. It is wild adventure."

Wild adventure indeed. When he learned his loan customer's identity and what he meant when he said "I just operate around here," L.E. was horrified. He felt like a cheese maker dealing with a rat. Frank didn't share his brother's concern. As always, he looked at the practical side.

"How much did he borrow?" asked Frank.

"Five hundred dollars," gasped L.E.

Frank looked at Starr's signature on the note and the names of the two references he gave. "Don't worry about it. He'll pay it back."

Frank was right again. The note was paid in full on the due date and Starr never robbed their bank. In fact, none of the Phillips banks were ever robbed. A remarkable record, considering that for many years there was at least one bank robbery per day in Indian Territory. Frank bragged that he never lost a penny in dealing with oil operators or outlaws. This gave rise to the popular theory that outlaws robbed other banks and deposited their money with the Phillips boys, something Frank and L.E. neither confirmed nor denied.

Frank had the West in his blood and was used to rough-and-tumble characters. He had met more than his share during the years he spent working in the silver-mining districts and in railroad camps. Many of the early roustabouts he hired to work on his drilling rigs were of the same cut —profane, salty, and tough as boot leather. To Frank, the men of the oil patch, and even many of the outlaws operating in and around Bartlesville, were colorful characters—pure entertainment. He relished their lust for life and took pleasure from their yarns about the women, and often the lawmen, they were trying to forget.

Frank knew many of the local outlaws maintained their own code of chivalry which included abiding by the basic tenet of "live and let live," particularly if that feeling of respect was mutual. The fact that he was comfortable around desperadoes, and at times could even enjoy their company, was clear from the start.

One of the first early-day outlaws to strike up an acquaintance with Frank was Henry Wells, a bank and train robber, born in Virginia but reared in Missouri—breeding ground for the worst bandits and killers alive. Wells moved to Indian Territory shortly before the turn of the century and made his home in the wild and rugged Osage Hills.

A top rider and crack marksman, Wells could head shoot a hawk with a rifle. At different times during his checkered career, he rode the outlaw trail with Henry Starr, Frank Nash, Joe Davis, and Al Spencer. It was Spencer who showed Wells the hideout he himself used out in the Osage. It was an ideal spot—a rock ledge surrounded by thick brush. Spencer's hideout wasn't far, as the buzzard flies, from a tiny community called Okesa, the Osage word for "halfway," because it lay midway between Pawhuska, the Osage capital, and the Cherokee Nation on the east.

Wells liked the lifestyle in Okesa, home to bootleggers and others who preferred that snooping marshals and sheriff's deputies keep their distance. In these parts, a thief or renegade could convalesce from a bullet wound in peace. A fruit jar brimming with "old bang head" fresh from a still cost under a half-dollar and there were whole tribes of folks scattered throughout the hills and thickets who had always lived their lives just a step across the line from the law. Wells would eventually choose a spot on Lost Creek, just southwest of Okesa, for his home. His cabin was built so one man could guard the only pass and from the vantage point on the top could see for twenty miles in any direction.

Although he was living less than ten miles from Bartlesville, Wells remained true to the local outlaw code and resisted killing his skunks too close to the house. That meant passing by the temptation of paying a call on the banks of Bartlesville, especially those controlled by the Phillips brothers. Word was that Frank Phillips gave everyone a fair shake. Why, he even lent money to Henry Starr. In Wells's book, Phillips had to be all right.

But more than Indian chiefs, outlaws, and wildcatters came to the Phillips bank. Anna Anderson, the Delaware girl who owned Frank's first successful oil lease, was a frequent visitor. Still living on the farm not far from Dewey, little Anna grew weary of being pestered by guardians who feared she'd become a target for kidnappers. She was also tired of going to town and hearing people whisper, "There goes that rich little Indian girl." What Anna did enjoy were her horses. She liked going to Bartlesville, by herself, in a buggy pulled by her favorite mount, for an afternoon of shopping. When she made her purchases, she always went to see one of her guardians, Frank Phillips.

"After I shopped I had to go see Mr. Phillips and get him to okay what I bought," said Anna. "I'd get a bunch of new clothes and one of the clerks would have to walk with me to the bank to get Mr. Phillips to sign for them. I liked him but I was afraid of him. And I didn't like walking through that bank and worrying because he'd scold me because I'd bought too much. I never got over that worry but it was silly. He never ever got angry. He was always kind."

Although he was gaining in popularity with most everyone in Bartlesville—from little Indian girls to known bank robbers—Frank still had a lone-wolf reputation when it came to the oil business. He played his cards as close to his chest as possible. A few critics said he was "too damned independent" for his own good.

"Frank didn't know an oil well from a badger fight," said Albert

Latta, the newspaper publisher, who met the banker years before through Brother Larrabee during one of Frank's early scouting trips to Bartlesville. "But in sixty days he was around the curve, and in six years he was out of the sticks and in the big time. For an Iowa product, that boy sure did travel!"

In 1906, Frank stayed close to home—rigging and drilling, opening and closing his string of successful oil wells, and helping out L.E. at the bank. In February, when his wife's brother John suddenly died in Creston, Frank accompanied Jane to Iowa to arrange for the funeral and burial in the Graceland Cemetery. Later that year, Lewis Phillips visited his sons in Bartlesville and returned to his farm feeling proud of their accomplishments. Always prepared to counsel his boys, Lew left his oldest son a bit of advice and his initials, L.F.P., written on the back of Frank's business card —"Enmity to none, justice to all." Words to live by, thought Frank, especially when doing business in Indian Territory. The year 1906 also brought another family member—young Waite—to Bartlesville. He arrived in the spring but came for more than a visit.

Since burying his twin three years before, Waite worked on the family farm until Frank and L.E. arranged for him to take a clerk's job with their sister Jennie's husband, R. W. Coan, owner of a grocery in Gravity, Iowa. Waite lived at his sister's home and worked at the store until 1903, when he left Gravity to take a short commercial course at a business college in Shenandoah. It wasn't until years later that Waite found out that his older brothers not only paid his tuition and living expenses while he went to school but had also paid for his wages when he worked at Coan's grocery. They figured it was a smart investment.

After completing his schoolwork, Waite moved to Knoxville and worked as a bookkeeper for L.E.'s floundering coal company until the spring of 1904. The World's Fair was beckoning and he moved to St. Louis for a brief stint as a clerk at the Southern Hotel. Satisfied that he had soaked in as much of the color and excitement of the Fair as possible, he returned to Iowa and continued working as a salesman for the coal company until it became insolvent.

Following a few preliminary visits, Waite was persuaded by Frank to move to Bartlesville and join his brothers in the oil business. He would be shown the ropes and could help manage the oil firms Frank had established. At first, Waite lived with Fred Spies, the young Iowa man working at the bank. They took rented rooms on the third floor of a boardinghouse across the street from Citizens Bank. But soon Frank had his kid brother

out in the oil patch working on the Anna Parady lease, a site maintained
by Phillips & Company, one of the brothers' many corporations. Sometimes Frank put Waite to work for the Anchor Oil and Gas Company's old
reliable, the Anna Anderson. It never mattered much to Waite where he
was assigned. He had much to learn.

When he first went out in the patch, Waite was as green as new corn.
One day, he became so involved with drilling a wildcat well that, while
laying pipe, he had a sunstroke and was found unconscious baking in the
sun. He recovered and chalked it up to experience. Like Frank, he was a
quick learner. He submerged himself in the business of oil. Waite lived
right on the lease property in the Caney River bottoms during the week,
and spent every Sunday at L.E.'s place. By working and living on the lease
premises, Waite quickly grasped the fundamentals and rapidly rose to field
superintendent of all the Phillipses' new oil properties.

As soon as winter blew in off the prairie, Waite retreated to the comfort of Frank's home and worked out of town. He joined Frank in his
buckboard on inspection tours of the leases. He watched every move Frank
made and listened to the old-timers working the rigs. In no time, Waite
became a first-class oil man. Frank told L.E. that their investment in
young Waite had already paid off.

As L.E. increased his role in the banking business, Frank and now
Waite were free to pursue their growing oil ventures. Before 1906 came to
a close, and Waite had put in a full year in Indian Territory, the brothers
decided it was time to found yet another oil firm. They named this one the
Lewcinda Oil & Gas Company, a name honoring both their parents, Lewis
and Lucinda Phillips.

Statehood talk was beginning to dominate the barbershops and saloons, and the Phillips brothers felt certain their luck and success would
continue with Lewcinda, especially if they broadened the scope of their oil
operations and moved into new lands.

But instead of chasing rainbows, Frank created a rainbow all his own.
It started in Bartlesville in his "dinky trust company" and ended out in the
Osage, where outlaws and Indians roamed free as deer and a mother lode
of rich oil slept beneath the tall bluestem grass. The land there was ripe.
The old Indian myths about treasures in the earth intrigued him. And
when there was just a sliver of new moon hanging in the night sky, Frank
thought he heard the siren sound of the Indian drums and flutes above the
steady noise of the town's pumping rigs. His dreams let him see the possibilities. The Osage, its fields of tallgrass waving against the wind, beck-
oned. Frank could not resist.

CHAPTER ELEVEN

Frank Phillips wasn't the first oil man to sniff out the oil fields of the Osage. Not by a long shot. That distinction belonged to others. But although he might not have been the first, Frank wasn't far behind the early Osage oil pioneers.

A family of bankers from back East, by way of Kansas, named Foster were the first ones to pursue the potential of the oil-rich Osage lands. The Fosters were active in the Osage when Frank was still cutting hair back in Iowa. Others had discovered oil in the Osage before them, but the Fosters were the first to develop the area's oil possibilities.

Patriarch of the clan was Henry Foster, a Quaker banker from Rhode Island. In 1882, Henry, joined by his brother, Edwin Bragg Foster, moved their families to Independence, Kansas, where they acquired a thriving bank and established themselves as prominent businessmen.

Henry and Edwin further boosted their fortunes when they successfully promoted and built a railroad line from Kansas City to Coffeyville. Henry was now in a position to buy several cattle ranches, including one spread which extended over the Indian Territory border into what had become the Osage Nation.

In 1891, after Congress passed legislation allowing Indian tribes to lease their lands for mineral purposes, John N. Florer, a trader and Indian agent at Gray Horse in the Osage Nation, took it upon himself to help the Osage tribe get their just rewards. At a time when Indians couldn't put much faith in the promises of whites, Florer was an exception. He was trusted by the Osage, who had shown their white confidant the oily scum floating on the streams and gurgling from seep holes. Florer realized the crude oil was worth more to the Indians than just as a grease for wagon axles.

After the Osage leaders gave him their consent to find the right party to lease the Indian land for oil exploration, Florer met Henry Foster through a mutual banking friend in Kansas. Already familiar with the Osage potential because of his ranching operations, Foster was interested in Florer's proposition and agreed to lease the Indian lands and search for oil.

On March 16, 1896, James Bigheart, the principal chief responsible for organizing the tribe under a written constitution, signed, on behalf of the Osage Nation, a $300,000 lease agreement with Foster. The contract allowed for prospecting and drilling oil and natural gas wells on the entire Osage reservation for the next decade and provided for a royalty to be paid to the Osage of one-tenth of all the petroleum taken from the earth and an additional $50 for each gas well discovered. By distributing 10 percent royalties equally among the tribe members, the Foster brothers gained exclusive rights to produce gas and oil. Wildcat wells had been drilled in Indian Territory since the early 1870s, but nothing on the scale the Fosters were proposing had ever been attempted.

With the groundwork laid, Henry Foster went to New York to arrange for financing. There he caught pneumonia and died at the Waldorf-Astoria in January 1896, two months before the lease was approved by the Secretary of the Interior. The Osage Council then granted Edwin Foster permission to substitute his name on the lease, and by April, Foster became the lessee of the entire Osage reservation, covering 1,470,559 acres—twice the size of his home state of Rhode Island.

Foster immediately organized the Phoenix Oil Company, under the laws of West Virginia. Phoenix would serve as Foster's operating company, with the late Henry's young son, Henry Vernon Foster, called H.V., acting from afar as treasurer. Fresh out of college, H.V. wasn't interested in coming to Indian Territory. For the time being, he was perfectly content to remain in Wisconsin, where he worked as a civil engineer.

The Fosters started drilling in the Osage during the summer of 1896. First results were less than encouraging—nothing but dry holes. Then, in the spring of 1897, after the Nellie Johnstone—the area's first commercially profitable well—started flowing in the Cherokee Nation, hopes were revived. The Phoenix drilling crews moved their rig to the banks of Butler Creek, only two miles from the Nellie Johnstone, just inside the Osage reservation. On October 28, 1897, the drillers brought in a well producing about 20 barrels a day. Not exactly a major gusher, it was, however, the first productive oil well in the Osage. Within a month, another well was flowing nearby. Osage production had begun.

But hopes for the future were short-lived. After seven dry holes out of the next eleven wells drilled, all oil operations in the Osage blanket lease came to a halt. Drilling was expensive and not very productive over such an enormous lease. Osage oil operators were also faced with the same problems as the drillers in the Cherokee nation—no storage facilities, no transportation to Kansas refineries, and no market. It appeared to be a stalemate.

The Fosters' interest in the Osage blanket lease turned their attention toward finding a means to ship the crude oil to the nearest refinery at Neodesha, Kansas. Valuable help came in 1899 when Bartlesville town boosters persuaded the railroad to extend a line to the south side of the Caney and the following year a two-inch pipeline was laid from the Phoenix wells to the depot in Bartlesville. The first oil from the Osage lease was shipped out of Bartlesville to the Kansas refineries in May 1900. Oil was selling for $1.25 per barrel, minus two bits for handling charges. This first commercial sale of crude oil started the flow of millions of dollars into the Osage tribe coffers. A total of 6,212 barrels of oil was shipped from the Osage in 1900 and another 10,536 in 1901.

Late that same year, the Phoenix Oil Company and its subsidiary, the Osage Oil Company, merged into the Indian Territory Illuminating Oil Company, better known for many years to come as ITIO. The new firm was named for kerosene used for home lighting—still the single most important petroleum product at the time. ITIO was incorporated in the state of New Jersey with a capital stock of three million shares at a par value of $1 per share and boasted nearly 400 wells making money in the vast acreage of Indian Territory.

Then, in 1902, Edwin Foster died. The heir to the huge Osage lease was twenty-seven-year-old H. V. Foster, still happy as a civil engineer in Wisconsin and very reluctant to get involved with all the problems surrounding the largest blanket lease for oil and gas exploration ever issued.

In the process of reorganization, a series of power plays and takeover attempts took its toll of the new company. There was dissension in the ranks and ITIO drifted in a state of disorder until finally litigation resulted in the company falling into receivership.

But in May 1903 the gloomy picture started to brighten for ITIO when H. V. Foster, the heir in absentia, was elected president of the company. Unable to sell off the lease, he knew he had to "fish or cut bait." And so, at last, H.V. took the reins of the firm and moved to Indian Territory, and Bartlesville, the town he had even misspelled in one of his business letters.

The oil company Foster presided over in 1903 was on the verge of financial ruin, but he stuck it out and used his native intelligence and honesty to rescue the huge lease. Old-timers from those early boom days, including Frank Phillips, described H. V. Foster as a man who could always be trusted. "You don't need a contract with him," they said. "He never breaks his word. If he says he'll do a thing, you can take it to the bank."

In little time ITIO and another early oil firm, the Almeda Oil Company, were selling oil out of the Osage lands. In 1903 the tribe received $3,800 in oil and gas royalties. But that was just the beginning. During the following year there were more than 100 producing wells in the Bartlesville area, and in the nearby Osage another 250 ITIO wells were pumping away. Petroleum exploration rapidly gained momentum.

By the summer of 1904, while Frank Phillips scouted the area, the Dawes Commission was nearing completion of its assignment of allotting the lands of the Five Civilized Tribes to tribal members. As a result, the Department of the Interior began to be more open to oil and gas lease applications from oil prospectors, and drilling activities were no longer limited as they had been in earlier years. This relaxed government policy concerning oil and gas leases would have a profound impact on drilling operations in the territory. In addition, there were now pipelines and transportation available to carry the crude to refineries at a reasonable cost. The lure for black gold was paying off. New finds—the Avant, Boston, Wiser, and Okesa pools—were opened northwest of Bartlesville and others were on the verge of discovery.

Prairie Oil and Gas Company maintained a six-inch pipeline running from Bartlesville and Ramona, in the Cherokee Nation, to Independence, Kansas. From there, the line stretched through Kansas City, to Indiana, and then on to outlets on the Atlantic seaboard. Larger pipelines were planned and pumping stations were under construction to funnel the crude from Osage wells into Bartlesville. It was an excellent time to be a banker or oil man in Indian Territory. Frank and L.E. had the wisdom, and good fortune, to be both.

Osage royalties had jumped to $108,934 by 1905 and were still climbing. Eventually, thirty-seven individual pools would be uncovered in the

Osage. The zealous Brother Larrabee and others were continuing to spread the gospel of oil as well as the area's success story far and wide. The stampede had begun.

Realizing that no one company could fully develop 1.5 million acres of the Osage during the balance of the ten-year lease, H.V. proposed to wait until the lease was up for renewal, and then divide up the land. ITIO would retain 680,000 acres in the eastern producing lands. Also, H.V.—in an effort to keep his word and gain more of their trust—would increase the Indians' royalty, reduce his firm's royalty, and give back the 717,000 acres of undeveloped land to the Osage tribe. He further suggested that his company then sublease to other oil men for a nominal bonus of one or two dollars an acre.

Foster made his bold proposal in March 1905, exactly one year before the blanket lease was due to expire. Despite much opposition, the new lease was signed on March 7, 1906. The Secretary of the Interior refused to approve the new agreement, but Congress voted to endorse it anyway. Foster had gained the Osage oil and gas rights for another ten years. The new agreement also covered subleases that had been made with other oil companies, including Barnsdall, Prairie Oil and Gas, and Uncle Sam. It now called for an increased royalty of one-eighth of the oil and gas produced and $100 a year for each gas well in operation to be paid to the Osage Nation. Congress had also passed what became known as the 1906 Allotment Act. It was supposedly a measure to protect the Osage, but many Osage came to believe its true purpose was to separate individual Indians from their land and mineral rights. The act gave each of the 2,229 enrolled members of the Osage tribe 659 acres of land and one share in the mineral trust. Individuals were free to sell their surface acreage but the mineral rights were held in perpetual trust by the tribe, a fact that would have great significance when the reservation became one of the great producing oil fields of the world. Only shareholders were allowed to vote in tribal elections, a concept that worked fine as long as all 2,229 members of the tribe were alive.

Because of their shrewd negotiations for the leasing of all their oil lands, the Osage became the wealthiest nation per capita in the world. In 1906 more than 5 million barrels were produced on the reservation. The "barren" Osage land was as special as the old soothsayers promised.

With the oil market opening up even more, many of the big names in oil and financial circles beat a path from the Bartlesville depot up Third Street to the Right Way Hotel. The Osage became the scene of renewed drilling activity, attracting hordes of eastern operators and speculators, all ready to sublease tracts from ITIO. Oil royalties to the Osage were increasing each year, with ITIO retaining a royalty of one-fifteenth interest plus the bonuses paid from oil actually produced.

To come up with a system for designating the subleases, ITIO divided the eastern portion of its blanket lease into three tiers, called "Lots" or "blocks." There were 348 of these Lots, each one a half mile wide north to south and about three and 3/4 miles long east to west, reaching from the Kansas border on the north to the chocolate-colored Arkansas River on the south. Lot 1 lay along the Kansas line and the rest of the blocks were numbered consecutively southward, so that Lots 33, 34, 35, and 36 fell on the western edge of Bartlesville and Lot 116 extended another forty miles south. Within the next few years, leases on the bulk of the Osage acreage would be sold by bids at public auction in Pawhuska.

Many of the leases on the initial lots around Bartlesville, and the ones in the Osage, were snatched up by the Phillips brothers and other independent operators. The Phillipses' Lewcinda Oil & Gas Company would be joined in exploring the Osage opportunities by the Patton Company, Sand Fork Gas and Petroleum, Carter Brothers Oil, Colonial Oil, Uncle Sam Oil, Minnehoma Oil Company, and many others. ITIO would grow in wealth, as would the Osage tribe, and, following more major oil-field discoveries, H. V. Foster would be called for the rest of his life "the richest man west of the Mississippi," a title he despised.

More new oil fields were being opened around Bartlesville and Dewey, and the Hogshooter Pool, only fifteen miles southeast of Bartlesville, was opened in 1907. It yielded as high as 500 barrels a day during initial production, and gas wells started off with a production of from 5 million to 15 million feet per day.

But the oil discoveries in the Cherokee and Osage Nations weren't the only good news in Indian Territory. Just to the south, in the Creek Nation, was Tulsa. This dusty little one-horse hamlet, for many years called "Tulsey Town," was perking to life. A sunbaked burg frequented by outlaws, Tulsa primarily served as a pit stop for the cattle trails and a meeting place for the Indians who traded there. It boasted nothing more than a dirty depot with a one-block-long dirt main street lined with a few ramshackle storefronts, and by the turn of the century supported a ragtag population of little more than 1,300. Tulsa didn't have much going for it. It was a town that looked to the past rather than the future.

But that was just about the time that Tulsa's quest for oil—which had been launched in other parts of Indian Territory thirty years before—began in earnest.

Thanks to Fred S. Clinton and John Bland—a pair of local physicians intent on bringing some attention to the cowtown they called home—Tulsa suddenly rocketed into the national limelight. It occurred shortly before midnight on June 24, 1901, in the Creek Nation just four miles west of Tulsa at Red Fork, known as "one of the vilest spots in the territory." There on the homestead of Bland's half-blood Creek wife, Sue, the two doctors' oil well—the Sue Bland No. 1—came in. "We decided on a rational development of the community and the state, with oil as the magic lure," said Clinton.

It was a good plan. For even though the well soon played out, word of the discovery swept across the land. The Tulsa *Democrat* headline proclaimed: "A Guyser [sic] Of Oil Spouts At Red Fork" and the story reported that the well "gushes gas and oil alternatively and the oil is unsurpassed west of Pennsylvania." All eyes turned to Tulsa and, within twenty-four hours, oil men from around the country were descending on the place.

During the next two years the Tulsa–Red Fork area grew rapidly. And to ensure that Tulsa would keep growing and would not be cut off from development by the broad Arkansas River, private citizens built a toll bridge with their own capital. Next they invited oil men from back East to pay a call. They also tried to attract some of the oil speculators and operators crowding into Beaumont, Texas, where in 1901 the wild Spindletop oil well was gushing 100,000 barrels a day. Tulsa city fathers invited the horde of oil men to "come and make your homes in a beautiful little city that is high and dry, peaceful and orderly. Where there are good churches, stores, and banks, and where our ordinances prevent the desolation of our homes and property by oil wells."

Oil men took Tulsa at its word. They came in packs and a building boom followed. Four years later, Tulsa really soared when two wildcatters, Bob Galbreath and Frank Chesley, leased the Ida Glenn farm about ten miles southwest of town and drilled the Ida E. Glenn No. 1, the discovery well of the fabled Glenn Pool—not just another major oil field but the richest the world had yet seen.

Galbreath and Chesley had drilled the well on a shoestring. They were playing it so close that they even sank their wildcat without using any casing. But their bean and corn bread meals were over. Oil came in with an explosion on November 22, 1905. Overnight, Galbreath and Chesley became millionaires and were forced to hire guards armed with shotguns to

keep prowlers off their claim. Black gold fever raged so intensely and the
oil flowed so fast and furious that storage tanks couldn't be built fast
enough to keep up. The Glenn Pool literally turned into a "lake of oil."
Tulsa's future was assured.

From the Osage in the north through the Creek lands and beyond, the
Indian Territory oil boom proved to be the last great mineral rush in the
American West—a stampede for riches that had never been rivaled before.
Soon pipelines were opened to the Gulf of Mexico and hotel and office
buildings were erected. Streets were paved. More banks were built. The
drillers, the roustabouts, and the other working stiffs of the oil patch lived
throughout the Creek Nation and up through the Osage into Kansas. But
the bosses—the men who gave the orders, paid the wages, and watched the
price of oil spiral upward—lived in Tulsa and Bartlesville and Ponca City,
a town near the sprawling 101 Ranch on the western edge of the Osage
where E. W. Marland, a Pennsylvania wildcatter with a "nose for oil and
the luck of the devil," managed his empire. Although oil was later discovered farther west, the real activity of the boom years was concentrated in
northeastern Oklahoma around these three important centers. They were
the cities where the successful oil firms established their headquarters;
where the operating money was banked and a lot of it was spent.

The big winners almost invariably came from humble beginnings. A
few, including H. V. Foster, got a head start in life, thanks to previous
family holdings. And still others were like Frank Phillips, who got a considerable boost from John Gibson's advice, not to mention a few critical
and healthy monetary transfusions. But for the most part, the most successful of the early oil operators accumulated their fortunes the old-fashioned way—by working hard and taking risks. Almost every one of
Frank's friends and associates from the oil patch were in this select company. To a man, they were all heavyweights.

There was big and burly William G. Skelly, who founded his own oil
company and became a prominent Tulsa citizen. Bill Skelly was a teamster
in the Pennsylvania oil fields in the early 1900s when he came West to try
his luck as a producer and struck it rich in Texas' famous Burkburnett
Pool. Within two years Skelly was independently wealthy and in seven
years he built one of the largest independent oil firms in the world.

In later years, the Phillips brothers were frequent guests at the oil
man's palatial Tulsa home, a twenty-five-room residence with five fireplaces of black-and-white Italian marble and a lush front lawn perfect for
mint julep parties. This was the domain of Skelly's wife, Gertrude. Sur-

rounded by servants and manicured gardens, she was the same woman who once lived in an oil-field shack and who never forgot the afternoon she was washing dishes in her worn dress and patched apron when her husband rushed in, grabbed her in his arms, and shouted: "Gertie, I've struck oil and now we're rich!"

Josh Cosden was another who tried his luck in the oil fields and won. Cosden had been a streetcar motorman in Baltimore when he tossed aside his uniform and came out to Indian Territory.

When he first arrived, he drove a tank wagon and his wife, like Skelly's, stayed home and cooked in the family's tent. But Cosden's wife didn't fair so well. After her husband became a successful producer and built a refinery at Tulsa and the money started to roll in, she was given the heave-ho so Cosden could marry a younger, socially prominent woman in Tulsa. He built a magnificent home with a $10,000 tennis court, using imported French clay, and threw lavish parties. Before long, he had a second home on Long Island, racing stables at Newport, and a private yacht at Palm Beach. He mixed with the Astors, the Guggenheims, and the Whitneys. The Prince of Wales was a lunch guest. Cosden built up a fortune of more than $20 million. Then his wealth began to fade. He eventually lost his oil company and had to sell off his mansions. By 1925 he was broke. Within two years, however, he made a recovery and accumulated another $15 million. He made a comeback and was able to pay off some debts, but Cosden was never the same. The Depression was the crowning blow. Josh Cosden—the generous and charming "Prince of Petroleum" who managed to lose not one, but two immense fortunes—died a broken man on a train in 1940 at age fifty-nine.

Cosden wasn't the only early oil baron who wasn't able to keep his riches. W. F. "Billy" Roeser was another. A classic poorboy-makes-good, Roeser made it big at the Glenn Pool and became a millionaire before he reached his thirtieth birthday. He soon found that the only way he could deal with his wealth was by spending as much money as he could in as little time as possible. In 1907 he purchased the finest house in Tulsa and threw elaborate parties. He bought a fleet of fancy automobiles, handed out money to relatives and friends, and backed countless wildcatters. After pouring most of his money into ventures that yielded nothing but dusters, Billy Roeser was busted. Like Cosden, he made a comeback, helped by other oil men who hadn't forgotten Billy's generosity, but before long he gambled everything on a wildcat lease that didn't pan out. Roeser had gone through fifty million dollars, and he was finished.

Harry Sinclair became another member of big-time oil's inner circle and an early associate of the Phillips brothers in Bartlesville. Following an uneventful childhood, Sinclair took over his father's drugstore in Independence, Kansas. He was not successful. By 1901 he had spent most of the money left him and was forced to close the store. His neighbors said Sinclair didn't have an ounce of business sense.

In 1903, still adrift before selecting a new career, Sinclair went to the prairies around Ochelata, a railroad community near Bartlesville, to do some rabbit hunting. While cleaning his shotgun, Sinclair accidentally discharged the weapon and his left foot was shredded by the blast. Some farmers loaded the young man on a handcar and rushed him to Bartlesville. After he was given opiates and returned to Kansas, part of his foot had to be amputated. Sinclair was at his lowest. He felt like he'd been snakebitten all his life.

But suddenly a silver lining appeared. It came in the form of a $5,000 check that Sinclair collected from his insurance policy. The hunting accident turned out to be a godsend. With the money, Sinclair moved to Bartlesville, set up housekeeping with his bride in a new house on Johnstone Avenue, and bought "mudsills," the big logs used as foundations for oil derricks. The venture quickly proved profitable and he began to negotiate leases on the oil-bearing lands in Indian Territory. He developed the Canary Field just south of Caney, Kansas, and had one hundred wells producing by 1905. Two years later he moved to Tulsa, invested his profits in the Glenn Pool, became an oil lease broker, and bought wells himself. Sinclair's success was phenomenal. Within several years he headed sixty-two oil companies and was a partner with his brother Earl in a Tulsa bank. Sinclair went on to serve as the chairman of a multimillion-dollar consolidation and own a Kentucky Derby winner, as well as a share of the St. Louis Browns baseball club.

In Bartlesville, Sinclair became acquainted not only with the Phillipses and H. V. Foster but with another local oil pioneer—T. N. Barnsdall, head of one of the earliest companies to drill in the Osage. Barnsdall had watched Drake drill the world's first oil well back at Titusville, Pennsylvania, and Barnsdall's father drilled the next well after Drake and launched the world's first refinery. Barnsdall moved to Indian Territory, formed the Barnsdall Oil Company, and led in the construction and operation of natural gas distribution systems.

Then there were the Gettys—George and his legendary son, Jean Paul. Just before Frank first arrived in Bartlesville, George Getty, a Minneapolis insurance executive and attorney-turned-oil-man, came to town and invested in extensive oil-field properties.

Captivated by the allurement of Osage oil, and pleased with the discoveries, Getty decided to move to Bartlesville. Although a near-fatal bout of typhoid caused him to reject Methodism and become a devout Christian Scientist, Getty, like the Phillips brothers he came to know and respect, was also from a strict, puritanical Methodist family. And like the Phillipses, he too was a frequent guest at the Right Way Hotel, where, in 1904, he moved his wife Sarah and son Jean Paul while the family residence on Osage Avenue was under construction.

Born in Minneapolis in 1892, Jean Paul, or J.P. as he came to be known, was only eleven, just a few years older than John Phillips, when he moved to Bartlesville. With his dog Jip by his side, J.P.'s first job was hawking *Saturday Evening Post* magazines to Bartlesville's oil men and their ladies. An inquisitive lad, fond of pulling neighborhood pranks, Jean Paul also had something in common with Frank Phillips—he loved to devour Horatio Alger novels and he longed to be an Alger hero. During an early trip to Bartlesville before leaving Minnesota, young Getty also was fascinated to watch his astute father buy $500 worth of oil leases from H. V. Foster in a new 1,100-acre tract west of town called Lot 50. Before his twelfth birthday, J.P.—subsidized by magazine sales and with some help from his father—made his first investment. He purchased one hundred shares of stock at five cents a share in the Minnehoma Oil Company, a name combining the Getty's former home state and the Indian name of their new business.

As 1903 came to a close, George Getty struck oil on Lot 50 at 1,400 feet. The well came in as a gusher and provided a steady production of one hundred barrels a day, making Getty an instant success. By 1905 Minnehoma had six wells producing 10,000 barrels per month on its Lot 50 lease in the Osage. The older Getty's drilling crew, which for a time on Lot 50 counted outlaw Henry Starr in its number, stayed busy. But petroleum prices were falling. Only two years before, crude sold for $1.03 per barrel. In the spring of 1905, the price dropped to 52 cents per barrel and was headed even lower. George Getty took advantage of the lull and in 1906 moved his family to the oil fields of California.

J.P. wouldn't stay away too long. A few years later, after he graduated from high school, young Getty went back to Minnehoma Oil's Lot 50 to work for several weeks as a roustabout. He hadn't forgotten his earlier trips to the lease with his father—a two-hour drive in the heat and dust over rough, potholed dirt roads which J.P. called "a great treat."

Lot 50 was important to the Getty family. Between 1903 and 1915, the leases there netted $426,000. George Getty finally sold the property in 1916 for $120,000 cash, just before the fifteen-year "golden age of Oklahoma oil" when prices ranged from $1.20 to $3.25 per barrel. J. P. Getty would eventually settle in Tulsa, but he'd always remember those early oil properties—especially Lot 50, where the original Getty fortune, which would one day make him the richest man in the world, first flowed from the land of the Osage.

But while some oil men—like the Gettys and the Fosters and the Phillipses—were winning vast fortunes, others—such as Cosden and Roeser—lost everything they had. Oil fortunes could vanish as quickly as they were made. Oil was a high-stakes gamble and the losers greatly outnumbered the winners. "Luck can always turn bad" became a basic truth in the oil patch. And not everyone had the luck, or the foresight, of Frank Phillips and his brothers. Those who did turned up big winners.

This band of independent oil barons, operating in the backwoods of Indian Territory, became known as history's best gamblers in the middle of the greatest gamble the world ever saw. They were rough and tough individuals involved in a business where heartbreak and ruin kept close company with wealth and success. They were men who would be kings, living in a time when anything was possible.

Frank was always proud to be part of this brotherhood of wildcatters. He and his brothers were pleased with their oil operations and their growing bank. Generally they liked Bartlesville even though the town was still untamed, as a newspaper story of the day noted: "It is to the shame of the city of Bartlesville that among the police force one ex-marshal is now in the Federal jail awaiting trial for murder; another, a deputy, so conducted himself as to cause his wife to blow out her brains, and still another, only a few days ago, tried to murder his wife, and failing, blew out his own light."

By the close of 1906, murder and mayhem still provided plenty of fodder for conversation, although both Jane and sister-in-law Node stuck to other topics as they attended endless piano recitals and social gatherings hosted in Bartlesville's best parlors, each one decorated with ferns and lace doilies. Besides socializing with the other ladies of Bartlesville and staying active in several local whist clubs, Jane still enjoyed getting out of town now and again. An ardent traveler all her life, Jane always found time for trips to California or Colorado and for visits with relatives in Nebraska

•

Frank Phillips, president of Phillips Petroleum Company. Official sixty-sixth-birthday portrait by Raymond P. R. Neilson. The November 28, 1939, celebration was one of the largest parties ever staged for an individual. *(Phillips Petroleum Company)*

TOP RIGHT:

Lucinda Phillips, pictured during a visit to Bartlesville just prior to her death in 1934. The mother of four girls and six boys, she lived to see three of her sons become successful Oklahoma oil men. *(John Gibson Phillips, Jr.)*

TOP LEFT:

Lewis Phillips, patriarch of the Phillips family, was seventeen years old when he enlisted as a private in the Union Army during the Civil War. He fought throughout the war and returned to Iowa, where he wed Lucinda J. Faucett on July 3, 1867. Frank Phillips kept this framed photograph of his father on a shelf near his bed at Woolaroc. *(John Gibson Phillips, Jr.)*

LOWER RIGHT:

John Gibson, successful Creston, Iowa, banker and businessman, was father-in-law, employer, business underwriter, friend, and mentor to Frank Phillips. *(John Gibson Phillips, Jr.)*

The Phillips clan at a family gathering at Lew Phillips' farm near Conway, Iowa, during the summer of 1899. Top row: Frank's father, Lewis; his sister, Lura; and his mother, Lucinda. Seated, center row: Frank's twin brothers, Waite and Wiate; Frank's brothers, Ed and Lee Eldas (L.E.); Frank; Frank's wife, Jane, holding their only son, John; and Frank's sister, Nellie. Seated, bottom row: Frank's brother, Fred, and his sister, Jennie. The family dog, Spot, is perched behind the twins. *(John Gibson Phillips, Jr.)*

Frank Phillips, far right, standing in one of his Creston, Iowa, barbershops in front of a rack holding customers' individual shaving mugs. Besides cornering the tonsorial market in town, Frank invented a hair treatment he dubbed Mountain Sage. Its principal ingredient was rainwater. *(Phillips Petroleum Company)*

Frank Phillips and Jane Gibson on their wedding day in Creston, Iowa, February 18, 1897. *(John Gibson Phillips, Jr.)*

Already successful as barber, entrepreneur, and business owner in Creston, Iowa, Frank Phillips journeyed to Bartlesville, Oklahoma, where he and brother L.E. became prosperous bankers and oil men. Frank is pictured here (top row, fourth from left) amid banking colleagues in 1911. *(John Gibson Phillips, Jr.)*

The skyline of Bartlesville in 1903 was punctuated by wooden oil derricks. One newspaper of the day called it "the dirtiest city in Indian Territory . . . the mecca for men who believe more in grease than grace." Oil drilling was outlawed by city ordinance in 1909. *(Phillips Petroleum Company)*

Blowing in a gusher on September 6, 1905, the Anna Anderson No. 1 was Frank and L.E.'s first big producer, leading a string of more than eighty-one drilling winners. Frank, Jane, L.E., Node, and Phil Phillips are barely visible in the far background of this dramatic portrait. *(Phillips Petroleum Company)*

RIGHT SIDE:

Checking operations at an early lease, Frank Phillips (center), one of many oil wildcatters in the Bartlesville area, became known as "Uncle Frank" to his hardworking roustabouts in the oil patch. *(Phillips Petroleum Company)*

BELOW:

Bartlesville's first tank cars transport Oklahoma crude from Indian Territory wells. By 1903, after the laws governing the sale of oil were changed, the railroad had started shipping tank cars of oil to Kansas refineries. *(Phillips Petroleum Company)*

This Phillips Petroleum Company office building in Bartlesville, Oklahoma, was constructed in 1925 and consisted of the eight-story corner structure. The tower, added a few years later, is the only portion remaining today. *(Phillips Petroleum Company)*

and Iowa. Sometimes she was able to coax Frank to take off a few days for holidays in Kansas City or Chicago. But for the most part he stuck to his hectic schedule and usually only took time off to tend to personal business, such as the funeral back in Iowa of his grandfather, Rev. T. L. Faucett, who died in 1906, just the day before he and his wife were to celebrate their sixtieth wedding anniversary.

Frank and L.E. were committed to making their bank and oil enterprises work. Deals had to be made, contacts maintained, and oil wells drilled. Time became as important as money. There were also more civic demands, including L.E.'s work with various banking associations and booster organizations, and Frank's election as a fire department captain and his volunteer work at the Methodist church. Frank found himself spending more and more time away from his family, a practice that would increase through the years and only end up proving costly, especially in terms of his relationship with young John, who was fast becoming a spoiled mama's boy.

Frank did pause for a few moments of celebration in 1907 when the twin territories were joined as a single state. The stage had been set to create the forty-sixth state when the work of the Dawes Commission was completed and the small tribes in the northeastern part of the territory divided their land and accepted the white man's system of land tenure. The federal government was finally convinced that Indian Territory was ready for statehood.

Frank Phillips couldn't have agreed more. On the eve of statehood, Oklahoma oil wells produced more than 40 million barrels annually. Frank realized that by combining the two territories as a single state—with uniform laws and representation in Washington—local economic development could emerge. He joined the other 7,855 citizens of Bartlesville in toasting the birth of the new state on November 16, 1907. At 10:16 A.M. that day, President Theodore Roosevelt, using a quill taken from an Oklahoma eagle, signed the official proclamation uniting Oklahoma Territory and Indian Territory into the state of Oklahoma. The tidings were telegraphed to Guthrie, Oklahoma City, Tulsa, and Bartlesville and soon every roustabout, farm boy, ranch wife, and Indian chief knew they had at long last become U.S. citizens. The new state, with an area of about forty-five million acres, was larger than any state east of the Mississippi and larger than all six of the New England states combined.

While Oklahomans were busy kicking up their heels over statehood, engineers and designers were working late into the night on the third floor of a cramped factory in Detroit. They were putting the finishing touches on a new automobile that would soon have an impact not only on the development of Oklahoma but on the rest of the nation. The man who inspired the new machine was Henry Ford and his mission was quite simple—build a durable, inexpensive "car for the great multitude." A car to help merchants and farmers do their work. A car to allow Americans see the rest of the country. A car to take roustabouts to oil wells.

Known by many names—Tin Lizzie, flivver, jalopy, Little Henry, and sometimes by some more peppery expletives—Ford's horseless carriage was called the Model T.

The practical development of the internal-combustion engine and the advent of mass-produced automobiles gave the petroleum industry its greatest push. Because of the Model T, it quickly became evident that neither steam nor electricity would be the best fuel for autos. Gasoline would be required and plenty of it. The oil industry now had a growing market for what used to be considered a useless by-product. Until the introduction of the standardized automobile, gasoline was thought to be a waste product in the distillation of crude oil. But not any longer. Frank Phillips had been right as rain years before in that Des Moines hotel room when he convinced L.E. to get in the oil business by telling him the motorcar was "the thing of the future." In 1910, there were fewer than a half million automobiles in the United States. By 1920, there would be nine million. Gasoline filling stations began to appear across the nation.

As the number of automobiles increased, so did the demand for fuel and petroleum products. It was necessary to find more crude oil and improve the refining process. Oklahoma was fast becoming the nation's leading producer of oil. The Phillips boys, and Bartlesville, were right in the thick of things. At statehood, the city was made the county seat of Washington County, named for the first President and, measuring only 425 square miles in size, the second-smallest of the state's seventy-seven counties. Although it was dwarfed by neighboring Osage County, at 2,293 square miles Oklahoma's largest, Washington County, by virtue of the oil businesses of Bartlesville, was one of the wealthiest per capita. Bartlesville was well on its way to becoming one of the principal cities in the new state. The population was swelling and more people were arriving each day. New oil and gas fields had been opened, and with gasoline proven to be a cheap fuel, the glass industry, zinc smelters, and a cement plant were attracted to the area.

Swarms of immigrants—mostly from Poland—came to Bartlesville to earn citizenship by sweating in the smelters and plants. Town-lot drilling was outlawed in 1909, and although the derricks vanished, the wells kept right on pumping. Bartians, as they called themselves, were pleased with their prosperity and proud of the new streetcar line—the Bartlesville Inter-Urban Railway. Frank was elected to the trolley line's board of directors. The streetcars ran back and forth from the smelters up Third Street, past Frank's bank building, carrying workers to the oil fields on the edge of town and to the cement plant in Dewey, where a local cowboy named Tom Mix worked for a short spell before riding off to become a hand at the 101 Ranch, a lawman, and, eventually, a flashy silent screen star.

Dewey was also the home of Jake Bartles' Dewey Hotel—a Victorian frame building where oil-field workers and cowhands bedded down. The hotel also reigned as the most fashionable place in the area for a family-style Sunday dinner. In 1908 the Dewey Hotel attracted a sizable crowd when Bartles invited his Civil War regiment, the 6th Kansas Cavalry, to the hotel to eat steaks and drink whiskey. Afterwards, Bartles and his son Joe arranged for some steer roping and bronc busting as entertainment. Although he died just a month after the reunion, that first Bartles rodeo party marked the birth of the annual Dewey Roundup, for many years one of the largest in the world. A shy half-breed Cherokee cowboy named Will Rogers, who met Tom Mix at the St. Louis World's Fair and introduced him to fourteen-year-old Olive Stokes—a local beauty who became one of Mix's several wives—was just one of the top rodeo performers who came to Dewey.

But the Dewey Hotel drew more than cowboys and roustabouts and trick ropers, or respectable families decked out in their church clothes and filled with the gospel but hungry for chicken and biscuits. A third-floor cupola, just above Mrs. Bartles' bedroom, proved to be a choice spot for marathon poker games, staged by the Phillips brothers and other oil men. The games would often last through the night and break up about the time the neighborhood newspaper boys started their rounds. Despite the Phillipses' busy schedule, there was always time to fit in some poker, particularly if it provided an opportunity to smoke cigars and drink a little whiskey and talk oil with some of the other important wildcatters of the day.

Not that nearby Bartlesville didn't have more than enough places for business entertainment. Bartlesville was, after all, a boom town. There were plenty of diversions, ranging from horse races, band concerts, and turkey shoots to rodeos, prizefights, and baseball games.

In the early years of the century Frank's favorite performers—the

Ringling Brothers' Circus—visited Bartlesville, the first of several appearances, and a new Phillips acquaintance, Gordon William Lillie, better known as Pawnee Bill, brought his Wild West Circus to town. Pawnee Bill and Frank developed a friendship which would last all their lives. Given his colorful stage name by a Pawnee chief after the white man saved the tribe from starvation during a harsh winter, Pawnee Bill had ridden with Buffalo Bill's Wild West extravaganza until Lillie decided to start his own show featuring his wife, May, originally a refined Philadelphian who could ride broncs sidesaddle and shoot bull's-eyes with the best of them, including Annie Oakley. And by 1910 yet another Wild West show was off and running when Jack Moore, a local Cherokee scout and short-story writer, formed "Mustang Jack's Wild West and Indian Congress," composed almost entirely of Bartians.

Besides family entertainment there were also ladies, with no visible means of support, ready and willing to serve a well-heeled oilman or a roustabout flush with pay. These "daughters of Jezebel" populated many of the downtown "rooming houses" and did a thriving business, as did the beer joints and bootleggers. Pool halls, smoke shops, and domino parlors were favorite haunts for local bankers and oil producers, and even an occasional outlaw, Al Spencer, Henry Starr, or Henry Wells, would drop by to shoot a rack or two of billiards. But pool sharks traveling the circuit of oil boom towns were usually no match for the sharp-eyed Fred Spies, the Iowan who worked for Frank Phillips' bank, and who could run off table after table without a miss.

Otto's German Band, featuring "Fritz the German Dog," was a popular group at the Jitney Bar on the corner of Second and Keeler, and earlier I. N. Nash, who had played violin and cornet for the famed John Phillip Sousa, settled in Bartlesville and organized a cornet band all his own.

Vaudeville shows and theatrical troupes came for a week at a time, setting up tents and drawing big crowds, and Bartlesville's first motion-picture theater—the Star Theatorium—opened and showed a constant stream of silent movies, including popular Westerns, many of which featured actual cowboys, and even outlaws, from Indian Territory.

After statehood the Oklah Opera House was built. On opening night —September 25, 1908—more than a thousand members of Bartlesville's high society, including the Frank and L. E. Phillips families wearing their very best duds, paid a steep five dollars a ticket and tiptoed on boards laid across the muddy ditches separating the carriages and the theater door, to see the premiere of *Princess Pocahontas,* described as an "elegant musical comedy." The blue-collar workers and their families enjoyed the open-air plays staged at the Air Dome, or weekend afternoons at the Dreamland Skating Pavilion, better known as the Coliseum, a huge skating rink with an outdoor roof garden large enough to seat 3,000. A brochure issued by the Bartlesville Commercial Club in 1908 stressed that "the restrictions are off on Indian lands," making the town "the most promising city of the new state . . . where wealth awaits you."

Bartlesville continued to grow. The city was quickly overflowing its boundaries and outgrowing the 334 acres originally platted. Annexation was difficult since the first townsite was surrounded by land allotted to the Cherokees and Delawares or intermarried whites. With the population swelling, many oil-field workers had to sleep on cots set up every night at the skating rink. Residential lots were becoming scarce.

Conscious of the city's growth, and also aware that his family required a more spacious home to entertain friends and business associates, Frank searched for land on which to build a new residence. In partnership with Harve Pemberton, a local developer, Frank shelled out $50,000 and purchased two sizable pieces of property. The first, a tract called Pemberton Heights, was announced in December 1907 with lots selling from $250 to $1,000, and the other was Johnstone Heights, a 40-acre tract. "This deal," said the local newspaper, "is one of the largest that has ever been pulled off in the local real estate market."

Frank decided to build a new residence for his family in the Pemberton addition and selected a site in the northwest portion of the tract on Cherokee Avenue between Eleventh and Twelfth streets. The rest of the property was subdivided and sold. Frank was well aware that by announcing that he had decided to build a substantial residence in this area, he was making an important statement. This would prove that he fully intended to stay in town and that the new tracts of land he was opening would be the most desirable places to live in Bartlesville. To ensure that his own homesite was even more valuable, Frank also saw to it that streets were laid out and paved, streetlights and fire hydrants were installed, and critical city ordinances were passed, including bans on streetcar tracks in the area. Although some of his detractors protested and said the Iowa upstart was getting preferential treatment, Frank hushed them by pointing out he had invested much of his own money in the development and had even paid for the street paving.

Frank acquired title to his ten-acre tract in Pemberton Heights from a Cherokee Indian named Fred Kerr on March 18, 1908, little more than a month before the plans for the new Phillips home were unveiled to the public. It was time for Frank to begin parting with some of the money he made from banking and the oil business. Walton Everman, an architect who had designed several of the town's more prominent residences and businesses, including the Phillipses' downtown bank building, was chosen to design the home. Frank wanted a mansion and Everman gave him just that. Construction, as always, took longer than originally planned, but in 1909 Frank, Jane, and eleven-year-old John left the family's first residence on Johnstone Avenue and moved into their new home.

The new Phillips residence was stunning. It was built seven years before Josh Cosden erected his Tulsa mansion and three years prior to the unveiling of the E. W. Marland's first estate in Ponca City. Frank and Jane's first "dream home," the mansion, or "town house" as they called it, was really one of the very first in the new state to be built as the result of big oil money.

The Phillips town house had four levels, including the basement, and there were servants' quarters, a gardener's cottage, and a gazebo. A limestone wall surrounded the property, and Frank had gardens and trees planted and a circular drive built at the north entrance to handle the carriages and automobiles bringing guests for the many parties and business dinners the family planned to host.

For the exterior, Everman conceived a stunning Greek Revival style —tall white wooden columns, mottled pink brick walls, spacious porches, and a tile roof with dormer windows. The corners of the brick walls were highlighted with flat blocks of white sawn limestone and the rest of the outside trim was wood, painted white.

Inside the main-floor entry the staircase and banisters leading to the upper floors were all fashioned of Filipino mahogany, shipped from John Gibson's timber operation. Jane decreed that the handsome grandfather clock Gibson also sent along as a housewarming gift would always stand in the entry hall like some sort of stately sentry of time.

Beyond the music room, in the imposing dining room was more mahogany paneling and woodwork and a dining-room table to seat eighteen. Upstairs Frank and Jane had their individual bedrooms and baths, and there were guest rooms and rooms for servants to supplement the living quarters built at the rear of the property. Jane kept a child's rocker, given to her by her father when she was two years old, next to her bed, and an oval photograph portrait of her mother hung directly over the headboard. Frank's bedroom was comfortable but understated. He preferred a single bed, and there were comfortable chairs, plenty of ashtrays, and a large cigar box. On the third floor was John's bedroom and "the ballroom," an expansive room for card games, entertainment, and John's exercise equipment. There was also an attic—a favorite hideaway for a little boy on rainy afternoons.

The Phillipses were delighted with their new residence, and before too long 1107 South Cherokee Avenue became the talk of the town.

But there was also more than talk about Frank Phillips' mansion. Most serious conversations in Bartlesville, especially in business circles, included discussions about the Phillipses' steady growth as both bankers and oil men. On November 7, 1908, just after Frank ventured into the real estate field, the Phillipses announced the purchase of one of their rivals— the Bartlesville National Bank.

It turned out to be more than a fiscal takeover. Overnight, the Phillipses merged the two institutions—still operating as separate banks—under the same roof at the Citizens Bank & Trust Building. They bought all the stock, kept some of the employees, and either retired or fired the new bank's directors and officers, replacing them with Citizens Bank officers, with Frank as president and L.E. as cashier.

The maneuver was applauded by many and criticized by some, including the other bankers in town. A Kansas newspaper called the Phillipses "tireless hustlers," and the Bartlesville *Examiner* praised the Phillips brothers, calling them "the kind of active financiers who are certain to keep the opposition awake at night." That was exactly what Frank wanted.

In 1911 the Phillipses decided it was time to consolidate their banking interests. Citizens Bank & Trust was absorbed into the Bartlesville National Bank, with the same slate of officers. The institution now boasted resources of $1 million and $150,000 in surplus profits, making it one of the strongest banks in the state. Bartlesville National advertised itself as "a bank for all people" and published a monthly newsletter offering financial advice and news about farming and the oil and gas business.

For the most part, as the second decade of the century began, the news about oil and gas was good. For the Phillipses it was excellent. All Phillips oil operations were still headquartered in the rear of what by now was called the Bartlesville National Bank Building on Third Avenue. Even though they were still crowded in the bank building, the Phillips brothers were expanding their holdings in the oil fields around town.

New faces were added to the payroll. The oil patch craved fresh blood, and it took lots of muscle and guts to get the job done. Many of the early employees hired by Frank would become key figures in the Phillipses' business plans for years to come. One was Henry E. Koopman, a hard worker with a business college diploma and some valuable experience with an oil-field supply firm in Tulsa. Koopman was hired as a stenographer and bookkeeper for the Lewcinda Oil & Gas Company and even devoted what spare time he had to helping out in the field.

Another early employee was Clyde Hamilton Alexander, a tough and profane oil man from Pennsylvania who was weaned on crude oil and started work as a teamster hauling oil-field equipment when he was thirteen years old. Alexander was stout enough to hunt bear with a switch. As a boy he was once knocked unconscious by a pulling rod on a well, but it didn't faze him a bit. He got up, brushed off the dust, and went back to his job. There was work to be done. By the time he was fourteen, Alexander was pumping his own lease and at age seventeen he was in charge of a crew of men.

In 1904, about the same time Frank was thinking about putting down roots in Indian Territory, Alexander left his job with the Mid-Continent Oil Company in Kansas and came to Bartlesville to drive a nitroglycerin wagon for the American Torpedo Company. It was one of the most dangerous jobs in the oil field, but Alexander had the nerves of a cat burglar. The next year he worked as a pumper and roustabout for Lahoma Oil Company before moving to the Douglas-Lacy Company, a firm absorbed in 1907 by the Phillipses, who promptly renamed it the Creston Oil Company, in honor of the Iowa city they loved. In 1910, when Frank sold Creston Oil, Alexander went to work for Wolverine Oil Company on Lot 169, first as a foreman and later as superintendent. Although Alexander went to another outfit, Frank kept in touch with him through the coming years. Like Frank, Alexander had the benefit of only a grade school education, but he possessed plenty of horse sense and solid oil-patch savvy. Frank knew that Clyde Alexander was a good oil man. He knew Alexander could smell crude oil and point to the spot on the ground where the drill bit should be placed. Frank also knew he would figure out a way to get Alexander back on the Phillips team. It would take another seven years, but Frank would get his wish.

As the Phillipses' oil operations grew, more companies were founded. Meredith Oil Company was organized to develop a new lease at Copan, and the Janenora Land & Development Company—a name combining the first names of Frank and L.E.'s wives—was started to prospect for oil near the ranching town of Vera. During the next few years, the Phillipses would launch other oil firms, including the Luddington Oil Company, Phillips and King, Phillips & Company, and the Standish Oil Company, owned 50 percent by H. V. Foster and named for Frank's illustrious puritan ancestor.

Frank and L.E. were also pleasantly surprised by younger brother Waite and his prowess as an oil man. In fact, Frank was so pleased with his kid brother's ability that he presented Waite with a Ford roadster to take the place of the team and buckboard he used to make inspection tours of the well sites. On top of that, Frank appointed Waite president of Creston Oil and sent him out to Lot 67 in the Osage. Waite was delighted to have earned Frank's confidence.

Waite was also contemplating settling down. Although both he and his pal, Fred Spies, had no shortage of young ladies to choose from to escort to the many dances and socials held in town, for some time Waite had his eye on a damsel living all the way back in Knoxville, Iowa. Her name was Genevieve Elliott, and like Jane Phillips, she was a banker's daughter. Waite and his Genevieve were married at the home of her father, J. B. Elliott, on March 30, 1909. Waite's mother, an aunt, Frank, Jane, and Node attended the wedding. L.E. was best man.

Waite brought his bride to Bartlesville and presented her with a new house he built on land purchased from Frank. Like Frank's and L.E.'s, Waite's new residence, with its large paneled oak front door, was also located on Cherokee Avenue. The address was fitting—711—gambler's numbers.

Times were good for the Phillips clan. Oklahoma had reached a population of 1,657,155 by 1910, and almost 7,000 of those inhabitants called Bartlesville home. It was becoming a flourishing city. Strong rumors about town suggested that Frank was about to accept the Republican nomination for Congress. The Bartlesville Enterprise went so far to say that "Phillips is the kind of man who would give this district a standing in Washington." Whether Frank was ever really serious about making the plunge into the political arena, the fact remains he didn't. He was making too much money to fool around with politics. Besides, he had more members of his family near him now than he had in a long time. He was happy to stay put.

During those years, on Saturday nights, at Frank and Jane's house, after the hired girls cleared the china and crystal from the dining-room table, and the Phillips brothers smoked their cigars and finished swapping lies and jokes, there was still time to gather in the music room with their wives and children to listen to the piano and sing a few songs. In 1910 a recently released song caught Frank's ear. A Kansas homesteader wrote the lyrics the year Frank was born in the wilds of Nebraska. The words and music were printed together in 1904 under the title "An Arizona Home." But it wasn't until 1910, when the song was published under a new title, that it became popular. Some called it the "cowboy's national anthem." Frank loved the words and the melancholy tune.

> Oh, give me a home where the buffalo roam
> And the deer and the antelope play,
> Where seldom is heard a discouraging word
> And the skies are not cloudy all day.

When he sang the chorus, Frank's deep voice became louder.

> Home, home on the range,
> Where the deer and the antelope play . . .

But despite the way he felt about "Home on the Range" there was still another hit song in 1910 that Frank—and Jane—liked even more. It was "Let Me Call You Sweetheart," a love song which sold five million copies that year. The sheet music on the Phillipses' grand piano was dog-eared after only a week. For the rest of their lives together, wherever they went, if there was a band or an orchestra, or just some people who knew the Phillipses and were willing to sing a cappella, that big hit from 1910 would be heard. "Let Me Call You Sweetheart" was Frank and Jane's song. And like the couple who cherished the music, the song was straightforward without a hint of pretentiousness.

> Let me call you sweetheart, I'm in love with you.
> Let me hear you whisper that you love me too.
> Keep the love light glowing in your eyes so true.
> Let me call you sweetheart, I'm in love with you.

On summer nights, when the doors and windows were open at 1107 South Cherokee, the music spilled outside and bounced down the avenue. Above all the voices was Frank's—singing poorly but doing his best to make Jane smile. It never failed.

CHAPTER TWELVE

Jane Gibson Phillips may have been the person whom Frank called sweetheart, but the oil patch was the best lover he ever knew. On those nights, after the music stopped on Cherokee Avenue and his brothers had carried their sleeping children home, Frank climbed the mahogany staircase with Jane, but his thoughts were on the rigs pumping out in the Indian lands. They were making him wealthy. He loved his wife and family, but he adored black gold.

Not that Frank didn't have plenty of chances to stray from both Jane and oil. He had more than his share. During the years, as he got out in the world, Frank had flings with other women, including a romance which rocked his marriage to its roots and persisted for decades. There were also times when it seemed like his smartest move would have been to get out of the oil business. Frank's head could always be turned by a pretty face or a solid financial opportunity, but, in the long run, he couldn't turn his back on the wife or the business that had been so good to him. Through it all, Frank and Jane—the determined banker's daughter—gutted it out and stayed together. Frank also stuck with the oil business. Frank found himself hopelessly caught up with both of these entities in a lifetime affair that he could never quite break off. He kept in mind the wisdom of the country saying: "Don't forget who brung ya to the dance."

Although he would have his dalliances, Frank ended up dancing the last dance with his original partners—Lady Jane and Oklahoma crude. Often the dance was interrupted by others trying to cut in, and there was a lot of stepping on toes, but through it all the band played on.

Drilling for petroleum, even in the oil-sodden Osage, was not easy. Frank learned that right off. The oil business depended on sheer luck and lots of muscle. Geology was a new science that wouldn't be developed and practiced on a large scale until the 1920s, when oil was found in huge quantities and the primitive exploration methods gave way to more scientific techniques. In those early days of the oil patch, practical oil men, such as the Phillips brothers and Getty and Skelly, liked to say that "geology has never filled an oil tank." Some old-timers were even more outspoken, and openly ridiculed anyone who suggested the use of modern scientific exploration methods.

Frank wore spectacles and looked more like a preacher than an oil man, but he was definitely a man's man; as macho as bootleg bourbon and barbed wire. He and his circle of cronies felt they had to be tough and rugged, or at least act like they were, in order to compete and survive in a business founded on the best principles of machismo and masculine dominance. By acting macho, they believed they kept the precious product they sought in its proper place. They built fancy homes and spoiled their children and cavorted like sultans, but they also lived and died at the whim of oil, the most fickle of all mistresses.

Smart operators, like Frank and his brothers, at least started to listen to the newfangled ideas about geology and engineering. Eventually they'd forsake the hocus-pocus school of oil prospecting, but for the time being they were content to sneer at those who hadn't soiled their hands on a rig site. They believed there was no time to waste on somebody who couldn't afford to take the heat in the smoke-filled poker room at Jake Bartles' hotel or who was afraid to gamble on wildcat wells on a hunk of untamed land in the Osage where coyotes and outlaws nested in the thick brush.

Frank did not avoid hard work, and he was never afraid to roll the dice. For instance, one evening, after dinner at the Phillips town house and before the singing and serious drinking commenced, Frank and H. V. Foster sat down at Jane's piano bench with a new deck to cut the cards for big money. Within seconds and a single turn of the cards Frank won ten thousand dollars. Foster peeled off the bills with a shrug. He had hit a "dry hole" in the Phillips music room this time, but he knew his luck would turn. Besides, what good was money if a man couldn't enjoy it. Frank fetched fresh toddies, and as the glasses clinked, he silently gave thanks that he was no longer shearing hair or peddling bonds but had become an Oklahoma oil man.

And not just any kind of oil man. Frank Phillips was pure wildcatter and proud of it. Wildcatting was always a high-stakes gamble, just like cutting cards. But Frank and the other wildcatters were courageous souls willing to take enormous risks and not afraid to wager the entire bankroll on the hope of discovering a new pool of oil. Without the early wildcatters and their willingness to defy the odds, the oil industry would have dried up like an old sow's teat. The wildcatter's fierce independence and audacity became the heart and soul of the oil patch. Oklahoma's prominence in the oil industry came as a direct result of persistent wildcatting which led to the discovery of a series of rich pools throughout the Indian lands and beyond.

Naturally, Frank, like any good Oklahoma wildcatter, had an arsenal of tricks to turn to when it came to finding pools of oil. Some were better than others, and a few were about as effective as rubbing rainwater on your head in order to cure baldness. Most wildcatters found oil "by guess or by God." They relied on their instincts and hunches.

High on the list of ways to find oil were divining rods. The belief was that a forked twig with an affinity for petroleum in the right hands would dip every time it was passed over oil. Although the divining rods were certainly as successful as random drilling, other equally remarkable devices and methods were constantly emerging, including the "doodlebugs," perfectly unscientific machines that were supposed to be foolproof when it came to locating hidden oil. One doodlebug inventor claimed his giant X-ray machine could focus on the ground and, through electrical current, excite the earth's molecules and separate them, causing the earth to be temporarily transparent so the doodlebugger could see the oil. The inventor's name never became a household word. Another sage said he could find gas fields by the tension in his neck, and still another man made a living by walking around with a jug of crude oil tied around his throat, swearing he received electrical jolts through his body each time he walked over deposits of oil.

"Creekology" was also a common method used to find oil sites for wildcat wells. The idea was to drill inside the curve or bend of a creek or stream, since past finds indicated that this was where the big discoveries were usually found. But most oil men worth their whiskey, including Frank, swore by the notion that "close is the best." They were convinced if they could get close enough to an existing oil well, they could actually smell the oil perking below the earth and know just where to drill. Charles Barnsdall, a big producer in the Osage, said "a good oil man can sniff it . . . he can smell it even if it's three thousand feet beneath the ground."

Once a site for a well was selected and the lease was secured, the rig builders—the most demanding of jobs in the oil field—came on the scene wielding two-headed axes and saws to cut timbers for the towering wooden derricks. Next came the drilling crews, working as quickly as possible to get the rig pumping oil out of the ground. For many years the industry was ruled by the "law of capture," which meant that oil and gas was the property of anyone who "captures" it at the surface, even though the oil and gas may have been withdrawn from a neighbor's adjoining land. Consequently, the early fields were drilled with great speed, with the drilling crews laboring under the idea that it was essential to get the well drilled and start production before a competing operator got his well down and tapped into the great streams of oil thought to be flowing below the earth. Some leaseholders thought the oil ran in a vast subterranean channel from Pennsylvania to Texas, and a few believed that drillers were removing the oil that kept the earth's axle greased. Prospectuses of some of the oil companies even talked about "lakes and rivers of oil."

To get the job done on his leases, Frank tried to hire the best men he could find. From the lease hounds to the roustabouts, they were a profane and salty bunch, especially the drilling crews.

The drillers and tool dressers entered the picture when the rig was up and in place. They came cussing and shouting, complete with their own lifestyles and cultures, including a language made up of hybrid words and phrases borrowed straight from the hobo camps and the outlaw elements which thrived in the Osage Hills. Words that were curious to most folks in Bartlesville and Tulsa were commonly used around the hundreds of rigs springing up in the oil fields. Words like "go-devil" and "pay dirt" and "goin' fishin'." The cry of "Headache!" became a warning that some object —usually heavy—was falling from the rig. And when a roustabout heard someone scream "Fire!" or "Fire in the hole!" it meant get the hell out of the way, an explosion was due.

A "boweevil," from "boll weevil," was a worthless fellow, or an oil-field novice who answered to a "tool pusher," the supervisor in charge of several drilling wells. The word "crumb" was used when a man infringed on the work rights of another, as in "Stop crumbing on me," and it also was the name given the swarms of pests and parasites which infested every oil-field tent and shanty. The "crumb boss," a person in charge of the tents or bunkhouses, had to be able to order around all the crumbs living in the roustabouts' bunks or he'd "drag up" his pay and quit.

Welders were "daubers," shiftless workers were "jerks," and construction hands who worked on the huge oil storage tanks proudly called themselves "tankies" and earned the reputation as the "meanest, roughest, hard-drinkingest, fightingest men" in the oil fields. Drillers, those specialized virtuosos with cables and bits, mostly were called "twisters" or "rope chokers." To the drillers, a "bindle" was a bundle, usually packed with cooking utensils and a change of clothing, and the migratory worker who carried it was known as a "bindlestiff." The sun was called "Bronze John," bread became "cake," beans were known as "cherries," and "jamoke," a word merging "java" and "mocha," was what the drillers called the stout coffee they swigged during infrequent rest breaks at their "knowledge bench," the driller's wooden stool.

But as important as the colorful drillers were to the oil-field operations, it was the "shooters"—the men who opened the oil sands at the bottom of the wells with "torpedoes" filled with nitroglycerin—who had the most dangerous job in the oil patch.

Many shooters started as helpers or drivers on the "soup wagons" which hauled the nitro in bouncing buckboards over rough roads to the field. The temperamental liquid was usually carried in square five-quart cans fitted into padded compartments in the wagon. Sometimes the padding didn't help. Not every trip out to the field was smooth. There were tales of search parties, looking for overdue shooters, who found only a smoking hole in the dirt road after a soup wagon exploded. Most of the shooters stayed drunk during their off time just so they could face their work. Their margin for error was very thin. They could make but one mistake. But if the shooters did their job properly the resulting explosion would break up the oil-bearing strata and increase the flow of crude. A good torpedo man always brought a big smile to a driller's face.

Frank's first drillers worked with tools that had been used for drilling water wells. At best these early drilling tools were as crude as the "fishing tools" used to retrieve drill bits lost in the well. Mostly cable tools drilled those first Oklahoma wells, based on a method of drilling invented by the Chinese in 500 B.C. This kind of drilling required a crew of only two men —a driller and a tool dresser. A cable-tool rig was made up of machinery and gear that would rise and drop a "string" of tools, consisting of a bit and stem on the end of a cable. The heavy bit would pound its way into the earth, pulverizing soil and rock. Every so often, the driller would pull out the string of tools, flush the hole, and remove the slurry of drilling cuttings. Drillers were critical to any outfit's success, and during a boom good drillers were always in great demand.

The tool dresser's job was no picnic. It was a twelve-hour workday spent sharpening the drill bits and hammering them into true gauge. One bit would barely be ready when the driller came out of the hole with the drill stem and left another one to be sharpened.

With the advent of rotary rigs, cable-tool drilling—for a long time known as the "standard" in the industry—took a back seat and was used mostly to drill shallow water wells. In rotary drilling, a cutting tool was attached to a length of pipe, and power to rotate the pipe was applied at the surface with additional lengths of pipe added as the hole deepened.

When a bit became dull, the entire length of drill pipe had to be removed, disconnected in stands of two or three joints and stacked in the derrick. After the new bit was attached, the crew reconnected the stands of pipe one at a time and ran the string down the hole once again. Rotary drillers had to take care not to allow their tools to drop to the bottom of the well or get stuck in the hole; nobody liked fishing for lost or broken tools.

During drilling, a mixture of water, clay, and chemicals, called drilling mud, was pumped under pressure down through the drill pipe into the slush pit and then returned to the surface. This constantly circulating fluid cooled and cleaned the bit and flushed out the cuttings. It was a demanding operation, requiring plenty of speed, precision, and skill. Rotary drilling called for a five-man crew—mostly itinerants who moved from rig to rig, independent to independent, wildcat site to wildcat site.

It usually took more than one crew to keep a rig running twenty-four hours a day. The crews, divided into shifts called "tours" and pronounced "towers," considered themselves fortunate if they could finagle a Sunday afternoon off. Most evenings they crawled back to their tent or boarding-house too tired to undress.

Drillers were a rugged lot. With no safety regulations and the extreme range of weather, they had to be. They worked hard, played hard, and always had their ears cocked for the deep underground rumble that told them they struck pay dirt. Even though good men for the drilling crews were in great demand, early on Frank made it clear that he would not hesitate to fire a man off the job if he couldn't pull his weight. There were no exceptions. Frank Phillips was always the boss. He even fired his kid brother, fresh off the farm, and didn't even blink.

It happened during the early rough-and-tumble years in Bartlesville, when the youngest Phillips brother, Fred, only a kid in his twenties, left Iowa and moved to Oklahoma. Frank immediately put him to work as a tool pusher of a cable-tool rig on a Phillips lease north of Bartlesville. During one of his routine inspection trips, Frank took a rented buckboard to the site and discovered that the rig was shut down. On the floor of the rig Frank found that the driller had lost his tools deep in the earth and was "fishing" them out of the hole.

"Where's Fred?" demanded Frank.

"Why, the last time I saw him he was headed toward that rock bluff by that little stream," answered the driller.

Frank made a beeline for the creek, where he found Fred, stick in hand, working as hard as he could to dig out a ground squirrel. Frank went berserk. "I let him have it right then and there," Frank said years later, long after he had reconciled with Fred. "I fired him on the spot and Fred went off to Tulsa and never came back to work on one of my rigs." Some time later, after he established his own independent oil firm, Waite also had the pleasure of firing Fred for a mindless blunder, but, just like Frank, he turned right around and helped grubstake the youngest Phillips brother in a new venture. Frank taught Waite to be tough as a driller's boot but to be fair and forgiving, especially when it came to family.

The Phillips clan was growing. By the time Frank built his fine town house, most of his brothers and sisters were married and raising broods of their own. In 1911, Genevieve and Waite became the parents of a daughter they named Helen Jane. L.E. and Node's two sons—Philip Rex and Lee Eldas, Jr.—were growing as fast as bluestem grass, and their daughter Martha Jane, born just before Oklahoma's statehood, blossomed into a petite beauty who enjoyed music and was a favorite of her Aunt Jane. But L.E.'s wife, Node, born with a heart condition that would keep her family hovering about her until she died at the age of ninety, was a cause for concern. It seemed Node's health fared better in cooler climes, and not in the sultry Oklahoma sunshine. After much prodding from L.E., she took the children and moved to Denver from 1913 through 1915.

Although Node was a sickly woman most of her life, L.E.'s health was not up to par either. Frail from youth and bothered with high blood pressure that wasn't helped by chain smoking cigarettes, in 1910 L.E. was forced to spend some time resting at the Battle Creek Sanatorium, the internationally known health resort in Battle Creek, Michigan, where Node was also a familiar guest. L.E. soon returned to help with the Phillipses' growing bank business. But his heart troubles continued, as did a plaguing gastrointestinal ailment. He seemed to always feel weak and wasted. Frank insisted on frequent checkups for L.E. at a clinic in Kansas City and periodic rests at Battle Creek. Besides the constant medical attention, L.E. brightened when he took a long vacation with his family or

attended one of the reunions of the Anchor Club, the group of former
Iowa farm boys who swore they'd never again have callused hands and to
a man kept their pledge.

When he was feeling healthy, L.E. concentrated his efforts on the
banking side of the Phillips family enterprise and left most of the oil-field
work to his capable brothers. L.E. might have been sickly, lacking Frank's
and Waite's oil-patch flamboyance and macho charm, but as a banker his
reputation was impeccable. The Bartlesville National Bank's balance sheet
in 1913 showed a surplus of $150,000 and deposits of $1.2 million. L.E.
had been elected a member of the executive committee of the Oklahoma
Bankers Association in 1910 and within two years he became president. In
1915 he was elected vice president of the American Bankers Association,
shortly after he narrowly missed being appointed to a directorship at the
new Federal Reserve Bank being established at Kansas City. Soon, L.E.
managed to work out a successful arrangement with the U.S. Indian
Agency at Muskogee making the Phillips bank acting guardians for the
Indian funds, a very profitable maneuver.

Although the banking business was thriving in Bartlesville, the town
was still on the short end of being civilized. In 1911, the town on the
Caney was having trouble with what the Bartlesville *Enterprise* claimed
were "canine ghouls"—packs of wild dogs which went on nightly raids in
the White Rose Cemetery to dig up fresh graves. A few sharpshooters and
a box of shotgun shells remedied that situation. But there were other
nuisances besides dogs despoiling graves. Even after statehood was just a
pleasant memory, as the century moved toward its teen years, a good
many outlaws still called the Osage Hills home.

Henry Starr, the notorious bandit who once worked for George Getty
on Lot 50 and bragged about robbing more banks than anyone else, was at
the height of his criminal career. On two occasions, Starr held up none
other than Jake Bartles, founder of Bartlesville. To his chagrin, Starr
learned that although Bartles was a wealthy man, he seldom carried any
cash. When Starr's third stickup of Bartles netted him only a few greenbacks, the outlaw lost his temper. He swore at Bartles and, according to
local legend, told him that "an important man like you ain't got no business runnin' around carrying nothing but pennies. It's a disgrace!" Then,
as he pushed the business end of his Colt revolver into Bartles' belly, Starr
declared that he planned to rob him again and that he had better be
carrying at least five thousand dollars or Starr would shoot him dead.
Bartles had no intention of winding up as a feast for some "canine ghoul"

at the White Rose Cemetery. The next morning he was at the bank before
the doors opened, waiting to withdraw five thousand dollars in large bills,
the sum he carried the rest of his life, even though Starr, for no known
reason, never bothered Bartles again.

Besides Starr, Al Spencer and Henry Wells—architects of the last
major train robbery not far from Okesa—and a slew of other desperadoes
were familiar sights on the streets of Bartlesville. Of course, the Phillips
boys continued to treat bank robbers and horse thieves just like any other
depositors and their luck held. They never had a masked visitor darken
their bank's door.

And there were other incentives to keep citizens on the right side of
the law. According to one newspaper report: "Bank robbers, as a rule,
fight shy of Washington County, preferring to take some other route in
making their escape after they have pulled off a robbery." The newspaper
pointed out that the "Anti-Horse Thief Association is more strongly organized than in any other county in the state, and there is probably no
county with better telephone service." Another deterrent was the decision
of local bankers to pay a bounty of $1,000 for each bank robber brought in
dead or alive. For the outlaws of the Osage it just seemed to make better
sense to rob banks elsewhere and stash the money in the hills or in Frank
Phillips' bank.

Content that between L.E. and Waite his growing oil and banking
empire was in good hands, in 1912 Frank allowed Jane to talk him into a
trip to Chicago to visit John Gibson and his wife, Bertha. Some years
earlier, Gibson had finally left the Philippines and returned to the States.
The Gibsons took up residence in Lincoln, where Gibson remained involved with the Nebraska Central Building and Loan Association. They
also spent considerable time back in Chicago so Gibson could help handle
the affairs of the Coliseum Company.

While visiting Chicago, Frank attended the International Air Exposition at Grant Park. He witnessed the aerial acrobatics of daredevil pilot
Lincoln Beachey and other pioneer aviators who threw caution to the
winds and left the spectators—including Frank Phillips—breathless. Frank
was so impressed by the stunt flying that he memorized every detail of the
air show and swore to himself that one day he'd become a part of the
fledgling aviation industry.

The trip to Chicago was important for another reason. After trying
for years to convince his daughter and son-in-law of the importance of
seeing the world, Gibson was at last able to talk them into accompanying

Bertha and himself on the first leg of what would become a trip around the globe.

Frank was thirty-nine years old, and although he and Jane had managed a few trips abroad, including one to England in 1910, he was again ready to experience new sights and sounds and taste adventure. He felt his sap rising just like when he was a teenaged barber setting out on his grand adventure through the American West. When friends reminded him about the sinking of the steamship *Titanic* and the loss of 1,502 lives less than a year before, Frank replied with a belly laugh. He was, after all, a direct descendant of Captain Miles Standish. If his distinguished ancestor could brave the ocean deep in a ship no larger than the *Mayflower,* then Frank could easily cruise around the earth aboard a modern liner.

After months of planning, and with son John—an overindulged fifteen-year-old—tucked away at Hotchkiss, the venerable Connecticut private school, Frank and Jane hosted a gathering at their Cherokee Avenue mansion the night before the start of their trip. Many neighbors and friends came, bringing their own wine and an array of bon voyage gifts. They all stood in a large circle in the dining room and proposed toast after toast. Then Frank and Jane said their goodbyes to Bartlesville and left to see the world.

On January 3, 1913, they departed Kansas City aboard the Union Pacific bound for California. After a brief stop in Denver, the Phillipses' Pullman car connected with the Overland Limited from Chicago carrying the Gibsons, who were to accompany Frank and Jane as far as Manila. After brief pauses in several western cities, including Ogden—one of Frank's old stomping grounds—the foursome reached the Palace Hotel in San Francisco, where they made final preparations to board the steamship *Siberia* on January 9 and head west through the Golden Gate into the Pacific.

From day one on the high seas, and at every port, Frank provided a personal account of the entire 120-day journey in a series of letters written to Lewis Phillips back on his farm in Iowa. The letters, meticulously dictated by Frank after he returned to Oklahoma, not only chronicle the entire trip but offer an interesting look into the mind of the writer. The letters also provide some clues to Frank's character as he reveals intimate impressions of the people and places he encountered.

And the letters also reveal some of the Phillipses' personal foibles. Like Jane's susceptibility to choppy waters. All the way to Honolulu she was seasick, a malaise she'd learn to live with the rest of her life. But once on solid turf in Hawaii, and after seeing the flowers and beaches, Jane was her old self and told Frank she felt "as if we had a trip through fairy land." The tropical paradise also agreed with Frank, who noted: "The view in leaving Honolulu on that delightful night has made an impression on me which I shall always remember. One interesting feature which is always to be seen in the Honolulu harbor is the natives swimming about the boat and diving for money which passengers throw in the ocean. They climbed all over our boat, and I saw one native dive from the highest deck into the water, a distance of more than fifty feet. I was so impressed at that time it was one of the most wonderful feats I had ever seen."

From Hawaii to Tokyo, Jane was in bed almost constantly because of near-typhoon conditions. Frank weathered the storm and put up with a contingent of missionaries who insisted on holding shipboard services even after the Sabbath was "lost" when the captain announced it was Monday as the ship crossed the 180th meridian before bedtime on Saturday night.

There were bright spots. All the way to Japan, and later when they sailed to Manila, John Gibson entertained his fellow passengers with eloquent recitations from his storehouse of poetry. Jane even managed to rally on several occasions to sit in a deck chair and listen to her father—his white mane of hair tousled by the salt spray—deliver from memory the best of Shelley, Poe, Longfellow, and Tennyson. Years later, Jane wrote: "I think that I never fully appreciated my father or realized what a really wonderful mind he had until I took a trip with him to Manila. On that voyage his recitations seemed to be unlimited."

When they arrived in Japan, the Gibsons and the Phillipses were hosted in traditional homes, attended both a native funeral and a wedding, and met many exotic people, including geisha girls, the wealthy widow of a war hero, and a tipsy Korean prince. They devoured everything and were hungry for more—from the local cuisine and bartering in the markets to visits to the sacred mountains and ricksha rides to the temples.

After Japan, the *Siberia* paused in Shanghai and then "turned her nose directly south" and proceeded through schools of flying fish to Manila Bay. They spent fifteen days in the Philippines. Winter carnival was at its height and the Gibsons' many friends gave parties every night. During the day Frank spent most of his time at Gibson's lumber operation. "Mr. Gibson had not been at the mill for two or three years and all the old employees anticipated his arrival with great rejoicing," wrote Frank. "They were glad to see him and extended a cordial welcome. We found enormous piles of logs and lumber. I saw fine tropical mahogany and other woods I had never seen before." On Sundays, Frank attended cockfights and then took Jane to the cemetery to leave tropical bouquets on her mother's grave.

To celebrate their sixteenth wedding anniversary, Frank bought two handcrafted chairs from the local prison and shipped them back to Bartlesville. Three days later, Frank and Jane boarded the steamship *Manchuria* and left the Philippines and the Gibsons and sped across the calm China Sea to Hong Kong for winter horse racing and a guided side trip to the city of Canton, where they "saw many dreadful sights . . . much distress and suffering."

The Orient was not at all what Frank had expected. "It seemed to me that half the people were crippled or blind. They have no sewers or waterworks and the filth and smell which we encountered was unbearable. When we refused to give the children money they called us 'gringos,' and once or twice I was not free from fear of attack by Chinamen who gathered in large crowds at places where we stopped."

For the most part China baffled Frank. "The Chinese do everything in a topsy-turvy manner and opposite our way of doing things. When you are introduced to a Chinaman he will shake his own hand instead of shaking yours. This sanitary method, however, was always satisfactory to me. If he wants to motion you away from him he makes the same motion of his hand which we make when we wish one to come toward us. He laughs when he announces the death of a relative, and the Chinese bride cries at her wedding. If you go into the office of a European and his hat is on his head, he will take it off. If you go where a Chinaman is sitting with the cap on the table in front of him, he will put on his cap. At dinner, contrary to our custom, the Chinese will eat his dessert first and finish up with rice and soup. The mourning color in China is white, not black. The men wear petticoats and the women wear trousers. A Chinese wears his vest over his coat. If an American boy is puzzled in doing a sum he scratches his head; a Chinese boy scratches his foot. We speak of being killed by lightning; the Chinese of being killed by thunder. The best places are occupied by the Chinese cemeteries, and the most filthy places are occupied by the living."

When they left Hong Kong, the Phillipses looked forward to new destinations, enjoying their first-class accommodations. There were stops at Singapore and the Malay Peninsula, before they sailed through the Indian Ocean and the Bay of Bengal to visit Ceylon. While staying at the Grand Oriental in the port city of Colombo, Jane busied herself with shopping for fine laces while Frank roamed the streets soaking up as much local color as possible. He was particularly struck by the colorful wardrobes of the strange and different people he met. "I observed that the people in other countries are more practical in lots of things than we are," wrote Frank. "I think that the European and American dress of both ladies and gentlemen is the most uncomfortable, if not impractical, of any nationality."

But even in exotic Ceylon, Frank couldn't forget the oil business. He noted in one of the letters to his father that gasoline was selling for about seventy-five cents per gallon. This caused him to observe that "the reason why it was so important for the Standard Oil Company to remain active in these foreign countries . . . is because of the vast amount of kerosene which is consumed. On this account the Orient is absorbing the surplus kerosene output of America today. Were it not for this outlet I am sure the oil industry would not be as prosperous in America."

The Phillipses boarded yet another boat, the *Marmora,* and before leaving port, Frank amused himself by "dickering" with natives who peddled curios from rowboats. The sailing was smooth across the Arabian Sea and the sunsets picturesque. But Arabia turned out to be a dismal disappointment for Frank. He was particularly upset with the people, whom he called "a dreadful annoyance" since "begging with them is a profession."

He was especially revolted by an encounter with a beggar in the town of Aden. "The little Negroes, as I would call them, are well trained in the art of begging. They could speak but a few words in English. One boy about ten years old held on to our carriage, whenever I did not drive him away with the buggy whip, for perhaps two or three miles, crying all the time: 'No fodder, no brudder, no sister, no mudder.' He would press in his stomach until it almost touched his backbone. I presume he thought he could fool us. As a matter of fact, I could duplicate his performance, as could anyone else who is as skinny as I am." Beggars didn't fare well with the industrious Mr. Phillips, who called Arabia "about the most destitute-looking land I had ever seen."

On Easter Sunday, the *Marmora* arrived at the mouth of the Suez Canal. In his letters Frank described his first view of the Red Sea, the voyage past the Sinai Peninsula, and the city of Suez. "One bit of scenery which attracted my attention and looked quite home-like and familiar to me was a number of large steel oil tanks and a refinery located near the entrance of the canal. There is a little oil down along the east coast of Africa, and an English concern bring the oil to Suez where it is refined."

The Phillipses, and some of their shipboard friends, spent ten days in

Egypt, exploring Cairo, the Nile Valley, the Pyramids and the Sphinx, and a village "alive with children and millions of flies." The museums, mosques, snake charmers, and ancient ruins fascinated Frank, and during a trip aboard a two-wheeled donkey cart to view a new excavation his guide presented Frank with a grisly souvenir—a human skull he found along the road. "If I had the ability," noted Frank, "I could easily write an interesting book in relating my experiences and telling about the people, their customs and the things we saw."

The *Osiris,* a sleek mail steamer and the boat which carried the Phillipses on the next portion of their odyssey, may have been the fastest on the Mediterranean, but it could not ride the waves worth a damn. The passage to Italy was so rough that Jane never left her berth and Frank himself spent a good deal of time leaning over the rails and chumming the seas, all within sight of the coasts of Greece and Albania. But the stormy voyage was worth every second of nausea. Of all the lands he visited, the traces of Roman civilization and the Renaissance cities he found in Italy affected Frank the most. "The most difficult part of our trip to describe is the time we spent in Naples and Rome," wrote Frank. "I have subject matter enough to write volumes of books. I cannot tell you about all the things we saw and the attractiveness of it all."

They toured the ancient cities of Pompeii and Herculaneum and peered at Mount Vesuvius. They took a train to Rome and checked into a large suite of rooms on the first floor of the Grand Hotel. The clerk told the Phillipses these were the last rooms left in the hotel, but he assured them they would be moved into new quarters the next day.

"We were more than delighted with the rooms and were wondering why they would not permit us to retain these quarters." Frank got his answer soon enough. "When we returned to the rooms after dinner my wife discovered they had been fumigated. I knew that J. Pierpont Morgan had died at this hotel only a few days before, so in a jesting manner I suggested that these were the rooms in which Mr. Morgan had died and that's why they had been fumigated." The next morning Frank learned he was correct. J. P. Morgan, the world's most influential banker, had indeed died in the Phillipses' suite. They quietly moved into fresh rooms and proceeded to visit every attraction Rome offered, including the Pantheon, St. Paul's, St. Peter's, and the original Coliseum, slightly older and more crusty than the one Frank helped build in Chicago.

Another highlight was the Vatican and the Phillipses' special audience with Pope Pius X, a kindly seventy-eight-year-old pontiff less than a year away from death. Jane wore a black gown and a veil and Frank was attired in a white linen suit. He also pulled on a long overcoat to cover his fancy duds, just in case they happened to bump into anyone from Oklahoma.

"We went to the Vatican and after waiting some little time were admitted to the private apartments of the Pope," wrote Frank. "There were about fifty admitted with us into a small throne-room where we all stood around the walls leaving the center of the room vacant for the Pope. The room was decorated in cardinal red, beautifully figured. There were many splendid gold hangings but there was no bric-a-brac worthy of mention. The Swiss Guards are a freakish-looking lot of soldiers and have been the royal guard of the Pope for centuries. On entering our room the Pope raised his hands and we all knelt, after which he blessed all present. He spoke in a foreign language and we did not understand what he said. The Pope was dressed in a long white robe trimmed in gold. He is a grand-looking old man."

Frank and Jane, the grandchildren of Methodist preachers from Iowa, were awed by the pageantry and splendor of the Vatican and the Sistine Chapel and the rest of the ancient city of Rome. Both of them were also charmed with the statuary and paintings and other art treasures, especially the Roman mosaics. "The Vatican process for making this mosaic work is a secret known only to them and is guarded very jealously," wrote Frank. "They also make fine mosaic pictures which are framed, and I have since regretted that I did not buy one for our home."

Following their tour of Italy, the Phillipses took a train for a whirlwind trip through Switzerland and France and then crossed the English Channel to spend a few days in London. They had visited London three years earlier for a month.

"Old London looked mighty good to us," wrote Frank. They attended a few plays and took in some vaudeville before boarding the steamship *New York,* a comfortable old boat that had served as a dispatch ship during the Spanish-American War and counted among its mariners most of the surviving members of the *Titanic* crew. The ship pulled out of Southampton bound for New York on April 12 and had reached mid-ocean on April 15, the first anniversary of the sinking of the *Titanic.* Frank and Jane watched as the crew gathered on the upper decks and, without a word, cast into the foamy sea wreaths and bouquets, many sent from people who had lost relatives or friends. Then the *New York* turned and took a southern course to the American shore.

"Only native Americans who have been abroad appreciate the feeling

Americans have upon returning to the native shore," wrote Frank. "I cannot describe the feeling which one experiences in coming up New York harbor in the face of the magnificent buildings which seem to rise up out cf the ocean. We passed the Statue of Liberty amidst all the turmoil of hundreds, if not thousands, of large and small boats which are floating in the harbor. Everyone is happy on such occasions. They are about to reach home and see loved ones and the things which are nearest and dearest to them in life."

John, on holiday from Hotchkiss, was on the dock to greet his parents. "I had a hard time to keep Jane from falling overboard in her efforts to get near to him," wrote Frank. But soon after the big trunks—stuffed with fancy clothes and fine silks and laces and mementos, including a human skull—were unloaded, the three Phillipses settled into the Waldorf-Astoria for a long visit.

The final letter of the series Frank sent to his father offers the most revealing look into the future, especially the course Frank Phillips would chart for himself in Bartlesville and the Osage.

My dear Father:

It has always been an ambition of mine to encircle the earth, and now that we were in New York I had a feeling of gratification at having accomplished something that would give me much to reflect over for many years to come.

We saw many wonderful sights and strange people, but after all doesn't Bartlesville, Oklahoma, afford equally as interesting things for the world traveler? We have oil and gas wells which to those who are not familiar with them should be an unusual attraction. We have the frontier life and the Indian history which, when openly explained to the traveler from abroad, should create an impression which he should carry home with him and prove equally as thrilling as any information which we might obtain from any foreign country. Distance lends enchantment and it is the things far away which seem to attract most people to a greater degree than those which we have in our midst.

Sincerely,
Frank Phillips

Itching to return to his banking and oil interests after such a long absence, Frank sent John back to his prep school in Connecticut and Jane hurried to Creston to attend a wedding celebration. Frank headed straight to Bartlesville. Despite the tiring voyage, he was refreshed and ready to roll up his sleeves and get back into the thick of things.

Even though the Phillips brothers and the other local oil outfits weren't directly involved, there was still big talk around town about the recent discovery of the rich Cushing field. Located on the northern edge of the Sac and Fox Indian lands below the Cimmaron River in north-central Oklahoma, Cushing was just another struggling farm community until 1912, when Tom Slick's big gusher blew in. By the time Frank completed his trip around the world and returned to Oklahoma, Cushing had become a brawling boom town. Other boom towns such as Drumright and Oilton sprouted out of the corn fields and offered haven to an army of roustabouts and whores. Fortunes were being made in a day and were squandered just as quickly. An excited newspaper reporter from Kansas City came down to see what all the fuss was about. "Any man with red blood gets oil fever when he gets to Drumright," the reporter wrote. "He sees things large and talks in terms of thousands, when a day earlier the best he could do was two bits." The hundreds of wagons loaded with pipe and derrick timbers, the teamsters snapping their bullwhips, and the shouts of the workers caused so much commotion that a blind man could find his way to Cushing and Drumright just by following all the noise.

Still, there was a shortage of drillers, despite the thousands of oil-field workers pouring into the state from Texas, Kansas, Indiana, and Pennsylvania. In 1915 there were 710 wells producing 72 million barrels of oil annually—almost 20 percent of all the oil in the United States—in the Cushing field and a virtual forest of wooden derricks surrounding the boisterous town.

The man who first "smelled out" the oil pools at Cushing—once called "Dry Hole" Slick by the old oil-field hands kibitzing in the fancy Tulsa hotel lobbies—was now the oil industry's youngest millionaire and would soon be known as "King of the Wildcatters." Cushing became the "Queen of the Oil Fields."

But most of the wealth from the Cushing field didn't stay in the little boom towns and makeshift camps. Oil money found its way to Tulsa. There the big oil men—Cosden, Skelly, Sinclair, Slick, and the Phillips boys from Bartlesville—hobnobbed in the Robinson Hotel or the brand-new Hotel Tulsa—"the state's finest." With its big leather chairs under the elegant lobby chandeliers, the Hotel Tulsa became the perfect setting for deals—and legends—to be born.

And there were more pockets of good luck and deep oil springing up across the prairies and ranchlands of Oklahoma. At Ponca City, oil men E. W. Marland and Lew Wentz were building their substantial oil empires. Marland would become known as "the patriarch of Ponca City," thanks to George Miller, founder of the 101 Ranch. It was Miller who interceded with White Eagle, the stern chief of the Ponca tribe, and convinced him to allow the wildcatter to drill for oil on Indian land near Miller's ranch. That move made Marland a multimillionaire and didn't do the Ponca Indians a bit of harm.

Frank relished day trips to Ponca City to visit Marland, but he liked going to the 101 Ranch, in the lush, rolling grasslands just south of the Salt Fork of the Arkansas River, even better.

Colonel George Washington Miller established the sprawling 101 Ranch in the late 1800s. His trio of brawny sons—Joe, George, and Zack —saw to it that the fences stayed mended, that the cattle grew fat, and that only the best cowboys rode herd over their 110,000 acre domain. Tom Mix, Buck Jones, and Hoot Gibson saddled cow ponies and drew pay-checks from the Millers. So did Bill Pickett, the black cowboy who in-vented bulldogging. During a ranch rodeo once he galloped into the arena, seized a wild steer by the nose with his teeth, and, allowing his mount to run from under him, wrestled the steer to the ground. Will Rogers prac-ticed some of his best rope tricks and sang cowboy songs all night long on the spread. Teddy Roosevelt, one of Frank Phillips' favorites from the Grand Old Party, was a guest. So was Geronimo. John Philip Sousa be-came a member of the Ponca Indian tribe while visiting the ranch, and Admiral Richard Byrd rode the Miller brothers' trained elephants. Paw-nee Bill, William S. Hart, Jack Dempsey, William Randolph Hearst, and Buffalo Bill Cody all ate steaks as big as doormats in the Millers' imposing stucco ranch headquarters, known as the "White House." While sheriff's posses hunted the hills looking for outlaws, early motion-picture makers made Western films on the ranch, featuring former bandits and 101 cow-pokes-turned-movie-stars.

The big-name guests and the motion pictures were exciting, but what impressed Frank the most about the glamorous Miller boys was their fa-mous 101 Ranch Wild West Show—an extravaganza which carried the spirit of the Old West across America and Europe. The show featured as many as one thousand performers—cowboys, Indians, Russian cossacks, sharpshooters, trick riders, ropers, and musicians. It played to standing-room-only audiences from 1908 until the Great Depression came along and clobbered the Millers. Through it all, one of their biggest fans was Frank Phillips. He found in the Millers men living the life he dreamed about as a boy. To Frank, they were genuine heroes of the American West who played cowboy and Indian for a living. He admired them for being true to their roots. Each time he paid a call at the 101, Frank felt like he did back in Iowa when he was chosen over all the other children to attend the circus.

Frank was so taken with the rough-and-ready Millers that he did his best to expose his son to them. Frank hoped some of their self-reliance would rub off on the youngster, who either was up to his hips in mischief or had his nose in one of those infernal books. Frank was determined to make a man out of him come hell or high water.

On occasion, when John was not away at school, Frank would break away from the hectic schedule in Bartlesville, coax the young man to leave his doting mother, and go see the Millers. Many times, Frank and John spent three or four days at a stretch on the 101 Ranch, riding lively cow ponies and listening to poetic cowboy yarns around a blazing fire. It was like really being "Home on the Range," with the worries of running a business left far behind. The times at the 101 Ranch were, without a doubt, the best days Frank and John would ever spend together.

But not all the clouds had silver linings. By 1914 sinister clouds of war were gathering in Western Europe. Despite Wilson's pledge of total neutrality on the part of the United States, talk of the impending world war dominated conversations from Wall Street to the oil-field camps.

Closer to Frank's home front, there was more bad news brewing. Sibling rivalry had reared its head. Little brother Waite, at thirty-one a competent and proven operator, was ready to withdraw from his older brother's oil interests and try his wings. By the summer of 1914, as World War I officially commenced in Europe, he made his move.

When Waite shocked Frank and L.E. by announcing his decision to leave, there were 157 oil companies headquartered in Bartlesville. Eleven of them were owned by the Phillipses. But what troubled Waite was that lately there had been some talk, between Frank and L.E., about liquidating some, if not all, of the assets of their oil companies in order to devote their time to other business interests, especially banking. Waite disagreed. He believed in the production of crude oil and gas in the midcontinent oil districts and he felt the future was in privately owned oil businesses.

Frank was not happy with the news that Waite was going to move out on his own. He tried to put himself in Waite's shoes, but it was difficult. As Frank would later say, "We Phillipses just can't get along with each other when it comes to business." Just the notion that one of his own brothers would leave the fold was distressing. Frank liked the idea of the Phillips brothers working together as oil men or bankers or whatever they chose to pursue. The trouble for Waite was that Frank liked to do all the choosing. L.E. was always perfectly content to play second fiddle. Not Waite. Despite his deep admiration for Frank, he had a mind all his own. His respect for Frank would remain forever, but a spirited competition between the brothers would soon develop.

Waite sold Frank and L.E. his share in their mutual property and their joint oil interests. As the guns of August roared in France, he and Genevieve left Bartlesville and their house on Cherokee Avenue with the lucky 711 address. The Saturday-night gatherings would continue at Frank's house, and some would be grand and glorious affairs, but they'd never be the same without Waite.

He moved to Arkansas, "The Land of Opportunity," where his first individual business venture was the purchase of the North Arkansas Oil Company, an oil-marketing firm headquartered in Fayetteville. Waite built up the company's assets quickly, and sold it for a profit after only one year. In the fall of 1915, he returned to Oklahoma and went into the oil-production business in the Okmulgee area. In Okmulgee, a Creek word for "bubbling water," Waite soon found that there was also oil bubbling. He took leases on several promising properties, including one just south of town near the cemetery. That well came in big. So did several others. Waite Phillips had struck it rich. Within a few years he managed to build up large and valuable oil-producing properties, pipelines, refineries, and marketing outlets throughout the midcontinent region. He was a millionaire.

But Waite didn't stop there. He was attracted by the unlimited possibilities which the oil industry offered to those who were able to take advantage of the opportunity. He bought more leases in Arkansas, Texas, and Kansas, where the Rainbow Bend lease near Arkansas City proved to be another big winner. Waite liked making money, and lots of it, but he enjoyed hunting a dollar more than possessing it.

Back in Bartlesville, Frank was still mulling over his choices—banking or oil. He and L.E. continued to talk—sometimes long into the night—about selling out of the "boom or bust" oil business, only now the conversations were getting serious. The brothers talked about the wisdom of getting out of the oil business while the getting seemed good. More and more wells were being drilled, and the crude oil production was creating a market glut. As always, there was the risk factor to consider. Only about one out of every ten wells drilled was successful, and as more oil was produced, even with the increased export demand brought about by the war raging in Europe, crude prices continued to plummet.

Finally, in 1915, during another of their lengthy conferences, Frank made up his mind. "Hell, L.E.," said Frank. "We're not oil men, we're bankers."

Within the hour, they decided to sell out their oil interests and leave Bartlesville. They'd bestow the job of hunting crude oil to Waite. He could have it. Frank and L.E. had made their mark in the oil patch. Now they were ready to move, lock, stock, and barrel, to Kansas City, Missouri, and open a bank. They even had a location in mind—the corner of Tenth and Walnut. It would be the biggest and best financial institution in the city and would serve as the parent of a chain of banks they planned to open throughout the Midwest. Frank figured that with a sizable banking empire he could become a national influence.

Starting in December 1915, Frank and L.E. began selling off their oil properties. They sold the bulk of their leases to several local oil outfits, including the Okla Oil Company, a subsidiary of the Tidewater Oil Company. They also sold off some properties to the Tidal Oil Company, except for a few leases out in the Osage which both Tidal and the Phillips brothers considered less than valuable. But besides the low worth of the mineral rights, the Phillips were also stymied by their Osage leases because of tribal regulations. The Osage Agency would not allow Frank to transfer or reassign any of the subleases originally assigned to H. V. Foster. Try as they may, it appeared, at least for the time being, that Frank and L.E. would have to put their Kansas City banking venture on a back burner, until they disposed of all of the oil property, including the leases in the Osage.

But while Frank and L.E. were plotting preliminary plans for their banking empire, events were transpiring thousands of miles away which would have a profound effect on the Phillipses' future. The ominous war clouds which already covered Europe were headed for American shores.

In March 1916, as a tune-up for the big war, Brigadier General John J. Pershing led 6,000 American troops across the border into Mexico to chase Pancho Villa and his 1,500 guerrillas after Villa's raid on Columbus, New Mexico. Oklahoma's National Guard was asked to help battle the border bandits. Most of the Okie farm boys and oil-field roustabouts riding

with "Black Jack" Pershing down Mexico way knew full well that they'd
soon be fixing bayonets and donning gas masks in France and Belgium
when they'd be asked to take on the Kaiser and his dreaded Huns.

With the threat of the U.S. involvement in the world war a real
possibility, Frank began to reconsider his decision to get out of the oil
business. He was having second thoughts. After all, it didn't take a genius
to realize that the days of the cavalry and horse-drawn wagons were numbered. Modern warfare would rely on motorized vehicles, tanks, and even
airplanes. That would create a tremendous demand for petroleum. The
price of Oklahoma crude was bound to skyrocket.

Frank's mind was turning faster than a handful of dice. He summoned L.E. to his office in the bank building on Third Street for another
conference. After Frank laid out his thoughts about the need for oil to
grease and fuel the machines of war, the room was silent. But just for a few
seconds. As soon as L.E. nodded his approval, Frank jumped up and
grabbed his brother's hand. "Hell, L.E.," shouted Frank, reversing his
proclamation, "we're not bankers, we're oil men."

In truth, L.E. wasn't joyous over the new scheme. He liked being a
banker. He also knew he'd be disappointed if the plan to launch a chain of
big-time banks fell through. Still, L.E., as was his custom, listened to his
big brother. Besides, he thought, Frank promised that all he wanted to do
was explore their few remaining leases out in the Osage. If they didn't turn
out to be worth much, as everybody predicted, then the Phillipses' oil days
would definitely be caput. There was still hope that they'd soon be up in
Kansas City overseeing their string of banks.

Frank immediately summoned his remaining top oil-field hands from
the Lewcinda company to a meeting. He gave them orders to proceed at
once with drilling on the last of the Phillips leases, located about ten miles
west of Bartlesville, past George Getty's Lot 50 in the Osage. Frank knew
the area pretty well. He had spent a good deal of time in those parts nosing
around. It was that outlaw country where Henry Wells and the others
found refuge, out around Okesa.

The Phillips Lot 185 lease was on a tract acquired from Foster some
years before—virgin land covered with scrub brush and wild blackberries.
Wolves and panthers were known to roam through the thickets stalking
their prey. Frank liked the Indian legends and the tall tales he heard about
the bandits who hid out in those parts.

Drilling commenced on these last Osage leases in 1916, and it looked
like all that Phillips had feared about the lack of oil on the lot was true—

nothing but dry holes or gas. Still, the crews kept working. Frank had a hunch about this land. He went there almost daily, driving a buggy up the steep "44 Hill," just to see how the drillers were doing. Same story every day—no luck.

But in February 1917—just a month after Buffalo Bill Cody died and two months before America declared war on Germany—all that changed. Frank sensed his luck was definitely improving. He told his boys to keep working.

Talk in town wasn't just about the war. It was about the Lewcinda crews in the Osage. The consensus was that the Phillipses had finally met their match. After taking the town by storm, it appeared their luck had run out. Six dry holes were all they had to show for all their hard work on Lot 185. It had to be worthless as far as oil production was concerned, good for nothing but growing berries and scrub trees.

Frank disagreed. He still had his hunch about the Osage lease. He also still had plenty of gumption and audacity. And lots of optimism. This was no time to become a pessimist and turn and run. Frank believed that no matter how many dry holes he drilled, the next one was going to hit.

The Lewcinda crews had drilled the first wells on Lot 185 back in 1914, about the time Waite departed, and had turned up only a little gas. But it was enough to tease them and keep the operation going. Also, when the H. V. Foster blanket lease on the Osage Nation lands was scheduled to expire once more in 1916, Frank was notified that he had to start testing the acreage in Lot 185 or wind up forfeiting the lease.

Then, on February 13, 1917, No. 7 came in as a modest producer, turning out 100 barrels a day. By no means a gusher, the flowing well at least inspired Frank and his men. It also entitled Frank to more land. A provision in Foster's Lot 185 lease made it clear that any leaseholder who discovered significant oil and gas earned the privilege of acquiring the adjacent acreage. As a result of uncovering oil in Well No. 7, Frank picked up the 320 acres in Lot 186. He then re-leased all but the proven 320 acres of the more than 1,000 acres he controlled in Lot 185.

Frank was ready to try again. The Creek Indians believed that wild onions were just the tonic for giving a person strength each year. That spring, Frank feasted on wild onions daily. He wasn't going to take any chances. After another supper of corn bread and fried onions, he returned to Lot 185 to confer with his drillers. They picked out a likely spot just west of the last well and, on February 28, started drilling No. 8.

Now, almost a month later, with the world around him in turmoil and a new season underway, there was still no oil in sight. But Frank knew they were close. And his wildcatter instinct told him that this well—surrounded by plugged "dusters" and lowly gas producers—would be the big one. Besides, it had to be nothing but lucky to be drilling on land leased from H. V. Foster, the richest man west of the Mississippi, not more than a half mile away from Lot 50, where the Getty fortune got its start.

Frank really wanted to be there when Lot 185 paid off. He got his wish. It was March 21, 1917—the date the vernal equinox marked the beginning of spring. The air was still chilly and damp, but the blackjack oaks were ready to sprout leaves and clumps of sumac and pokeweed and the other brush had started to come back to life. That afternoon Frank was standing on the rig platform when the Bartlesville sand was struck at 1,790 feet. After drilling only seventeen feet into the sand, the well was shut down so the crew could prepare for the anticipated flow of crude. The No. 7 well took an ace shooter and a hundred quarts of nitroglycerin to get it going. Well No. 8 was a different story. Not one drop of nitro would be needed—it flowed naturally and never had to be shot.

The next morning Frank returned to the site in his buckboard pulled by a team of bay ponies. He gave the order to commence drilling, and after a little more than half a screw had been turned, or about four feet, the wooden platform began to tremble and a low rumbling came from deep in the earth. Bill Noland, a driller for Lewcinda since 1915, was standing near Frank when the oil, straining to break loose, started to growl. Noland pounced on Frank, like a bobcat on a jackrabbit. He grabbed his surprised boss in his arms and the two men dove off the platform and fell to the ground clear of the rig just as the oil and debris burst from the well. It was a tremendous flow and oil was shooting over the top of the derrick before the tools could be pulled out.

Dripping with crude oil, Frank hugged Noland and shook the hand of each member of his excited drilling crew—their happy shouts drowned out by the roar of the gusher. Working as quickly as they could, the crew connected the well with a 250-barrel tank, but in less than an hour the tank was filled and oil was spilling over the countryside. Five more lines had to be connected to the well and a pump installed to force the oil into earthen tanks. Frank remained at the lease all night, swigging his tin cup of "jamoke" laced with Scotch whiskey, as gangs of workmen were summoned to help throw up more tanks and stem the flow of crude oil.

There was so much commotion that some of the people from Okesa wandered out to see the gushing well. Some of them had kin or friends working for Lewcinda Oil. Henry Wells and a few of his cohorts also heard the racket and paid a call. They tied their horses' reins to a stack of old oil-field pipe and stood off—away from the crowd—under a cluster of black-jack oaks and watched.

The next day, March 23, Bartians awoke to find the Frank Phillips critics were wrong again. The brash Iowan hadn't lost his Midas touch. He had found another Anna Anderson. On the front page of the newspapers, alongside stories telling of the Germans and British battling in the trenches of France and the latest dispatches about the abdication of the Russian Czar, was the news of Frank and his big gusher out on Lot 185 in the Osage.

Nothing but dry holes on Lot 185, and then they hit a gusher. It dashed forever any idea of Frank Phillips leaving Bartlesville and the oil patch. Try as he might, Frank was not able to wipe the Oklahoma crude from his hands. Now—and forever—the black gold was in his heart and in his soul.

CHAPTER THIRTEEN

"A Wonder Well," screamed the headline in the *Washington County Sentinel*. Although, in truth, the well more than likely was good for at least 1,000 barrels a day, the newspaper stories boosted the output to more than 3,500 barrels a day. No matter—there was more than enough production to earn the Lewcinda's well the enviable claim of "biggest oiler in local field." The whole town celebrated.

"After drilling six dry holes on Lot 185 in the Osage, Frank Phillips, president and owner of Lewcinda Oil Company, yesterday startled the local oil fraternity by bringing in the biggest well yet brought in around Bartlesville," said the *Sentinel* story. "There are no other completed wells in this section of the Osage, for the territory has always been considered dangerous."

In the Bartlesville *Morning Examiner,* another reporter—who liked using the word "immense"—was obviously impressed with the amazing Mr. Phillips. This story claimed the new discovery would "open up a pool of immense richness." The newspaper account went on to explain that Frank Phillips had "a 'hunch' that the well would be a big one," and predicted that its discovery "may mean some immense development in that portion of the Osage country and will mean an immense sum of money coming into Bartlesville."

Crowds of townspeople went out to the Lewcinda well on Lot 185 and found that everything the newspaper reports said was true. The five gravity-flow lines were carrying excess oil to earthen tanks and a gasoline pump was working overtime trying to force the remainder of the oil to the surface line in the prairie.

Later that same spring, the other shoe dropped on the proposed bank venture. Just when the oak leaves were in full bud in the Osage hills, Woodrow Wilson—in his second term as President—convinced Congress to declare war on Germany. Plans were made to send tens of thousands of American doughboys overseas. "The world must be made safe for democracy," Wilson told Congress. George M. Cohan put it another way: "The Yanks are coming." And come they did, in every imaginable motorized vehicle in existence. Just as Frank had figured, the market for oil and gas exploded. The buried treasure the Lewcinda drilling crews tapped on Lot 185 was part of the thousands of barrels of oil that helped stoke the fires of war.

There were more oil and gas wells drilled by the Lewcinda crews during the spring and summer of 1917. Most showed an initial production of anywhere between 30 to 150 barrels a day, and No. 16 came in at 1,200 barrels per day.

By this time, the banking-chain plan was totally discarded. Frank and L.E. decided to keep their bank in Bartlesville and consolidate their remaining oil properties into the Lewcinda Oil & Gas Company. They also set about expanding the firm's holdings. Soon Lewcinda was valued at more than $1 million. Frank and L.E. owned almost the entire interest. Money was rolling in, and the prospects looked even better. With the war in Europe going full throttle, the price of oil skyrocketed from less than forty cents a barrel to more than a dollar a barrel.

Frank and L.E. realized they were fortunate to own their own bank to help underwrite many of their oil-field operations. But they also knew that because the industry was on the brink of major change, due to continuous technological developments, the cost of competing required major capital investment. This financing could only be found in the large financial centers back East.

Frank became more certain that the time was right to organize a public company, sell stock, retain a good share of it, and keep right on going. There were no limits. All they needed was a name.

Through the early years in Bartlesville, the Phillipses had formed many private oil companies, each with its own identity. They named them for their wives, their parents, a revered ancestor, and even a favorite town from back in Iowa. Now that they were going to form a public company, another name would be needed. It would have to be a name of distinction that could instantly tell the world what kind of business Frank and his brother L.E. were running.

Frank had a name in mind right along. He thought it was straight to the point and no-nonsense. He tried it on Jane and John. They liked it. He tried it on his best hands from Lewcinda. They liked it too. He wrote the name out and looked at it. He repeated the words out loud. Frank liked the way it looked and sounded. It was the one name that said it all.

"Phillips Petroleum Company."

It was a natural.

CHAPTER FOURTEEN

Frank Phillips was forty-three years old when he signed his name to the sheaf of papers incorporating Phillips Petroleum Company. The date was June 13, 1917. It was a Wednesday morning—bright and sunny—and Frank was in New York working out last-minute details and taking care of the financial arrangements. The company was incorporated for tax reasons under a Delaware charter, with assets listed at $3 million.

After weeks of planning and wheeling and dealing, and a barrage of wires and correspondence between Bartlesville and eastern lawyers, Phillips Petroleum was a reality. Frank was ready and eager to take command and start doing business under the new company name.

But before the impatient Mr. Phillips could assume control of the new firm, he had to endure at least one strategic maneuver. During a meeting in New York conducted shortly after Frank signed the incorporation documents, the firm's first president was chosen. It was not, as expected, Frank Phillips. That was the twist. Instead, Robert S. Sloan, member of a New York law firm, was elected the first president of Phillips Petroleum.

Sloan's election was actually just a technicality. He was, in fact, a professional incorporator whom the real founders of the new oil company —Frank and L.E.—hired to get the firm off the ground. The technique of using incorporating experts, such as Sloan, to serve as temporary corporate officers was a common practice in order to legally establish a new company and to be sure that the corporation was properly organized. But from the start, even during the brief period when Sloan acted as president, there was really never any doubt who controlled Phillips Petroleum.

Watching from the wings all the time were Frank and L.E. Their assets consisted of the Lewcinda Oil & Gas Company and Standish Oil, comprised of twenty-seven employees who worked for the two firms, and an array of leases, equipment, and rigs scattered over the oil patch.

The first Phillips Petroleum board of directors meeting was held in Bartlesville on August 8, 1917—Jane Phillips' fortieth birthday. As the first order of business during that initial meeting, the new company absorbed all the assets of Lewcinda Oil. At a second director's session, held just two days later, half the capital stock of Standish Oil was purchased. Lewcinda, which supplied most of the first Phillips Petroleum employees, held only a dozen leases, covering 15,500 acres in Kansas, Oklahoma, and Kentucky. At the time the firm was absorbed by Phillips Petroleum, Lewcinda had only four gas wells and sixty-eight oil wells producing an average of 380 barrels daily, mostly from properties in the Osage and throughout Washington County.

At the first board meeting, the interim company directors and officers, including temporary president Sloan, resigned. This allowed the real Phillips Petroleum players to be unveiled.

Frank Phillips was elected president and a director; L. E. Phillips was chosen vice president and a director; and Henry E. Koopman, the Kansan who first joined the Phillips brothers in 1911 as a stenographer and book-keeper, was named secretary-treasurer and a director. Koopman, a trusted officer of the firm who would eventually become a vice president and serve as a director for almost thirty-seven years, also found himself one of only three people to own original shares of stock in the fledgling $3 million company. Frank was issued 59,925 shares of common stock; L.E., 15,000 shares; and Koopman, 75 shares.

After the permanent officers were chosen, the company's first board of directors was selected. Besides the two Phillips brothers and Koopman, two others were named as directors—George S. Marshall, for a brief time the manager of Phillips Petroleum's land and lease department until he resigned and L.E. took the job, and Frank's son, John Gibson Phillips, only nineteen years old.

Besides being named a director, young John was also made assistant secretary-treasurer with the company, and was given a desk in the crowded Phillips Petroleum offices in the bank building on Third Street in Bartlesville.

John was known around town as a "spoiled rich kid," and his deportment problems, brought on by a growing affinity for strong drink, led to his expulsion from Hotchkiss. He returned to Bartlesville, where he and a band of cronies—mostly other local boys with fathers grown wealthy off the oil business—quickly earned reputations as hell-raisers.

When John and Mildred Beattie, the young daughter and only child of a prominent Bartlesville banker, announced their engagement, every gossip in town labeled it a union of convenience, arranged by Jane Phillips to keep her son out of the Army. Whether for love or safety's sake, John and Mildred were married in Kansas City, Missouri, on March 1, 1917, less than three months after John's nineteenth birthday.

Mildred, an attractive young lady, was educated at Monticello School for Girls in Alton, Illinois, just upstream from St. Louis. Her father, Robert Lee Beattie, started his banking career in Missouri and later moved to Kansas, where he met his wife, Betty Claypool of Bowling Green, Kentucky. Beattie served with a Kansas City bank before moving to Bartlesville and eventually he became president of the Union National Bank, a venerable institution which rivaled Frank and L.E.'s bank and counted H. V. Foster as an officer and member of the board of directors.

The newlyweds, with ample help from their parents, set up housekeeping in Bartlesville. John was prepared to put his best foot forward. Aware of his "Peck's Bad Boy" image, he wanted to make something of himself and please his bride, and especially Frank and Jane.

But in Jane's eyes her dearest John, given to spells of petulance, could do no wrong. He was the darling of her life. Frank, however, had a different opinion about his son. He was forever exasperated by his antics. But, mindful that John was his only child, and resigned to the fact that Jane could bear no more children, Frank held out in hopes that this marriage to the Beattie girl would help. He continued to groom his sole heir for his rightful place in the growing Phillips empire and thought that if John was given responsibility to shoulder, as both an officer and a director of their new oil company, there would be a change for the better.

John duly accepted the challenge. He was diligent and he tried to mend his ways. He was going to do his best to show Mildred and his parents that when it came to matters of business, he was a chip off the old block. In truth, John Phillips' mind—and certainly his heart—were never in the oil business. It would take a few more years for that fact to become clear to everyone.

In the beginning, John and the other Phillips directors dutifully carried out their corporate chores. Also, they took all their cues from the watchful Frank, who frequently summoned them to his office for board meetings during the first few months after the founding of the company.

On September 27—a little more than three months after the firm was incorporated—Frank called one of his special meetings of the board of directors. A decision needed to be made concerning the offering of shares of preferred stock. The board reviewed the reports, listened to Frank's recommendations, and took swift action.

The Phillips directors decided that from October 1, 1917, to August 16, 1918, a total of 15,000 shares of Phillips preferred stock would be sold, with an equal number of common shares given as a bonus. Frank and L.E., naturally, received the lion's share. The rest of the shares of preferred stock were purchased by eighty-three people, including Zack Miller of the 101 Ranch; Arthur B. Eisenhower, older brother of Dwight D. Eisenhower; H. H. Westinghouse, founder of the Westinghouse companies; and Wirt Franklin, a pioneer oil man. By September 1, 1919, all of the Phillips Petroleum preferred stock was retired and the company never again authorized or issued any more.

Phillips Petroleum kept Frank and his officers and directors busy day and night, but for most people in Bartlesville the founding of the oil company hardly caused a stir. It was business as usual for the townsfolk. After all, oil companies came and went. Besides, the Phillipses had started many other firms in the past. This was just another.

Even the Lewcinda crews—the roustabouts and tool dressers who made up the majority of those first Phillips employees—didn't notice any changes in operating procedures. It didn't matter to them if they were working for a private or a public company. All they knew were two things —that they were expected to work their butts off six days a week and Frank Phillips was still the boss.

To ensure instant success for his company, Frank saw to it that the best hands in the oil patch came to Phillips Petroleum. Many already worked for Frank and there were others who were lured to the new company because of the Phillipses' reputation for treating their workers fair and square.

On June 1, two weeks before the company was incorporated, Clyde Alexander—the former teamster and soup wagon driver who could smell oil better than anyone else—reported for duty. Frank wanted Clyde back working for him ever since the two men parted company in 1910, when the Creston Oil Company was sold and Clyde joined the new owners. Now he had returned. Frank immediately made Alexander the general superintendent of production and turned him loose to find more pools of oil.

John Stewart Dewar, a native Canadian who came to Bartlesville in

1911, was hired to assist Alexander. Although Dewar soon left for two years of active duty in the infantry, he rejoined Phillips in 1919 after he was discharged as a first lieutenant.

There were other bright and shining stars added to the company roster. Oberon K. Wing, better known as O. K. Wing, or simply "Obie," came aboard in September to assist Koopman and help run the accounting department. Wing was a Missourian and only five years older than John Phillips. He had moved to Bartlesville the year before to work for the Empire companies, but when he heard that Frank was going to stay in town, Wing switched to Phillips. He soon became one of Frank's favorites and rose rapidly through the corporate ranks.

John H. Kane and his former law partner, Judge R. H. Hudson, provided legal services for Phillips Petroleum; A. C. Branson was the first purchasing agent; Earl Beard, former superintendent of production for Lewcinda, transferred to Phillips; Paul J. McIntyre joined the land and geological department; William N. Davis, eventually a vice president, handled any governmental matters that arose; H. A. Trower was put in charge of natural gasoline sales and transportation; F. E. Rice, bringing experience in design and construction, became chief engineer; and J. M. Sands, employed as a consulting geologist, quickly became head of the geology department and by the early 1920s took the reins of the newly created economics department.

The first female employee at Phillips Petroleum was Sydney Fern Butler. A former schoolteacher, Miss Butler was hired to serve as Frank's secretary as well as operate the company's primitive telephone switchboard.

Ralph Stewart was the company's first truck driver before he went off to war. As a boy, Stewart shot marbles and raided neighborhood orchards with John Phillips. Now he worked for his former playmate. He hauled supplies and pipe and equipment in an old truck with solid tires. The truck could only go about twelve miles an hour, but it was still the best way to get up the steep "44 Hill" on that two-rut trail they called a road between Lot 185 and Bartlesville. Stewart used up an entire day just to make the round trip, but at least he didn't need six teams of horses to pull him, like the wagons did, when the road became muddy. Rain or shine, Stewart managed to keep the rigs supplied with fresh pipe and tools.

Out in the field, Frank's crews were bossed by oil-patch veterans Bill Noland and Benjamin Gray Barbee. Noland was the alert driller who pushed Frank off the drilling platform when the Lot 185 well was getting ready to blow, and Ben Barbee, Noland's father-in-law, had been a Phillips roustabout since 1914, when he started on Lot 195.

Patriarch of the Barbee family—early settlers in the Okesa area—Ben Barbee had served as a U.S. marshal and kept the peace in Dewey, Oklahoma, with Tom Mix, before Mix hung up his star and six-shooter and rode off to film stardom in Hollywood. Barbee also knew men on the other side of the law and on occasion shared a plug of tobacco with the outlaw Henry Wells. As soon as Phillips Petroleum was formed, Ben—a strapping man at six feet two inches and two hundred and forty pounds—saw to it that his son, William "Mark" Barbee, landed a job on Lot 185. The Barbees pulled down seventy-five dollars a month each in wages. Just for the hell of it, Ben's younger boys worked on the rigs alongside their papa and big brother for no pay.

When any of his crew got hurt, Frank had Dr. O. S. Somerville tend to them. Somerville came to Bartlesville in 1905 to open a medical practice. He first met Frank Phillips at his bank when the young doctor needed to borrow money to pay his office rent.

"He was a smooth-looking, clean-faced fellow with a banker's attitude," Somerville recalled of that first encounter. "He looked you over real good before he started any business relations with you."

Somerville got his loan, and years later he got Frank's business when the oil man started sending him injured or sick workers. After Phillips Petroleum was organized, Somerville went on salary with the firm, but Frank allowed him to treat private patients in his spare time. That all changed in only a few months when operations out at Lot 185 picked up. There were no safety regulations in the oil field at that time and Somerville was on call day and night. He'd take his horse and buggy over the rough roads to set bones or sew up gashes. One night he even delivered a baby in the worker's camp on the Lot 185 lease. Bill Noland was the proud father, and he and his wife named their son James Phillip, in honor of Frank. When he reached manhood, James Phillip Noland, grandson of old Ben Barbee himself, joined Phillips Petroleum as a roustabout in the Texas panhandle and worked for the firm for forty-four years.

There were more and more babies named for Frank. Robert Jeffers, an early Phillips employee and the husband of Ben Barbee's daughter, Tennie May, pumped leases for Frank Phillips for close to forty-three years and was described by one of his sons as "a rowdy who raced horses, roped goats, called square dances, raised Shelton Roundhead cockfighters, and fought at the drop of a hat." Jeffers also liked Frank so much that all five of his boys went to work for Phillips and he named one Frank Phillips Jeffers. News of the Jeffers baby pleased Frank and he sent the family a twenty-five-dollar War Bond in behalf of his namesake.

But there was a lot more for Frank Phillips to keep track of besides namesakes. All the bright young men Frank was hiring were coming up with ideas to improve business and expand the bottom line. For instance, some of Frank's hotshots convinced him there was actually a good use for natural gas.

At the time Phillips Petroleum was organized, most oil men didn't consider the natural gas discovered along with crude oil an important source of energy. They thought it was nothing but a nuisance. In most places natural gas was simply "flared off" at the wellhead because, without pipelines and a distribution system, there were few ways the gas could be used. But that was changing. Even before the company was organized, Frank committed his resources to building a "plant to extract liquids from natural gas." By October 1917, construction of the Hamilton Gasoline Plant was completed. Located between Bartlesville and Dewey, it was Phillips' first facility for manufacturing motor fuel by transforming waste gases into a marketable product. After the waste product was blended with naphtha, Frank's workers sold the casinghead gasoline to passersby from a tank alongside the road in front of the plant. The sales didn't bring much profit but did provide Frank and his young bucks with some early insight into the worlds of processing and marketing.

After the Hamilton plant was off and running, Frank turned his attention back to Lot 185, the lease that was responsible for the birth of his growing company. Things were popping out on the Osage lease, which, by now, was Frank's favorite hunk of real estate. The workers' camp had been expanded to include a boardinghouse and a school, and new oil and gas wells were being drilled. Ben Barbee's son Calip, who had come to work for Phillips at the age of fifteen to clear land at Sand Creek for four dollars a day, moved his wife, Anna, and their two children into one of the company houses on Lot 185. "The house was made with one-by-twelves," recalled Calip. "It had a bedroom, a sitting room, a kitchen, and a path out back leading to the little house. No running water. No electricity. We put linoleum on the floor, and if you didn't nail it down, the wind would seep in under the door and make the linoleum flap like a flag in the breeze."

By the summer of 1918, a second casinghead gasoline plant was opened near the Lot 185 lease. Called the Osage plant, it was able to produce more than two thousand gallons of natural gasoline a day, several times as much as the Hamilton plant turned out.

In order to supply water to the gasoline extraction plant and the steam boilers running the many wells springing up on the lease, Clyde Alexander saw to it that his men hooked up a pump to a nearby spring-fed lake. Located about a mile and a half from the Lot 185 camp, the small lake—actually an overgrown stock pond—was surrounded by steep rocky bluffs and woods. The locals called it Rock Lake.

Frank admired the natural beauty of the area, especially the huge rock cliffs and the lake below. He liked going out to the lease in his buckboard and then going over to check on the pump at Rock Lake. There were many other small lakes scattered about his land, but the rustic setting of this natural lake, among the rocks and trees, appealed to him the most. Frank had his boys build a modest cabin on the bluff overlooking the lake. He called it his "ranch house." And whenever the tension of running a multi-million-dollar oil business got to him, Frank went to his sanctuary in the Osage and hid out, just like the outlaws.

A local character named "Daddy" Miller ran the water pump and, whenever he could find the time, caught a few bass and perch in Rock Lake. Earnest Traywick, an eighteen-year-old roustabout, helped operate the pump and also acted as Frank's official driver after the horse-drawn buckboard was replaced with a Model T Ford.

Ironically, Frank Phillips—the man who early on predicted the success of the internal-combustion engine and would spend most of his adult life involved in the oil and gasoline business—never learned to drive. He remained an accomplished horseman but left driving to others. "I'm a thinking machine. I don't have time to drive," is how Frank explained his aversion to operating a car.

When young Traywick came from Arkansas to teach school in Oklahoma, he saw there was more money to be earned in the oil field, so he took up the life of a roustabout. One of his main jobs with Phillips was to drive Frank, sometimes joined by L.E., on surveying trips through the countryside near Bartlesville. They mostly searched for more oil and gas wells. Traywick would drive until he was told to stop and then Frank and L.E. would get out and walk around the land. When they found a likely place they'd have Traywick, always armed with a hatchet and a bundle of stakes, mark the site. Next the rig builders and drillers would come with their timbers and cable tools and go to work.

By the close of the first year of operation, Phillips Petroleum was beginning to take shape. The handful of employees were still jammed into the second floor of the bank building in Bartlesville, but the future was as bright as an Osage sunrise. Thanks to the war, petroleum sales spiraled. Tin lizzies were chugging down the roads and airplanes were no longer novelties. Under Frank's guidance, the company was expanding right along with the growing market.

Frank paid out $12,963.67 in executive salaries in 1917. George Marshall and H. E. Koopman each brought home $1,535. Bill Noland earned $940. And Clyde Alexander, the trusty Phillips superintendent of production, received the biggest salary—a whopping $1,950.

L.E., in a fit of patriotic fervor, devoted much of 1917 and a good part of 1918 to the war effort. "Laying aside all excuses," as he put it, L.E. temporarily left the banking and oil and gas business and, until the close of the war in November 1918, spent his time diligently serving on the home front.

L.E. put his heart and soul into raising money for the war cause. He was so successful in his efforts in Bartlesville that Governor Robert L. Williams made him a member of the State Council of Defense and chairman of the entire Red Cross and Liberty Loan campaigns for Oklahoma and the panhandle of Texas. L.E. set up his headquarters in Oklahoma City and traveled throughout the Southwest delivering speeches and preaching the doctrines of loyalty and patriotism. One of L.E.'s coined phrases— "All that is needed for membership in the Red Cross is a heart and a dollar"—was adopted by the national Red Cross for fund-raising campaigns.

Meanwhile, back in Bartlesville, in an effort to keep the dollars coming in, Frank was putting his heart—and a good deal of his soul—into the fledgling Phillips Petroleum Company. The hours were long and the work demanding, but Frank worked right alongside his men in the field and stayed at the office every night until he couldn't keep his eyes open. Frank's devotion to both his business and his workers didn't go unnoticed. Traywick and the other Phillips employees would have fought a panther if Frank asked. "Firm but fair," is how Traywick remembered Frank Phillips. "If you had a problem of any kind you could go to him and talk. Even before there was an employee benefits program, I was never docked or lost a day's pay. If you were sick you stayed off work until you were well and that was that. He'd give us pep talks just like we were his own boys. We all liked him."

Frank treated the Phillips employees—from roustabout to vice president—as if they were members of his family. They all learned quickly that Frank was stern but evenhanded. They also learned that he lived by one steadfast rule. "Work hard and demonstrate loyalty, and I'm a great guy to work for," said Frank. "Do neither and there is no one worse."

Out in the oil field, the word got around that Frank Phillips was a square shooter and his Phillips Petroleum Company was a good outfit with a bright future. Although he could be bombastic as hell and as short-tempered as a cornered skunk, if a worker out at the Hamilton plant or a roustabout breaking his back on Lot 185 ever had a problem, he could always go to Mr. Phillips and talk.

"Why, it's just like visitin' with one of your uncles," is what the oil-patch boys said about Frank. And pretty soon that's the name the workers hung on him. Whenever they saw him coming in the Model T, with Traywick or somebody else driving—maps and wooden stakes going every which way—the boys would call out: "Howdy, Uncle Frank, how are you today? And how's Aunt Jane?"

Frank would wave and flash his best grin. Then he'd motion to the driver and the Model T would chug on and disappear in a cloud of dust. The crews would return to their work knowing full well that Uncle Frank —his face buried in a map—was intent on locating another good site for his next oil well before nightfall.

CHAPTER FIFTEEN

Not even the old Osage chiefs knew for sure the age of the big elm tree shading the lawn on Agency Hill in Pawhuska. Songbirds nested in the broad spreading branches each spring. In the fall, when cold rains washed across the Osage prairies, the pointed oval leaves with toothed edges turned yellow and brown and gathered on the ground like clumps of wet tea bags.

The tree was tall and stately, but most of the Indians and whites who climbed the steep hill to transact business or attend a meeting in the Osage Agency paid the tree no mind. It was simply there, like the grass and the stone building and the sky.

About the time Phillips Petroleum was getting off the ground, all that changed. The big elm was destined to become well known, especially among the oil men. Anxious to grab up Osage land, oil operators and wildcatters flocked to Pawhuska at the invitation of the tribe and, when the crowds were too large for the local theater to hold, they stood beneath the old elm tree to bid on Osage leases.

The public sales of oil leases on the Osage reservation started at the Indian agency grounds in Pawhuska years before when the Interior Department and the Osage tribal council came to terms and agreed that a competitive auction system was the best method for selling future leases to oil companies. The first leases were sold by sealed bids on April 11, 1912, and several months later the first public auction of oil leases on the reservation took place.

In March 1916, the H. V. Foster blanket lease expired and ITIO's hold was finally broken on the Osage lands. This meant that all oil leases in the Osage could be acquired only through the public auctions. Tracts with wells producing an average greater than 25 barrels of oil per day were sold at auction within a month. Public lease auctions at Pawhuska became a way of life for oil men interested in the Osage.

The early lease sales back in 1912 had been quiet affairs, attracting limited interest since only unexplored tracts of land were on the block. But after the blanket lease expired and all of the Osage tracts were up for grabs, the number of bidders increased and many of the auctions lasted until past midnight.

When the weather was bad the auctions were held in the Constantine Theater and later the Kihekah Theater, but usually the crowds attending the auctions were so large that the sales were held beneath the big elm—the solitary tree just north of the Osage Council House. Because of the vast fortunes paid for the leases the tree became known as the "Million Dollar Elm." It was a fitting name. It was estimated that during three-quarters of a century of oil development in the Osage, more than $300 million was paid to the tribe.

Since 1906, the Osage—thanks to the oil-lease auctions under the elm—had become the richest Indians in the world. The tribe's mineral rights were held in common, with all oil and gas revenues divided equally among the 2,229 Osage on the tribal rolls. Each share was called a headright. In 1914 more than $1 million was deposited in the Osage coffers. By the time the ITIO blanket lease was broken up two years later and the region was opened to outside oil men, the increase in competition raised the Osage headright payments to $3,672.33 for every man, woman, and child and by 1923 leaped to $12,400 per headright.

Before Phillips Petroleum Company could celebrate its first anniversary, the average Osage family received a yearly income ranging from $5,000 to $10,000, with many families getting much more. News of the Osage's wealth made the tribe famous around the world. The biggest oil men of the day and lots of curiosity seekers came to Pawhuska.

Crowds attending the auctions gathered under the elm tree in the early morning, when the air was crisp and dry. On Saturdays, ranchers and farmers from throughout the Osage came to Pawhuska to buy feed and supplies and gawk at each other on the crowded streets. Many would find their way up the hill to the agency and the elm tree. They'd stare at the Indians and they'd point at the oil men's fancy automobiles parked off to the side and the chauffeurs who polished the chrome trim with their coat sleeves.

In the years following the formation of Phillips Petroleum there were many such Saturdays in Pawhuska. And Frank Phillips was usually there. Even in the autumn of 1918, when his company was only a year old and just days before the Great War came to a close, Frank came to Pawhuska. He could be found on the hill above the business district, watching the Indians and waiting.

Frank liked coming to Agency Hill to bid on the oil leases. He enjoyed the company of the other oil men and he especially liked seeing the Osage. On auction day in Pawhuska, he'd arrive early, as was his custom. He'd quickly find a good perch in the bleachers or else sit against the elm and study the faces of his competitors and the Indians and the spectators. Old full-blood Osage and their squaws squatted on the grass near the tree. They dressed in colorful blankets, buckskin leggings, and beaded moccasins. Their hair was in long braids, and most wore traditional beaver-skin caps or Stetsons with eagle feathers tucked in the bands. Their earrings were gold and some of the old men had bear-claw necklaces around their necks. A few, even when the weather was cool, clutched eagle-plume fans. They sat silent and dignified and talked only with their eyes.

All around the old full-bloods gathered groups of younger Osage and clusters of white men dressed in fine business suits and neckties. Tending to ignore the old Indians, they babbled away about the big news of the day, which in the summer and fall of 1918 meant talk of the end of World War I.

In early November, the Germans were about ready to throw in the towel and admit defeat. The German high command, close to signing an armistice treaty, conferred with the Allies in a dining car in a French forest. The crowd surrounding the Million Dollar Elm in Oklahoma knew that soon the boys—including a good many local cowpokes, oil workers, and Osage sons—would be coming home. Some had already returned. They lay at rest in the Osage graveyard, where American flags waved on tall poles at each plot.

The gossip swapping and talk of the soldiers' return eventually stopped. There was business to tend to, and the chatter about the end of the war could wait until lunch. Almost in unison, the chauffeurs leaned back against the shiny automobiles and folded their arms across their chests. A few latecomers took the last open spots in the wooden bleachers set up around the elm tree. It was time for the auction to start.

Colonel Ellsworth E. Walters, a man with a booming voice from Skeedee, Oklahoma, conducted the auctions. Walters was a showman and used humor, dramatic gestures, and a modulating voice to keep these affairs exciting and the bidding lively. He was smooth as corn silk and could charm a crowd of anxious oil tycoons better than a show girl. His fee for an auction was only $10. Later the tribe decided Walters was worth more and they raised his fee to $100 and provided him with $50 to cover his two-day trip from Skeedee to Pawhuska. In his first eight years of peddling Osage leases Walters took home less than $140, while the tribe earned more than $27 million. The stoic Osages were so pleased with Walters and his rapid-fire gavel that they presented him with a diamond-studded badge and a giant diamond ring, which he proudly wore to every lease sale. Not to be outdone, the Miller brothers, owners of the nearby 101 Ranch, gave Walters a jeweled tiepin to go with his fancy Osage gifts.

Prone to using homespun wit to get his audience involved in the auction, Walters would poke fun, cajole, and chatter like a squirrel in heat just to get the crowd going. "Come on, boys, this old wildcat is liable to have a mess of kittens," was one of the flamboyant Walters' favorite lines. He started the auctions at about ten-thirty in the morning and sold leases until noon. After lunch, the auction resumed. Walters "cried" the sales until a supper break, catered by the Christian church, followed by more auctioning of leases until late into the night. Wearing a celluloid collar and striped shirt, Walters liked to scan the crowd of prospective bidders, looking for even the slightest nod—a movement which usually boosted the price by tens of thousands of dollars. When a lease was sold, the bidder sent a runner to the closest telephone to alert the company's drilling crews. As soon as the call came, roustabouts, horses, and equipment rushed to the lease site so drilling could commence. The "laws of capture" left little time to get the crews in the field and drain the black gold from the earth.

At one auction, Frank and L.E. sat under the elm listening to Walters try his best to drive up a bid for a lease by another $100,000. There were no takers. Walters was close to slamming down his gavel when he noticed L.E. flick his fingers across his nose. Thinking the movement an apparent gesture to up the ante, Walters accepted L.E.'s offer and closed out the bidding.

Frank later claimed his brother hadn't really bid on the lease, but was just swatting a fly. "L.E. cost the company $100,000 because he couldn't stand the flies," laughed Frank.

At another Osage auction, Walters mistook the sleeping nods of W. J. Knupp, an independent driller from Pennsylvania with a reputation as a tightwad, as signals that he was bidding on some rather uninteresting lots. Knupp awoke to find he owned the properties. He was horrified, but all his ranting and raving fell on deaf ears. Still, the luck of a wildcatter prevailed and before the year ended, Knupp's unproven property was supporting five producing wells and he ended up selling the lease for almost $1 million.

There was a lot of money to be made in the oil business and that meant lots of big money changed hands beneath the elm at Pawhuska. Any oil man worth anything was there. Harry Sinclair came from Tulsa in his private railroad car. Bill Skelly, E. W. Marland, J. P. Getty, and the Phillipses were always represented.

Waite Phillips came too. By 1918, Waite had moved his headquarters from Okmulgee to Tulsa, now acclaimed as the "Oil Capital of the World." He made millions of dollars with a daily output of 10,000 barrels when oil sold for $3.50 per barrel. It was said that by the time Waite and his family arrived in Tulsa his income was $30,000 a day. Frank's kid brother was a full-fledged tycoon.

But it took tycoon-style money in order to stay in the running beneath the Million Dollar Elm. Often the bidding became heated. During one auction, the Phillipses, Skelly, and Sinclair were all vying for the same tract. The stakes were high and tempers were short. After several minutes, the bidders became so agitated that Frank and big Bill Skelly, eventually Osage partners, started a shoving match with each other which quickly ended up with the two of them wrestling on the lawn. While they were battling, Sinclair gave Walters a nod and a wink and secured the lease.

A single Osage lease earned the tribe at least hundreds of thousands of dollars, usually a million. The tribe grew richer every day. So did the individual Indians. In 1919, an Osage couple with nine children received a total of almost $55,000 in cash bonuses and royalties. In June 1921, fourteen Osage leases brought $3.25 million, and a few months later a group of eighteen leases sold for $6.25 million. The first sale when a 160-acre tract brought a bonus of $1 million or more was on March 2, 1922. On March 18, 1924, the opening bid on a single tract was $1 million, and on March 19 the largest bonus ever paid for a tract was shelled out—a cool $1,990,000. In a single day in 1924, Walters more than earned his $10 fee when he auctioned off more than $10 million worth of Osage oil properties.

Beyond the huge prices paid as bonuses for the right to drill, the Osage received a royalty of one-sixth of the oil on leases producing less than 100 barrels a day, and one-fifth on wells bringing in more than 100 barrels a day.

Some of the tribal members used their newfound fortunes wisely.

They spent the headright earnings to educate their children or tucked large sums away for the time when the white men would no longer buy huge quantities of oil. But for many Osage, the oil-boom money ultimately proved to be a curse. They squandered their newfound wealth as if there was no tomorrow. A party of Osage—dressed in traditional moccasins, leggings, and jewelry—being chauffeured along a dusty reservation back road became a common sight. It was reported that in a single day in the 1920s one Osage woman spent more than $40,000, including $12,000 for a fur coat, $3,000 for a diamond ring, $5,000 for a new automobile, $7,000 for furniture, $600 to ship the furniture to California, and, finally, $12,800 for some swampland in Florida. During one month in the 1920s, when prime cuts of steak went for only a quarter a pound, one Osage forked over $2,000 just for food.

Oil-rich Osage thought nothing of hauling pigs in the back seats of their new autos. There were stories of those who left brand-new cars on the side of the road if the slightest thing went wrong. No need to bother with repairs when there was plenty of money to buy a new motorcar. One Osage bought ten new cars in less than a year, and another Indian, wishing to impress his friends, bought a hearse and hired a chauffeur to drive him around the countryside. Other Osage built large homes, filled with modern conveniences and lavish furnishings, yet they continued to live in huts in the front yard. A historian of the time noted: "Grand pianos often stood out on the lawns year round; priceless china and silverware sat on shelves while the Indians ate with their fingers . . . expensive vases were used to keep vegetables in or as corn bins."

Ho-tah-moie, or Rolling Thunder, a full-blood Osage and the grandson of Big Elk, one of the early Osage chiefs, was one of those least affected by the curse of oil. He was not only one of the most colorful members of the tribe; he was the most misunderstood. As a young man living near Pawhuska, Ho-tah-moie became ill during an epidemic and was thought to have died. He was prepared for burial in the best Osage traditions. His face was painted and he was clad in his fanciest costume. His body was placed against a tree so that he faced the east and he was covered with rocks.

Two days later, Ho-tah-moie, for no explainable reason, rose from the dead and walked into his camp, only to find that all of his possessions had been given away. His friends and relatives were terrified. They thought he was a ghost, and they shunned him. Among the whites, he became known as the Indian who died and came back to life. Ho-tah-moie left the camp and lived as a hermit in the hills with only a pack of dogs for companions.

While Ho-tah-moie was living as a pariah, more bad luck followed. He was stricken with "King's Evil," or scrofula, a tuberculosis of the lymph glands which produces a disagreeable odor. Ho-tah-moie soon became known as John Stink.

Then it seemed his luck changed. The reclusive Indian became very wealthy when oil was found in the Osage and he was awarded his share of the royalties. Frank Phillips drilled producing oil wells on the Indian's land and bags of letters poured in, many containing offers of marriage. But Stink ignored the money and the attention.

When his guardian built Stink a comfortable house, the old Osage moved his pack of dogs in and fed them steak while he continued to sleep in his crude log cabin. He lived to be seventy-five. He slipped on a rock and broke his leg, lingered for six months, and died. John Stink—the Indian without a tribe who managed to ignore the intoxicating boom years—was eulogized in *Time* magazine and was laid to rest for the second, and final, time. The legendary ghost of the Osage left behind his cur dogs and a sizable estate to be squabbled over by those who claimed kinship.

Not all of the stories of Osage wealth during the oil-boom years were as colorful as the tale of John Stink. The black gold that lurked below the tall bluestem grass brought a mixed blessing.

Many Osage, unsophisticated and unable to handle their newfound wealth, fell easy prey to the unscrupulous—hucksters, swindlers, corrupt guardians, shyster lawyers, or whites who married into the tribe just to gain the headrights. Embezzlement of Indian funds, political bribes, and fraudulent land sales were commonly used to cheat the Indians. Most of the Osage would have been better off if they had become hermits like Stink.

The oil madness became known as the "Osage reign of terror," climaxed by twenty-four unsolved murders in a three-year period. Indians were shot, poisoned, and blown up. Some just disappeared. By the early 1920s the situation in the Osage became so critical that tribal leaders appealed to the federal government for help. The government answered by sending teams of FBI agents into Osage County. The agents helped, but many Osage still lived in constant fear.

Osage Chief Fred Lookout, principal leader of the tribe for many years, summed up the situation: "My people are not happy. Someday this oil will go and there will be no more checks every few months from the Great White Father. There'll be no more fine motorcars and new clothes. Then I know my people will be happier." Despite his somber prophecy,

Chief Lookout enjoyed his chauffeur-driven limousine as well as his relationship with some of the whites who drilled for oil on his tribe's land. One of those Lookout liked the best was Frank Phillips.

Frank revered the proud and haughty Osage Indian and was a frequent guest in the home of Fred and Julia Lookout at Pawhuska. Any abuse of his Osage friends troubled Frank, who was one of the good whites old Baconrind referred to when he spoke before the tribal council and said: "There are men amongst the whites, honest men, but they are mighty scarce—mighty few."

Frank did everything in his power to see that his Osage friends were treated with the respect he believed they deserved. He would never forget that the early success of Phillips Petroleum Company came during the peak years of the Osage's great production.

Throughout 1918, Frank's four-wheel-drive Jeffry Quad trucks rumbled up and down the Osage Hills, hauling men and drilling tools to new well sites. By year's end, Phillips sold its "wild" gas from two sites in Bartlesville, while the directors peddled the preferred stock at $95 a share. Phillips Petroleum now had 227 oil and gas wells, and the new gasoline plants produced from 75 to 100 million cubic feet of gas a day.

It was still only the beginning. Frank knew there would be more Osage oil discoveries and his hunches hadn't let him down so far. He told Jane that as they entertained at the town house, singing their favorite songs, bidding farewell to 1918, and welcoming the new year.

But Frank and Jane had more to celebrate than the growth of the oil company. Their family, after twenty-one years, was growing too. Just before the holidays, Frank, forty-five, and Jane, forty-one, became the legal guardians of two young sisters from New York City.

They had wanted additional children for some time, and when it became clear that Jane could not bear any more, they began searching for suitable candidates.

On a trip to Colorado Springs with Minnie Hall, Jane's childhood friend from Creston, the Phillipses heard about "adorable twin girls" available for adoption at a local orphanage. Frank made discreet inquiries and arrangements were made for the Phillipses and Minnie to see the babies that afternoon.

Their chauffeur-driven car took them to the orphanage, where they were ushered into a room and told to wait. Frank, impatient as always, paced the floor, and Jane and Minnie sat on the edge of their chairs. Soon the door opened and the nurses brought in the twin baby girls. Frank's mouth fell open and Jane and Minnie gasped. The infants were black. After a moment, the Phillipses and Minnie broke into shrieks and laughed till they cried. Frank was finally able to regain his composure and explain to the nurses that the babies were indeed adorable, but they were not suitable for the Phillipses. It would be impossible to explain two black babies to the folks back in Bartlesville. The Phillipses' search continued.

In 1918, during one of their lengthy stays in New York, Frank and Jane heard about a five-year-old orphan girl living with foster parents. The girl was white, in fairly good health, and a prime candidate for a permanent home. She sounded ideal, but when the Phillipses went to the tiny apartment where the little girl was staying, they found out that she was not alone. She had a three-year-old sister. Jane was charmed with the little girls, but Frank was dismayed. He had not bargained for two orphans.

If Frank was disturbed, the little girls were also not immediately impressed with the Phillipses. Dressed in their fancy clothes, the strangers frightened the two sisters and they clung to the long skirts of their foster mother, a large woman they called Aunt Bella.

The Phillipses asked about the girls' background and found the details were sketchy. No one knew much about the real parents, including their first names. They did learn that the girls' father was a man named Canfield who reportedly fell into the sea and drowned while working on a ship during the war. He left behind a young wife and a sizable brood, including the two girls and several children from his wife's previous marriage. Their mother was described as a "flighty southern girl" who remarried and then died quite young. The girls were first sent to an orphanage and then Aunt Bella and her husband took them in until a permanent home could be located.

Frank still wasn't sure about taking both of the girls. He thought perhaps it would be best just to take the older sister. But Jane watched the girls hold on to one another and she saw the terror in their eyes as Aunt Bella explained their brief but unfortunate lives. In the end, Jane's will prevailed. Both sisters went with the Phillipses.

They renamed the older girl, born October 3, 1913, Mary Francis. Her little sister, born January 2, 1915, was renamed Sara Jane.

Years later, Mary remembered two men coming to Aunt Bella's to take her and her sister away. "I think it was my father and Uncle L.E.," she said. "I climbed up into my father's lap and said, 'I hear you're going to be our new daddy.' I believe when he heard that, Father had no more thoughts about separating us. The next thing I remember was this lovely lady taking us to a hospital to get checkups and have our tonsils and adenoids removed. That was my mother."

Frank, advised by his lawyers and others against legally adopting the girls in case other relatives suddenly appeared on the scene, instead legally arranged for both of them to become their foster daughters. Frank and Jane would be their legal guardians and would give the girls the Phillips name and treat them like their own flesh and blood.

The girls were sent to a doctor for thorough medical examinations and then moved into the Phillipses' suite at the Plaza Hotel, where the family stayed during their long stays in New York. Next they were enrolled in a preschool and taken to Central Park for afternoon romps. Sara Jane also had memories of those first days with the Phillipses in New York. "We were all living at the Plaza and I was having trouble with my ears and Mother was always taking me to the doctor because she was afraid I'd go deaf. And then I remember going on the train to take us to Oklahoma. It was all very exciting and I recall looking out the window and seeing animals I'd never seen before. I said, 'Oh, look at all those little dogs!' And Mother laughed and said, "Those aren't dogs, they're pigs.'"

On the way back to Oklahoma from New York, Jane stopped in St. Louis, where she had advertised for a governess to take care of her two new daughters. After interviewing several candidates in her hotel room, Jane settled on an attractive young woman named Matilda Barnhart, who had been raised on a farm near Jefferson City, Missouri. Tillie's kinfolk told her she was crazy to leave a civilized city like St. Louis and her good position at a shoe factory to move to Oklahoma and take care of some oil man's kids. Tillie paid them no mind. She accepted Jane's offer and boarded the train with the two little girls in tow.

In Bartlesville, the orphan sisters were given a huge second-floor bedroom suite, brimming with new dresses and toys. Their bedroom was part of a major renovation of the Phillipses' town house, completed just before the girls were brought home. Four stories were added to the southwest wing, the downstairs sunroom was expanded, and a servants' dining room was added. Tillie moved into the servants' quarters at the rear of the residence.

As soon as the little girls arrived in Bartlesville, the rumors about them began circulating. Few people got the story straight. Most stuck to the popular notion that the girls were French war orphans the Phillipses had found. Nobody stopped to think if that was the case why the girls couldn't speak a word of French. Other Bartians, especially those who were puzzled about Frank's not legally adopting the girls, spread the story that the two sisters were actually bastard children sired by Frank. A man can't adopt children who are already his by blood, was the reasoning of those who subscribed to this theory. And there were still others who believed that if the girls weren't Frank's real daughters, they were the illegitimate children of one of his brothers.

Despite all the attention they received and their new plush lifestyle, the little girls were miserable. Bartlesville and the big rambling town house with all its servants were strange and different. Jane did her best to make them feel loved and wanted. Before dinner parties, she and Tillie dressed the girls in bright new outfits and paraded them before the guests. On many occasions, Jane would ask her daughters to show everyone what it was like when they were poor little girls living in a tenement and had to hang out their clothes to dry. The girls would face the adults and do their carefully rehearsed pantomime of pulling clotheslines inside the windows. The amused guests would applaud, and Mary and Sara would curtsy and then scamper up the staircase with Tillie.

After they were safely tucked into their beds, the girls would listen to the party downstairs until their minds took them back to their past. They'd lie there in the darkened room and think about New York, and they could see Aunt Bella and her big bathtub with the claw feet. Then they'd think about their real parents—the couple with no names. Sara could faintly recall being lifted up on a man's shoulders to see a parade of soldiers march past. And Mary could vaguely remember a man taking her and little Sara to a church. Then she'd remember being on a busy New York street with her sister and seeing a crude image of the Kaiser hanging from a streetlamp and how the angry crowd gathered around and burned the effigy. It was still so real and frightening. The sisters would hold each other in the darkness and cry and cry until they fell asleep.

After a short adjustment period, the house on Cherokee Avenue became familiar and comfortable to Mary and Sara Jane. They made friends and played with their dolls in the gardens and served tea to imaginary guests and kittens in the sunroom. Frank and Jane doted on their daughters and showered them with attention. The girls soaked up their love and blossomed. There was also Tillie, with her sweet-smelling hair the color of sunlight. She had a gentle touch and treated the children as though they were her own. Before long, the girls stopped crying themselves to sleep and their nightmares disappeared.

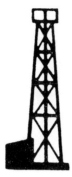

PART THREE

OIL BOOM
1917–1927

"So little done—so much to do."
—Last words of Sir Cecil John Rhodes

CHAPTER SIXTEEN

In 1919 Frank Phillips was forty-six years old and just reaching his stride. Family, friends, and fortune were all firmly fixed in his mind. He was sure of himself and ready to take on the world and the Roaring Twenties. Phillips Petroleum was growing like a prairie flower. The company needed leadership and required lots of nurturing and a guiding hand. That was Frank's job and he loved it.

As the company surged ahead toward the new decade, Frank waited —with impatience—for approval of the application made to list the Phillips shares on the New York Stock Exchange. It was just a matter of time. Already that year Phillips Petroleum had purchased Mountain States Oil Company—lock, stock, and oil barrel—and opened up twenty-two new properties around the oil patch. And although oil production was still less than 6,000 barrels a day, the company now owned close to 450 producing wells and had a third gasoline plant under construction. Phillips Petroleum—with Frank at the helm—was attracting lots of attention.

The 1918 annual report for Phillips Petroleum and Standish Oil, its subsidiary company—a slim document issued at the annual meeting on March 4, 1919—touted the potential of the properties and Phillips preferred stock. The report predicted a steady and substantial increase in earnings for those who stuck with Phillips.

"At the present time the company's properties offer splendid promise," said the report. "In many districts we are developing territory where large wells are the rule, rather than the exception. We are seeking to take the fullest advantage of the law of average, which we feel assured will in the end reward us, and further maintain our position by the purchase of the highest class of drillable acreage available in the best known proven areas."

Prentiss Moore, a veteran newspaper reporter used to reading upstart oil firms' annual reports, heard about Frank Phillips for the first time in 1919. Moore was in Texas, and E. R. Brown, president of the Magnolia Petroleum Company, advised him that "up in a small town north of Tulsa is a man whom a good oil news writer should keep his eyes on."

At first, Moore didn't think anything of it. But when Brown mentioned the Phillips name again, Moore sat up and listened. He was puzzled by Brown's strong endorsement of Frank Phillips. When he found out that Magnolia Oil lent Phillips Petroleum a cool $12 million, to be paid back in crude oil, Moore realized that Frank was no ordinary wildcatter.

"The woods and Wall Street were full of coming oil men in those days," wrote Moore. "Sinclair and Cosden, with their earlier start in the Cushing pool, were soaring higher. Marland was rising in Oklahoma. John Gilliland, Pat White, and Simms of Simms Petroleum each had companies which guaranteed them a cordial reception in New York City."

Wherever he went Moore heard more about Phillips Petroleum and its founder. "With so many great names flying through the air, Mr. Brown's remark about the man 'in the little town north of Tulsa,' stuck in my mind." Moore decided to see for himself, so he followed Brown's advice and went to Bartlesville. There he found Frank, peering from behind wire-rim glasses, in his office at the rear of the bank. Moore recorded his first impressions:

"The party line—'long, lean, and lank'—fitted him well. Tall, of easy carriage and suave frank speech, he greeted me in a way to put one at ease. Had he been a politician, he might have been a very good one. Of what he said I only recall one remark, his ironically deprecatory 'Why, I'm only a small-town banker here in the prairies of Oklahoma, trying to get together a few oil properties as a kind of sideline to the bank."

As Moore was headed home, he couldn't get Frank out of his mind. "All questions seemed, at the immediate time, to be fully answered, but when I was on the train going south again I had a nagging impression that somehow I had failed to get all the data I wanted. I felt that Mr. Phillips had been too smart for me—which he probably had been." Frank's earthy charm had struck again.

Throughout his life, Frank purposely worked at keeping his "down home" image. It usually worked like a charm, especially with reporters. Although he was often egotistical to the point of being pompous and could behave like a raging tyrant if something or someone displeased him, Frank was never able to get his beginnings out of his bones. He stayed as common as corn bread and it showed.

That attachment to his roots became a major ingredient in Frank's success and the subsequent good fortune of Phillips Petroleum Company. Frank had the ability to chew the fat with a nosy reporter, a roustabout, or a president. He could attend a fancy-dress ball and be as charming as a prince, and then hunker down with a grizzled outlaw and talk weather, women, and the price of beef.

Frank learned by listening to the cowboys and Indians, and the salt-of-the-earth men and women who lived on the land. Country people and Indians who staked their claims in the Osage Hills and had their own beliefs and customs. Frank respected their horse sense and valued their wisdom. He knew they could tell exactly when to plant crops and how to treat a rattlesnake bite. He believed them when they said they could forecast storms or sunshine just by looking at the clouds and the sun and the moon or by watching the movements of the stock and the wild animals. When an Osage cowboy told him that cattle and horses bunched in tight groups was a sign that a storm was coming, Frank was smart enough to head for cover. If an old-timer from Lot 185 pointed out mules and ponies rolling in the dust and not shaking it off, Frank cussed and chewed his cigar. That meant—sure as Satan—more dry weather was in store. A dog that constantly sniffed the air warned everyone of a change in weather, and if a hard Oklahoma winter was coming there were three sure signals—a hound that buried its food and bones in the autumn, a heavy coat of fur on a dog by early fall, and a dog that curled up to sleep by the fireplace before the leaves fell.

But Frank didn't spend all of his time listening to the denizens of the Osage talk about folk medicine and the omens of future storms. He had places to go. Places far beyond the Osage and Bartlesville, Oklahoma. The man with the West in his blood was turning to the East. In order to raise badly needed capital for the company, Frank knew he had to make as many contacts as possible in established business centers. He made frequent trips to Kansas City, St. Louis, Chicago, and New York. He began to spend more time in bankers' offices than he did in the oil field.

The hard work and travel paid off. Phillips Petroleum began to attract heavy hitters as directors. A pair of Kansas City bankers, Fernando P. Neal and R. P. Brewer, were added to the board of directors, followed by J. L. Johnston, a prominent businessman; Eugene E. Du Pont, part of the

Du Pont industrial dynasty; and John Markle and R. H. Higgins, both financial big shots from New York.

But even with a Du Pont on the Phillips board, Frank was faced with an uphill battle. He needed all the help he could muster. The oil business was still considered by many to be as risky as shooting dice. Conservative eastern bankers weren't too interested in taking a chance on a relatively new oil company from Oklahoma. Frank was resolved to change their minds.

In 1919 alone, Frank spent 116 days in New York, and in 1920 he missed that mark by only a day, spending another 115 days in the city. He stayed at the Plaza Hotel and camped out in the offices of Dominick and Dominick, an investment banking firm known as "one of the aristocrats of the New York Stock Exchange."

It was while Frank was temporarily headquartered at Dominick and Dominick that Prentiss Moore caught up with him for the second time. The journalist, out of Texas, was in New York to develop a series of articles about the oil industry for *The Wall Street Journal.* He made an appointment and went to see how Frank was faring with the big-city bankers.

"He met me, as in Bartlesville, with the same unassuming frank manner, discussed the times, the oil business, his company as unreservedly, apparently, as if the newspaperman were a partner in his business," wrote Moore. "This manner of his, in fact, might have always been one of his strongest points."

Moore also noticed that despite Frank's easy style and obvious confidence, there didn't appear to be a trace of pretension. "He talks with men levelly, without apparent reservation, with none of the assumption of greatness which is affected by some important men. It is probable that he is without pretense of any kind—an intrinsically superior man, so superior that it has never occurred to him to try to impose that superiority on others."

Not inclined to be a follower, Frank left New York in mid-April 1920 to return to Bartlesville and preside over the acquisition of another bank. On April 30, he summoned his closest Bartlesville banking lieutenants, including L.E., Kane, Holm, and Spies, for a final conference and that afternoon the announcement was made.

The Phillipses, already owners of the Bartlesville National Bank, bought the First National Bank of Bartlesville at a cost of $425,000. The newspapers called the merger—which put a combined $6.5 million under the Phillipses' control— "the most important financial transaction in local banking history." Not only did the Phillipses assume the First National Bank name but they also took over the spacious bank building at Fourth and Johnstone. Frank was named president of the combined banks and held that post until 1928, when he became chairman of the board. L.E., who had retired from his banking duties the year before in order to help Frank manage oil operations, was named vice president—in title only— and Holm became the active vice president.

With Bartlesville under control, Frank returned to New York and continued his politicking with the Wall Street influences. He met daily with bankers and brokers, and lunched at the better clubs and restaurants with a steady flow of oil-field personalities—Skelly, Cosden, Sinclair, and others.

Frank always found time to visit his aging parents back on the farm near Gravity, Iowa, and he made sure his schedule allowed him to attend key political get-togethers, including the Republican Party's National Convention held at Frank's old stomping grounds—the Chicago Coliseum, where in June 1920 Frank helped support Warren G. Harding's presidential nomination.

Trips to Oklahoma required two days by train with stops at the Muehlebach Hotel in Kansas City or the Statler in St. Louis, where Frank would meet with local directors. He returned to Bartlesville for special occasions such as H. V. Foster's reception for ex-President Taft or to meet with other oil men, such as Alf Landon, an independent oil operator from Kansas who worked oil leases not far from Frank's Lot 185 property.

Whenever he came home to Oklahoma, Frank liked to bring along a group of eastern directors. They were always anxious to tour the Phillips field operations, and after he wined and dined them at his town house or the Maire Hotel, Frank would haul the whole bunch to a ripe lease. Once they were in the oil patch, Frank didn't let them down. He'd "arrange" for a gusher to come blowing in—a dramatic sight that never failed to impress the Easterners.

Crude oil was impressing many people in 1920. The price of oil, which had lingered at $2.25 a barrel in 1918, soared to $3.50, and Oklahoma became the leader in production, turning out 106,206,000 barrels— more than either California or Texas.

By summer's end, with the price of oil rising, Frank felt confident enough to open a permanent location in New York. He selected the company's first female employee—Sydney Fern Butler—to serve as the New York office manager.

Together, Frank and Fern Butler looked for just the right spot. They finally settled on a location at 115 Broadway in Manhattan. The offices weren't anything special and they still were not as close to Wall Street as Frank wanted, but they would do for the time being. Besides, anything was better than using a spare desk in someone else's bank. On September 29 the new offices were officially opened. There was little fanfare. A few of the directors paid calls and a couple of them took Frank to lunch. Frank left late that afternoon to prepare for a trip to Pittsburgh, and Fern stayed to tidy up the new little office. She was proud of her new post with the company. In only three years she climbed from answering the telephone in the bank building in Bartlesville to running the first New York office for Phillips Petroleum.

She never thought she'd go this far. There had been more than a few strained moments in her tenure with Phillips Petroleum. Frank Phillips was not the easiest man to work for. He could be demanding and unreasonable. He worked as hard as a roustabout himself to get the job done. That counted for something. Besides, she rather liked his style—firm and direct, always in command.

The air was warm and moist that night, and Fern decided to walk home from the new office. Above the clatter of the traffic, she thought about her job. She felt as lucky as an Osage wildcatter. Even when an easy rain started to fall she didn't mind. She pulled up her collar and walked all the way home through the streets of Manhattan with a smile on her lips and Frank Phillips on her mind.

CHAPTER SEVENTEEN

The 1920s bloomed sweet and delicious for Frank Phillips. It was an exciting period to be alive, especially for a millionaire oil man, like Frank, with the confidence of a cat burglar and the bearing of a sergeant major. "The uncertainties of 1919 were over," wrote F. Scott Fitzgerald. "America was going on the greatest, gaudiest spree in history." Frank was right in the middle of it.

The temperament for the decade was set on March 4, 1921, in Washington, D.C., when a broken and bent Woodrow Wilson limped out of the White House, and the silver-haired Warren Gamaliel Harding was sworn in as President of the United States. By 1923 Harding would be dead after serving less than three years of a term marked by scandal and corruption. He'd be remembered as one of the worst Presidents ever.

For the time being, though, Frank and his Republican cohorts were content. So was most of the country. Wilson, the scholarly and incisive Democrat, was literally history. Harding, who they said at least looked and sounded like a President, and his cronies—known as the "Ohio Gang" —were in power. The age of normalcy—whatever that meant—had begun.

Not everyone was ready and willing to be normal. Not everyone wanted to conform. The quest for normalcy was popular in parts of the nation—especially the rural areas—but not in the rambunctious big cities. And not in the life of Frank Phillips. Or at least in every part of his life.

As the 1920s got underway, Frank began leading two lives, dividing his time between the Oklahoma oil fields and his office in Bartlesville, and the financial circles and his office in New York. Frank considered himself as having the best of both worlds—the normalcy of mid-America in Bartlesville and the unconventional urban attitudes of Manhattan. "In both places, I lived like a spoiled city dude—nothing but stiff collars and expensive Havana cigars," said Frank. Whichever world he chose to be in, it was a topsy-turvy time for Frank, and the nation. Old values were tossed aside. A revolution of manners and morals seized the land.

In the 1920s, women won the right to vote. Some became flappers and wore bobbed hair. Many even smoked cigarettes in public. Mah-Jongg was the rage. So were Eskimo Pies, yo-yos, flagpole sitters, marathon dance contests, raccoon coats, the Charleston, and ukuleles. Will Rogers, the cowboy philosopher, still had Oklahoma on his mind, but now he held court in New York and made Flo Ziegfeld's Follies famous. Isadora Duncan twirled in her tunics and silk scarves in support of the Bolsheviks. John Barrymore electrified the London public with his Hamlet. D. H. Lawrence titillated the world with his stories of spiritual and sexual passion. Chaplin, Fairbanks, Valentino, and the Gish sisters kept the theaters filled. Ruth, Dempsey, and Cobb kept the fans hoarse.

In the 1920s, motion pictures were on the verge of finding their voice, while Sheiks and their Shebas whispered secret passwords at one of the thousands of speakeasies operating in New York City alone. Bootleggers did a booming trade, but bathtub gin and near beer couldn't quench America's thirst. The nation was between wars and on the wagon. Jolson, in blackface and voice at full throttle, was on his knees belting out songs. The Ku Klux Klan was marching; the Teapot Dome scandal was brewing. Political shenanigans, race riots, rum runners, and Al Capone captured the headlines.

And in the 1920s, Fern Butler captured Frank Phillips' heart.

The precise and dedicated secretary and her flamboyant and imperious boss became lovers. From the start theirs was a star-crossed romance. Fern, a career-minded young woman, was not inclined to marriage. Frank, a middle-aged tycoon, was a partner in a marriage of more than twenty years that had grown stale.

Jane was still vivacious and winsome. Her dark eyes never lost their sparkle and flash. She was animated and lively, a gracious hostess, a loyal friend, and a relentless but charming gossip. Her manner captivated everyone she met. She was sophisticated yet, like Frank, rooted to the Iowa soil. Her zest for life was evident in everything she did—from throwing a tea party on the town-house lawn in Bartlesville to hosting a luncheon at the Plaza to leading a throng of revelers on the rounds of the choicest speakeasies in Manhattan. Wherever she went, wrapped in ermine with her ever-present cigarette and holder stylishly tucked between two fingers, she gathered a crowd. Her wit could be biting or cheery; her throaty laughter contagious.

The years did take their toll, though. Jane was no longer the beguiling banker's daughter sending innocent notes to an upstart barber. Her dark hair had become strikingly white, just like her father's. Lack of regular exercise and a healthy appetite for life—and food—took their toll. She installed an exercise machine—the motorized kind with a belt that wraps around the torso—in her dressing closet at the town house. It did little good. There were too many trips, too many dinner parties, too many cocktail receptions. Jane was heavy, overweight. Her girlish figure was gone. She became what people referred to as "a handsome woman." Her voice, tempered by a steady stream of cigarettes, had deepened. It was husky and rich. Scott Fitzgerald would have called it a voice full of money.

Jane and Frank still cared a great deal for each other. Their love was never an issue. She was concerned about him and his business, and Frank considered her an important sounding board when he struggled with decisions. Jane was focused on her foster daughters, her son John and his wife, Mildred. Then there was her circle of friends and the growing number of social obligations to attend to both in Bartlesville and, when she visited Frank, in New York. Jane would be Frank Phillips' wife until the day she died and Fern would be his lover for almost a quarter of a century. But oil remained his favorite mistress. Jane and Fern understood this about Frank more than anyone else. Sometimes it was hard to admit. Often, their circumstances with Frank were tested. He could be sentimental and caring, then suddenly become abrupt, impatient, and tyrannical. But both Jane and Fern learned to deal with Frank. And both women endured. They were extraordinary persons living in extraordinary times.

No one ever accused Jane Phillips of being an innocent little girl from the backwoods of Iowa. She was bright, had a good sense of humor, and was not easily fooled. In many ways, she was more intelligent than her husband. Frank was inquisitive and had plenty of horse sense, but Jane possessed an inner intelligence and, with few exceptions, was an excellent judge of character.

Fern Butler was no ordinary woman herself. She was neither a fly-by-night lover nor some ambitious little secretary who thought she could sleep her way to the top. Born Sydney Fern to Edwin and Minnie Butler in Weaubleau, Missouri, on May 24, 1893—the same month a panic started on the New York Stock Exchange when securities fell dramatically. The Butlers had little way of knowing that their baby girl would spend much of her life on Wall Street wheeling and dealing with brokers and bankers and appeasing the stockholders and directors of one of the largest independent oil firms in the world.

Her real name, Sydney, was for one of her uncles. When she entered the business world, she simply signed her name S. F. Butler. Her friends and family—and Frank—always called her Fern. She was one of three sisters, wedged between Monta Lura and Una Zee. The Butler girls attended Weaubleau Christian College, founded by their grandfather in 1842, and all three became schoolteachers.

After her mother died, Fern moved to Colorado to keep house for her father. She taught in a crowded one-room school and she hated it. The long prairie winter and her strict father's relentless interrogation of her suitors became intolerable. Fern lasted barely a year. She went back to Missouri to attend secretarial school in Springfield and then moved to Bartlesville to join little sister Zee and big sister Monta, who was married to Cleve Locke, owner of a local feed store.

Fern and Zee moved into Monta and Cleve's stucco house with its dormer windows and screened porch on Santa Fe Avenue, not far from one of the old residences of Emmet Dalton, survivor of the ill-fated Dalton Brothers gang. When she completed her education, Zee started teaching second grade at Jefferson School. A few months later Fern accepted her position with Phillips Petroleum. Frank took notice of Miss Butler the first day he met her. She was hired on the spot.

Fern was intelligent, had a fiery temper, and was fastidiously neat. Her hair was light brown and her eyes were China blue. She was not extraordinarily attractive, but she had a stunning figure—firm bustline, small waist and hips, shapely legs. She wasn't flashy, but had great style and poise. At the office every move she made counted. There was no lost motion.

In Bartlesville, she caught the eye of every man who came to the Phillips office to see Uncle Frank. When she moved to New York, where people were prone to pass each other on the street without a glance, Fern turned even more heads. When Frank told her he didn't like her hairstyle, Fern had it cut, bleached, and curled. People stopped on the busy streets and watched the smart-looking young woman march down Broadway to her office.

Frank's attraction to Fern Butler was more than physical. He genuinely liked the gutsy Missourian—almost twenty years his junior—who shared his love for the Republican Party, sleek riding horses, and the oil

business, yet wasn't afraid to speak her mind if she disagreed with something Frank did or said. He respected her business ability and he quickly
grasped that, like him, Fern was married to her work. He liked her devotion to the company and her personal loyalty to him. Most of all, Frank
admired her independence.

Fern dearly loved and respected Frank Phillips. She came to know his
likes and dislikes and his personal quirks. She knew he liked wearing
expensive silk underwear. She knew he hated going to restaurants where
the lights were turned low. She knew he couldn't stand seeing a postage
stamp with even a slight tear or jagged edge—every letter that left the
office had to be perfect. She knew he liked the color red. She knew about
his compulsion for promptness.

She also knew too well that Frank would always stay married to Jane
and to black gold, but that she'd get her share of the dances.

In Bartlesville, the tales of Fern Butler and Uncle Frank fit nicely in
the latest rounds of gossip. Although both Fern and Frank were discreet
and went so far as to take separate trains, and later airplanes, when traveling, spicy accounts of their love nest back in New York were whispered at
cocktail parties and at the country club. A popular story was that the
Phillipses' foster girls were actually Fern and Frank's daughters. A few
even suggested that Fern, with Jane's blessing, served as a surrogate
mother for the two girls.

Jane didn't learn of the affair between her husband and his trusted
secretary for many years. If she had suspicions, she shared them only with
her closest friends who gathered in her bedroom for regular confessions.
For the most part, Fern and Jane treated each other cordially and with
respect. From time to time, they exchanged gifts and traveled together.

When Fern first arrived in New York she rented a room from a stockbroker and his family for a while before moving to her own apartment on
Lexington Avenue, just on the fringe of the Grand Central area. Frank was
a frequent visitor but maintained his own residence at the Plaza Hotel—a
New York institution since 1907, the year Oklahoma became a state. Designed in elegant French Renaissance style, the Plaza hosted endless com-
ing-out parties and charity balls. The cream of New York society—including many prominent financiers and bankers—came to the hotel. During
the 1920s, the Plaza was celebrated as a popular rendezvous of the flaming
youth described in *The Great Gatsby*.

Frank relished his morning taxi trips through Gotham's canyons to
his office. Fern, his diligent office manager, ran a tight ship. She and her

staff kept things running smoothly, and Fern made daily entries in bound diaries, carefully listing all of Frank's business and social activities. Fern also maintained contact with the firm's Oklahoma headquarters through Marjorie Loos, the young woman Frank hired as his Bartlesville secretary.

Frank spent 158 days in New York and only 120 in Bartlesville during 1921. He spent another 87 days traveling to other places, including trips to Washington, D.C., where he based himself in the Willard Hotel while he sought final government approval of the most recent Osage leases purchased at Pawhuska. In Washington, Frank also socialized with a bevy of influential contacts—especially congressmen and senators who could effect legislation governing the oil business. On several occasions he conferred with President Harding at the White House.

In 1921, Frank also had to deal with his aging father back in Iowa. In early February, Frank went to Gravity to be with Lew Phillips and to accompany him to the Research Hospital in Kansas City for medical treatment. L.E. joined Frank, and the brothers waited by their father's hospital bed for several days. It was a sad vigil. The old man's body had played out. There was nothing the doctors could do. At four-thirty in the afternoon on February 15, with his two oldest sons at his bedside, Lew Phillips—the Iowa farm boy who survived Rebel bullets, rampaging Indians, prairie fires, blizzards, and swarms of grasshoppers—died. He was a little more than one month into his seventy-seventh year. Lew's boys took his body back to Gravity and they laid him to rest beneath the rich Iowa soil he had farmed for so long.

Aside from family obligations, Frank spent as much time as he could with his board of directors, especially in New York, Kansas City, and St. Louis, where Frank conferred with contacts such as Charles Lemp, the brewery magnate, and served on the board of a local bank.

In Bartlesville, at the early directors' gatherings and annual meetings of the stockholders, hardly anyone showed up. But a few locals, who bought as much Phillips Petroleum stock as possible, always came, including Bill Leonard, owner of a printing shop, and his mother, for a few years the company's only female stockholder. Those early shareholders who attended the meetings would be ushered into Frank's office and told to pull up chairs around his desk. Afterward, Frank handed out cigars and took the directors to his town house on Cherokee Avenue for a big meal.

Frank had another good reason to pass out his best cigars in the fall of 1921. He and Jane became grandparents for the first time on October 26, when John's wife, Mildred, gave birth to Elizabeth Jane Phillips. Frank was impressed with anybody who was punctual and he was especially pleased if they arrived ahead of time. John and Mildred's baby did just that. She made her appearance at seven-thirty in the morning, a full two and half hours before Frank got back to Bartlesville following one of his whirlwind tours of oil properties in Texas, Louisiana, and Arkansas. The baby instantly became a favorite of both Jane and Frank. The day she was born Frank shortened her name to Betty Jane and he made sure his diary noted the hour of her birth.

There were other diversions, such as annual company picnics, either at L.E.'s farm or else out at Pershing, an oil camp named for the famous general where, in the early 1920s, a young man from Ohio named Clark Gable worked with his father for a brief stint as an oil-field roustabout for seventy-five dollars a week.

In April 1922, when an explosion at the Osage plant, on Lot 185, killed a young worker named Steve Haynes, Frank was beside himself with grief. He met with the dead man's brother and several times went to the hospital to visit the other Phillips workers burned in the accident. He worried Dr. Somerville with questions and advice about his injured men until he was certain they were out of danger.

Whenever he was in Bartlesville, Frank also tried to spend time with his son John. Although he wasn't as dynamic and forceful as Frank wished, he still hadn't given up on his son. John accompanied his father on business trips and sat silently during directors' meetings. Frank persisted. He took his son to the country club for rounds of golf, to prizefights and baseball games. John was a frequent companion in New York, where Frank enjoyed going to shows and plays. He took John up to Rye, New York, to play golf and arranged for weekend yachting excursions. He worked at becoming the father he couldn't be when John was growing up and Frank was either in the oil field or glued to his desk at the office.

Sometimes John, but often just Fern, went with Frank on long Saturday drives to look at Long Island or Connecticut properties. Frank was itching to entertain his major eastern contacts and directors in style, and longed for a palatial country estate. It was a frustrating process—Frank found nothing that pleased him.

But at least back in the city he was able to expand his operations and make some moves. In February 1922, Frank canceled his lease at the Plaza and looked for a new residence. The following month, he signed a lease for new and larger Phillips Petroleum offices in the Equitable Building at 120 Broadway—the only street that runs through Manhattan from one end to the other. Frank approved of the larger offices on the twenty-third floor of the forty-one-story building located between Cedar and Liberty streets. Its total of 1.2 million square feet of rentable office space made it one of the largest buildings in floor area in the city and one of the most formidable structures in the Wall Street district. It was a prestigious address. The move also meant that, despite the fluctuating price of oil, his company was growing.

Frank recalled that back in 1917, during the first six months of its existence, Phillips Petroleum earned $3,114 on a gross income of $218,000. Five years later the firm was earning $3 million net on an annual gross of $12,500,000 and total assets had climbed to $50,155,000.

On June 28, 1922, Frank stood under the Million Dollar Elm at the Pawhuska sale and spent $4.5 million for new leases. The company's daily average oil production had reached 21,014 barrels from 1,160 producing wells. Seven Phillips Natural gasoline plants produced an average of 75,000 gallons a day, and Frank bought 300 new insulated tank cars to ship the gasoline.

Even after he established a New York office and despite the string of successes in the Osage, Frank still had his share of detractors. Some people believed that Phillips Petroleum was just another flash-in-the-pan operation. Skelly Oil, a partner with Phillips in some of the Osage fields, experienced similar criticism.

"We are recommending only those companies which we consider safe and whose stock is likely to show an appreciation," said one brokerage firm. "Our chief objection to Phillips Petroleum is that it is not a complete unit in the oil industry. We cannot recommend Skelly, as the stock is not a well-seasoned one and fundamentally sound. We think Mr. Phillips and management are sound and conservative men. We would like to be able to recommend Phillips Petroleum as we regard Mr. Phillips very highly, but until the company adopts standards which are suited to our rules for protecting the clients' interests and money, we cannot be sure that the stock of Phillips Petroleum Company is safe."

Despite the criticism, Phillips stockholders remained loyal. One of them, L. T. Crutcher, even became a spy. As bullish about Phillips as anyone, Crutcher was concerned when he heard Wall Street rumors that a brokerage house was coming out with nothing but bearish statements about Phillips stock. The indignant stockholder paid an incognito call on the firm, which was located at 111 Broadway, just down the street from the Phillips Petroleum offices.

Crutcher listened to idle conversation around the ticker-tape machines. The more he heard, the madder he got. "During that period of my first visit an employee of the firm . . . made several radical statements about the Phillips Company as the quotations would come out of the tape, and every indication of weakness in the stock seemed to derive considerable glee," Crutcher wrote in a letter to Frank. "He made such comments that the Phillips Company was a 'lousy company'; that they had the most remarkable ability of any of the oil companies of kidding themselves into the belief that they were making money; that they did not have the slightest idea how to keep books; that he thoroughly agreed with the recent statement put out by some financial writer that the company was in perilous condition."

Crutcher was indignant, but he bit his tongue and returned to the brokerage house the following afternoon. This time he talked at length, still undercover, with the office manager. Again, Crutcher heard nothing good about Phillips. Fed up with all the negative commentary about a company he had heavily invested in, Crutcher finally asked the broker why Phillips stock price remained so firm if it was such a bad outfit. The broker had a ready reply: "He said it was simply on account of the fact that Frank Phillips was the most remarkable salesman in the world and can make anyone believe that black is white," Crutcher wrote. "He wanted to take off his hat to you for being able to so completely pull the wool over the eyes of your followers. He also said that because of your double ability of knowing the oil game as well as the financial game, that it made it possible for you to become the richest oil man in Oklahoma, and that you and Sinclair were really responsible for taking oil to Wall Street."

Frank studied Crutcher's letter. He especially savored the broker's backhanded compliments about Frank's sales ability. The other barbs stung, but he shrugged them off. Frank had grown thick-skinned and wasn't worried about the sharp criticism. Besides, he knew most of it could be overcome. He was sitting on the boards of banks in New York and St. Louis, his own bank in Bartlesville was healthy, and his oil firm was bucking all odds and making handsome profits. Company earnings for 1923 were close to $12.5 million.

It was about this period that Frank began traveling with a retinue of personal attendants which often included chauffeurs, secretaries, geologists, and barbers. Frank's regular traveling companions were Dr. C. W. Hammond, a Bartlesville osteopath, and Dan Mitani, the Japanese servant who served as Frank's valet. Doc Hammond, as everyone called him, frequently gave Frank an osteopathic treatment, a manipulative technique thought to be helpful for correcting aches and pains and keeping bodies healthy. Mitani's job was to help dress Frank, tend to his wardrobe, and assist with all of his needs. Mitani was provided with a room in the town house but went with Frank almost everywhere he traveled—from New York to the Texas panhandle.

In the 1920s, Dan and Doc Hammond were usually with Frank when he met his family for vacations at Excelsior Springs, outside Kansas City, or at the Broadmoor Hotel in Colorado Springs. Both Dan and Doc were at the Broadmoor with Frank and members of the Phillips family in the summer and fall of 1922.

That September, shortly after returning from his Rocky Mountain holiday, Frank was in the midst of heavy bidding during a lease auction at Pawhuska when he felt a sharp pain in his lower abdomen. John Kane and Stewart Dewar helped him to the car and they raced back to Bartlesville and summoned Dr. Somerville to the town house. Frank still didn't feel well but recovered enough to go to his office the next morning.

Two days later the pain returned and persisted. Frank, accompanied by Dan, Doc Hammond, Kane, L.E., and several others, took a train to Kansas City and checked into the Research Hospital, where Lew Phillips died the previous year. Jane, beside herself with worry, arrived from Colorado in time to hear part of Frank's consultation with the physicians attending him. John Phillips pulled into town the next afternoon and became part of the watch committee crowded into his father's room.

After examining Frank, the team of doctors formed their diagnosis—appendicitis. An operation was scheduled for the following day. Frank hurriedly dictated a telegram for Fern, anxiously waiting in the New York office.

"In answer to any possible inquiry you may say that I will be operated on [at] Research Hospital Tuesday morning for appendicitis which has troubled me for years. Physicians announce my physical condition otherwise perfect, therefore I approach the situation with extreme confidence. Everything in business and otherwise so favorable now I consider this a good time to have it over with."

At Frank's insistence, Doc Hammond and John Kane witnessed the entire procedure. It was a complete success. Immediately afterward, still a bit groggy from the anesthesia, Frank sent word to New York that everything went according to plan. Happy there were no complications, but eager to get back to work, Frank convalesced in the hospital and at the

Muehlebach Hotel for several days but was able to get back to Bartlesville in time to see Babe Ruth clobber a baseball a country mile at an exhibition baseball game. A few days later Frank and Dan returned to New York, where Fern had the Phillipses' newest Broadway office running as smoothly as Frank's silk skivvies.

On November 9, with Dan overseeing the movers, Frank bid the Plaza farewell and moved into a large suite of rooms at the Ambassador, one of the city's newest and most luxurious hotels.

Located on the east side of Park Avenue between Fifty-first and Fifty-second streets, the Ambassador was opened in 1921 with the Father Knickerbocker Ball, attended by the crème de la crème of New York society. The following morning the New York *Times* reported that the Ambassador's ballroom had "all the intimacy of the ballrooms in the large houses on Park and Fifth Avenues," and that "a successful attempt has been made to provide individuality and the spirit of a fine home rather than the usual hotel atmosphere."

Primarily a residential hotel, the Ambassador rose eighteen stories tall and had six hundred rooms. It was so popular in diplomatic circles that it was called the "social embassy of two continents." No two apartments in the hotel were furnished alike and each suite featured "an electrically operated alarm clock . . . which awakens the sleeper by chimes." Many of the interior walls were paneled with embossed leather. There were vaulted ceilings and the foyers on the main floor had green stone walls and marble columns.

Frank, ensconced in his twelfth-floor suite, considered the Ambassador an ideal location. It was close to Fern's Lexington Avenue apartment and Grand Central Station. In fact, the railroad tracks were buried under a mall which ran the length of the avenue. Across the street from the hotel was the fashionable Racquet and Tennis Club, and next door was St. Bartholomew's. A Byzantine design with a multicolored dome, salmon-pink bricks, and gray Indiana limestone walls, this Episcopal church was filled with art treasures and was the home of one of the city's wealthiest congregations.

Best of all, Frank knew that Park Avenue—Manhattan's street of dreams with its posh clubs, fine apartments, uniformed doormen, powerful and privileged people—was the correct address for both those who had always had it and those, like himself, who had just made it. Park Avenue meant wealth, social standing, and achievement. Even a few years later when the park malls—where nurses and nannies congregated with their babies—were reduced to narrow islands in order to give motor vehicles more room, *The New Republic* was able to report: "If America has a heaven, this is it."

In Bartlesville to celebrate his forty-ninth birthday on November 28, less than three weeks after moving to the Ambassador, Frank described the Phillipses' new Park Avenue residence to Jane and the family, gathered around the dining-room table for cake and hand-cranked ice cream.

A few feet away, in the help's kitchen, Dan Mitani, Frank's valet, gave his impressions of the Ambassador and New York to the rest of the servants as they devoured their own bowls of ice cream, scraped from the wooden bucket. Dan told them about the Phillipses' suite at the Ambassador and about how he had his very own room. He told them about how he got Mr. Phillips ready for work every morning and about taking long walks with Mr. Phillips in the evenings, when the color and the light and the mood of the city changed. He told them about the skyscrapers and the subways and the crowds of people. He told them about all the important tycoons and bankers and executives—Marshall Field, Bill Skelly, Amos Beatty, the head of the Texas Company, and the rest—who came to see Mr. Phillips.

Dan finally stopped talking and went back to the dining room to watch Mr. Phillips cut the cake. They all sang "Happy Birthday"—Dan, the servants, the little girls, John and Mildred, Jane, and the other guests. Then they sang more songs, popular tunes of the day. Somebody started in with "Let Me Call You Sweetheart." Soon the whole house was full of music, just like years before when Waite was still there and the promise of oil in the Osage was as new and fresh as a prairie spring.

CHAPTER EIGHTEEN

Frank had a knack for hiring, firing, and delegating—special talents he developed when he was just a fresh-faced kid bottling Mountain Sage and bossing barbers back in Creston. The oil business was rough-and-tumble, not nearly as glamorous as most thought. An independent operator had to be able to look a man square in the eye and know in an instant if he was worth a try on a rig or behind a desk. Once the man was hired on, he was given his tasks and left alone. If he did a good job, he drew his pay and, every once in a while, even got a day or two off. If the quality of his work stayed high he could look forward to someday getting a promotion. But if the man didn't pull his share of the load right from the start, he was run off quicker than the bookkeeper could count out his wages. Oil-field bosses were quick to fire workers who didn't suit them. It had to be that way. There was no time for second or third chances.

Many of the early executives hired at Phillips Petroleum had worked for Frank before he founded the company. Some got their start out on an oil rig. Others were brought aboard during the first few years, while Frank developed his New York connections and built a team of future company leaders back in Bartlesville. Clyde Alexander, John Kane, Stewart Dewar, Paul McIntyre, William Davis, H. E. Koopman, F. E. Rice, J. M. Sands, H. A. Trower, C. R. Musgrave, and O. K. Wing were all on the early fast track at Phillips. An ambitious and confident lineup, some of them were members of the influential executive committee headed by Frank and L.E. Several of these executives were confident they'd one day step into the senior vice president's position, or perhaps even inherit the president's desk if and when Uncle Frank ever decided he had made enough money and stepped down.

It didn't take a genius to figure out that L.E. was in no physical shape to remain a viable force for long and that none of his sons was slotted for greatness with the company. It was also clear that Frank's own son, John, while a director with the firm, didn't march to the same drummer as his dynamic father. Although he was a friend and drinking pal for some of the younger Phillips executives, and he appeared at board meetings and accompanied directors on tours of the oil field, John was never considered a serious contender for a position of authority. He was more interested in drinking Scotch than drilling for oil.

As the 1920s proceeded, other candidates for leadership at Phillips Petroleum emerged. One of the most eager—some said ruthlessly ambitious—was a young man who joined Phillips in the company's third year. He was born August 31, 1899, in the farm town of Horton, Kansas, to a locomotive engineer on the Rock Island Railroad and the daughter of a stockman. His name was Kenneth Stanley Adams.

When the boy was only one year old, the Adamses moved to Kansas City. By the time Kenneth was two, he was chattering so much that they predicted he'd grow up to become a preacher or a lawyer or maybe even an important business tycoon.

Kenneth Adams had just celebrated his third birthday and had just received a new pair of boots when torrential rains flooded the city. His parents took in about two dozen railroaders and their families who had been driven from their homes. Kenneth's pair of shiny black leather boots with fancy red tops were his proudest possession and he wore them day and night. He wouldn't even take them off when he went to bed. The flood refugees were so amused with the little boy and his prized footwear that they started calling him "Boots." The name stuck forever. Some folks never knew his real name.

Boots Adams was full of ideas when he was young. He thought a lot about the future. "I'm not sure why or how, but I can just see in my mind the plan for things," Boots told his mother once. He was also fearless. Even after a big draft horse pitched him through the air, Boots dusted himself off and climbed on for another round. When he was six he almost drowned but was rescued just as he went down for the third time. He was back at the swimming hole the next day.

Popular in school, Boots was also a natural athlete, and he was a good-looking cuss—all the girls at Wyandotte High School were crazy about him. Boots, never a shrinking violet, admired their good taste.

After graduating from high school in 1917, the year Phillips Petroleum was founded, he went door to door peddling cooking utensils to earn tuition money for premedical school. Later that summer, he went to Oklahoma to visit his uncle George McClintock, whose brother Harry owned the Crystal ice plant in Dewey. Boots, eager to make more money, hired on as a deliveryman on one of McClintock's ice wagons.

That autumn, Boots enrolled as a freshman at the University of Kansas in Lawrence. He also enlisted in the Student's Army Training Corps, played intramural sports, joined the Sigma Chi fraternity, and stoked the furnace at the frat house to earn extra cash. Following a brief stint in the Army, cut short by the armistice which ended World War I, Boots went back to college. This time he entered the School of Business, for he realized he could not afford the years of education it took to become a doctor.

During the summer, he returned to Oklahoma and his job lugging ice to housewives in Dewey and Bartlesville. Every evening, Boots relaxed by playing pickup basketball at the local YMCA gymnasium. One of the players there was Bill Feist, an employee of Phillips Petroleum Company. Feist told Boots there were several men at Phillips who liked to play basketball, and he encouraged Adams to quit school and join them. Boots told him no, thanks. He was going back to college.

But that fall, before Boots returned to school, Feist told him there was a job opening at Phillips that Adams could have. It paid $125 a month. That was a lot of money. Boots figured he could work for a while and then go back to complete his education. He dropped out of college and on November 17, 1920, he reported to work at Phillips Petroleum. He stayed for forty-five years.

Told he would start in the warehouse, Boots arrived for his first day on the job wearing overalls and heavy work shoes. His boss, L. E. Fitzjarrald, was waiting for him.

"Where are you goin' in that getup?" barked Fitzjarrald.

"To work in the warehouse," answered Boots.

"Well, get your clothes changed," said Fitzjarrald. "We don't have a warehouse like that here. We have an office."

Boots dashed home and returned in his Sunday suit and necktie and went to work in the Phillips offices in the First National Bank Building.

"He was an unusual fellow," recalled Fitzjarrald. "Seemed he never could learn enough. During the years he was working for me he'd never take a vacation—always said he'd rather put in the time working in some other department so he could learn how it operated. He liked to get out into the drilling fields—roustabouting, ditch digging, tool dressing, pumping."

Boots, with his good looks and roving eye, also liked to get out among the young ladies of Bartlesville. He soon settled on one of the choicest of the lot—Blanche Keeler. Raised as a Cherokee because of the bloodline of her Indian grandmothers, Blanche was a member of a prominent family and the granddaughter of two of Bartlesville's pioneer settlers—Nelson Carr and George B. Keeler, one of those who drilled Bartlesville's first big gushers. Smitten with the debonair Kansan and his easy chatter, and impressed by his obvious desire to climb the corporate ladder at Phillips, Blanche—a demure, dark haired beauty—accepted Boots' proposal of marriage.

After the wedding, Blanche and Boots moved into one of the more respectable neighborhoods in town, where they raised a son and daughter —Mary Louise and Kenneth Adams, Jr., better known around Bartlesville as Bud.

Boots worked long hours at Phillips, quickly moving from the warehouse-materials department to the production department to the accounting department, where he became assistant chief clerk. Besides starting his family and earning his keep at Phillips, Boots also managed to fit in time for another passion—basketball. For some time, the same group of Phillips employees Boots had met when he was hauling ice gathered at the YMCA to work off excess steam on the basketball court. Shortly before Boots took the job at Phillips, an official company team was organized.

The brains behind the first Phillips team were two of Uncle Frank's most valued young executives—Stewart Dewar and O. K. Wing. Dewar was the Canadian destined to become vice president of operations; "Obie" Wing, the irrepressible and energetic executive in charge of accounting, and Frank's personal favorite of all his young lieutenants.

Wing coaxed Frank into coming out to see the Phillips quintet play rival teams, representing everything from a local furniture store to other oil firms. Frank was impressed with the caliber of play and the excitement of the crowds. He quickly realized that sponsoring an official company team wouldn't do the Phillips name any harm. Besides, the young athletes were developing desirable executive skills—learning teamwork and how to make quick decisions under pressure. Frank decided Wing's idea was sound. A basketball team would make a good breeding ground for future company leaders. He consented to spring for uniforms with "Phillips" emblazoned on the jerseys and even shelled out a little extra to buy his boys hamburgers after the games.

In the first game, with Wing as coach, the Phillips team manhandled the "Town team" from Collinsville, Oklahoma, 86–24. They were off and running. Original team members included Boots Adams, Melvin Heine, Kenneth Beall, Horace Allen, Kenneth Slater, and V. T. Broaddus. Bill Feist, the man who convinced Boots to join Phillips Petroleum, was the captain.

Boots fit right in with the basketball program, and Obie Wing saw to it that the college dropout got plenty of playing time as the Phillips team took on all comers.

In 1923 the Phillips Petroleum team added a new player, another young man from Kansas, who could shoot a basketball better than Henry Wells could shoot fish in a rain barrel, and his name was Paul Endacott.

He was born to Frank and Rebecca Endacott on a scorching July day in 1902 in Lawrence, the home of the University of Kansas, and when he was a boy, young Endacott had his heart set on going to school there.

Every day after school, he worked as a hired hand on a farm and at a horse-collar factory, but he also made time for sports. When he was eight years old, Endacott was known for his fearless dives off a twenty-foot-high platform into Potter Lake on the university campus. "Paul Endacott is probably the bravest little swimmer in Lawrence," reported the local newspaper. "The lad dives in perfect form with a lithe movement. It is doubtful if there is another lad in town of equal age who would dare to dive from this height."

He also played basketball every chance he got and became a playground student of the sport under James Naismith, the legendary originator of the game, who at the time was coaching at the university in Lawrence.

After graduating from Lawrence High School as an all-state basketball sensation, Endacott enrolled at the University of Kansas. He entered the School of Engineering and naturally went out for the basketball team.

Endacott was outstanding. A sharpshooting guard with ice water in his veins and a defensive standout, he never held back; he played like a man possessed. His coach, the renowned Phog Allen—"Basketball's Winningest Coach"—and the rest of the Kansas team always wanted Endacott to have the ball if there were only a few seconds left on the clock and the game was on the line. In one battle to the wire against a nationally acclaimed University of Missouri team on their home court, Endacott played with such intensity that some of the fans thought he was in a trance. By the time the contest ended he had to be carried off the floor.

Endacott played basketball at the University of Kansas for three years and was the team captain as a senior. He was an all-conference selection three years running, including 1922 and 1923, when Kansas was designated the national collegiate champion. He was an All-American guard twice, national collegiate player of the year in 1923, and years later was named to the Helm's Foundation's all-time All-American team and was enshrined in the National Basketball Hall of Fame.

Endacott was not just a super jock. He also excelled academically, and when the university initiated its honor award for scholarship and leadership, Endacott was the first recipient. Although he was modest almost to a fault, Endacott never worried about finding a job.

He first learned about Phillips Petroleum during his senior year when L.E. Phillips came to Lawrence to speak at the annual Engineer's Day banquet on St. Patrick's Day.

"In those days Phillips Petroleum was a small company," Endacott recalled. "I can't remember what all L.E. said at the banquet except for one thing. It stuck in my mind. He said: 'I'm in the oil *business,* not the oil *game.*' That sounded pretty good to me."

Anxious to go to work, and ready to put his basketball playing days behind him, Endacott turned down job offers with several firms, including Gulf Oil. "They wanted me to come down to Port Arthur, Texas, and work, but they also wanted me to play on their company basketball team. I didn't want any part of that. So I remembered what L.E. had said and I decided Phillips was a serious oil company." Two days after he graduated, Endacott was on a train bound for Bartlesville. On June 11, 1923, he reported for work. Like Boots, he was surprised by what he found. The first thing Endacott was told was that a spot was waiting for him on the Phillips Petroleum basketball squad. Endacott couldn't believe his ears. Nobody told him Phillips had a team. He thought he had made a big mistake.

But it was too late. Endacott had already committed himself to the job and to Phillips. For a young man whose word was his bond and who had the heart of a lion, there was no such option as quitting. The word wasn't in his vocabulary.

Endacott suited up and played for the company team for a couple of years, alongside Boots Adams and Ken Beall, Slick Slator, Horace Allen, Tommy Sears, Ray Parker, Mel Heine, Bill Feist, and the others. His skillful dribbling and passing, his whirling pivots and deadly two-hand set shots, dazzled everyone who crowded into the high school gymnasium to see the Phillips team run up an impressive string of victories.

During the years that followed, especially after the company began marketing to the public in 1927, the Phillips squads were regarded as some of the best amateur basketball teams of all time. The nation's basketball elite was drawn to Bartlesville and the Phillips teams played in some of the biggest arenas in the world, including Madison Square Garden. Phillips teams became the talk of the industrial leagues. They piled up 1,543 victories against only 271 losses, won a pair of Olympic Trial championships, captured eleven Amateur Athletic Union (AAU) crowns, produced thirty-nine AAU All-American players and fourteen Olympic players, and helped train scores of top-flight managers and executives, including four company presidents.

But Boots Adams and Paul Endacott, and the other ambitious young men who joined the company team, had more on their minds than making their marks on the basketball court. They knew that Uncle Frank was pleased with Obie Wing's fast-breaking athletes, and that the team provided good exposure for the company, but they also knew that payrolls, stock dividends, and bonuses didn't come just from their antics in a stuffy high school gym. Phillips Petroleum was still in the oil business, and that meant discovering new pools of crude, getting men and tools to the sites, taking the oil out of the earth and transporting it to market.

In the 1920s, as the company's early basketball teams kicked into high gear and Frank shuttled back and forth between his growing headquarters operation in Bartlesville and the bustling New York office, giant strides were made which propelled Phillips Petroleum to the forefront of independent oil firms.

A place called Burbank was a big help.

Burbank was big. Bigger than the Glenn pool, bigger than most other oil fields, and located fifty miles west of Bartlesville, the Burbank field covered more than 20,000 acres. It was named for a nearby Osage settlement, started in 1903, which, according to legend, took its name from the profusion of cockleburs in the area.

Burbank sprang to life on May 14, 1920, when Frank's old friend E. W. Marland, the Ponca City baron, struck oil. During the first twenty-four hours, Marland's discovery well, the Bertha Hickman No. 1, produced 680 barrels of oil. The following year, Marland opened up the Tonkawa field, named for the Indians who onced owned the land. Marland became enormously wealthy. By 1923 he was selling Standard Oil of New Jersey 20,000 barrels a day. He also lived in princely style. Marland gave money and blocks of stock to everyone he could think of, built a fifty-five-room mansion, and bought strings of expensive polo ponies. But Marland wasn't the only oil man who made lots of money at Burbank. Once the pool was proved, other companies rushed to the Osage.

The Burbank field was Phillips Petroleum's first step into the big leagues. To Frank, the opening of the great Burbank oil pool in the Osage was an open invitation to even more wealth. He and his representatives practically lived beneath the Million Dollar Elm in Pawhuska, snapping up oil leases at fabulous prices during Colonel Walters' colorful auctions. At the time the field opened up for exploration the sale of leases brought less than $10 an acre, but after production was well underway, leases went for as much as $10,000 an acre. In June 1921, the sale of fourteen leases brought $3,256,000, and by 1922, a pair of Burbank leases sold for $1,335,000 and $1,160,000. The prices didn't really matter. The payoff below the earth did. The first year's production of Oklahoma crude totaled well over 20 million barrels. Frank and his frequent Osage partner Bill Skelly were prepared to spend as much as it took to get their share of the new field. They felt it was well worth every penny.

In 1923, Frank and Jane, accompanied by John and Mildred, returned to New York refreshed and rested from a European vacation, only to find Fern, Wing, Dewar, and two bankers waiting at the docks to meet them. Their business was most urgent. More oil storage tanks were needed at Burbank to hold the thousands of barrels of oil Frank's crews were bringing to the surface. A conference that same day at the Broadway office lasted until the early hours of the next morning. An agreement over future financing was reached. Within several weeks seventy-five new steel tanks were under construction at Burbank. In the meantime, "that little Oklahoma oil company" had become better known in New York financial circles because of the large-scale borrowing it took to develop the field.

Phillips Petroleum soon became the largest single operator in the Burbank field. In 1923, its peak year of production, Burbank yielded nearly 32 million barrels of crude.

That summer Paul Endacott arrived on the scene. The booming Burbank field was his first real assignment. The young civil engineer, with his survey instruments and grip full of clothes, rode out to the job on a company mail and freight truck that was already loaded down with tires, equipment, and supplies. It took all day and well into the evening to reach the Phillips camp. Because the camp's bunkhouse caretaker was off duty, it appeared Endacott would have to spend his first night wrapped up in a blanket on the prairie. Finally a "crumb boss" took pity and issued the new man a mattress.

"I was shown to a steel cot in a bunkhouse full of laborers—mostly ditchdiggers and pipeliners," Endacott recalled. "I hadn't been there long when things got rowdy. Seems that each night these guys would get a little playful and start throwing cots and knocking each other around."

Endacott survived his first night in the Burbank field, but life didn't get much easier. His primary job—when he wasn't hustling back to Bartlesville for a basketball game—was to survey the field and find suitable well and pipeline sites. To win friends, Endacott played down his university degree and dressed like a rodman on the survey crew. That helped earn the confidence of the men of the oil patch, where the old-school way of doing things still prevailed and engineers and geologists weren't ever going to win any popularity contests. Endacott quickly found out that tough characters thrived at Burbank, and as in most boom towns, fistfights were common occurrences.

Oil-field workers, or roustabouts, were a profane and rough lot. They came from all over, but they were mostly big strapping Okies and half-breed farm boys—rawboned and hard-knuckled. Some people thought roustabouts breathed fire and ate their own young. A few of them may have done just that. Many were oil-field tramps. Many were men on the run. They were escaping their past, escaping women, escaping the law. Foremen hired men on the spot. The two main requirements for working in the oil patch were a strong back and lots of endurance. There were no examinations, no social security numbers, no insurance policies, no nothing. The timekeeper got the new man's name and he went to work.

Some roustabouts changed their names like most men changed socks. Out in the oil patch, they found new identities. They worked hard, but they were restless as they moved from site to site. They turned black under the summer sun. Their hands were thick and their muscles hard from pulling icy tools in fields covered with snow in the wintertime. Out on the rigs and in the bunkhouses or in the boom towns, physical strength ruled and a belief in a man's independence was everyone's credo. There were workers who drank on the job. Others who took days off when they felt like it or mouthed off to the boss if something didn't suit them. Jobs didn't last that long anyway. Some men worked until the oil played out or they got fired. Theirs was a nomadic life and followed the oil fields' boom-and-bust cycles.

Boom towns were noisy and crowded, dirty and rowdy, twenty-four hours a day, seven days a week. They were violent, but they gave the men some place to go. Burbank was no exception. Like most boom towns, Burbank was black-eyed and split-lipped; filled with outlaws, whores, pimps, con artists, and plenty of roustabouts, most of them with a roll of greenbacks burning a hole in their back pocket. By the mid-1920s, when the field was famous and was producing 125,000 barrels a day from more than 1,000 wells, there were an estimated 45,000 people living in the Burbank area.

Most lived within the town limits of Burbank or in one of the other little hamlets which sprung up like mushrooms in a manure pile. Towns with names like Webb City, Carter Nine, and Lyman. At the height of the boom, more than 10,000 people got their mail at Shidler, a boom town that popped up on the prairie and was known for a time as the "Oil Capital of the Osage." Another was Whizbang. One of the most colorful crossroads in the Burbank field, some people thought that the town was named for Whiz Bang Red, a notorious Kansas City whore who worked the Oklahoma oil fields. Others said the town was so named because it "whizzed all day and banged all night." It was actually named after *Captain Billy's Whiz Bang,* a bawdy magazine of the day.

Most evenings, after the sun sank in the hills and the potent bootleg whiskey was flowing, the action began in Whizbang. One wild Saturday night a Phillips welder named Jim Vernon danced once too often with a young woman attached to Seth Lewis, an Osage County deputy sheriff. In a rage, Lewis jumped up and pistol-whipped Vernon. Word got back to the bunkhouses and the Phillips bunch got boiling mad. The next night about 150 Phillips roustabouts, armed with guns, knives, and rocks, went in to Whizbang to pay a social call. When they arrived, there wasn't a man to be found in the entire town. Someone had tipped them off. Only women and children were left.

Most of the time the workers didn't have to resort to mob tactics to protect themselves. Uncle Frank wasn't interested in having his equipment stolen or paying Dr. Somerville to patch up drillers who were hit over the head or knifed by an oil-patch outlaw. Frank believed in fighting fire with fire. He hired his own lawmen to patrol the rig sites and the boom towns and take care of his workers, especially on payday, when every cardsharp, bootlegger, whore, and thief suddenly appeared.

Some of the lawmen were nothing but hired killers. One of the most controversial was Bert Bryant, alias José Alvarado, a law enforcement officer described as running the gamut "from a cold-blooded killer to a Robin Hood." Tonkawa, Oklahoma, was the domain of "Hooky" Miller— a tough-as-a-boot lawman—who had a hook for a right hand. And the town of Shidler was where Joe Gonzales, a convicted bank robber, defended Frank's mén with a pair of revolvers.

As rough and profane as the boom towns could be, there were also pleasant times for the Phillips crews and the others working in the Osage. Moving across the oil patch in her fancy chauffeur-driven automobile was a sensuous show girl—as free as the Oklahoma breeze itself—who melted even the most hard-bitten roustabout's heart. Ruby Darby was her name and entertainment was her business. Nobody could top her. She was known as "the toast of the oil-field workers."

The daughter of a jolly old teamster and mule skinner who labored in the Burbank field, Ruby began her professional career at the tender age of fifteen when she was plucked from a chorus line in Dallas and given her big break. Ruby was said to have been the first—and possibly one of the greatest—blues singers. Certainly, she was the first person to sing the blues in the boom towns of the Southwest. "Memphis Blues," a W. C. Handy song, was her most famous number. Besides captivating audiences with her sultry voice, Ruby knew how to move her eyes, hands, hips, and torso until every man watching her was as excited as a wild colt. She was a good singer, but she was a hell of a good stripper.

The Ruby Darby stories were legend and most of them were true. It was said that Ruby "stripped at the drop of a driller's hat," that she had "ridden a hoss completely nekkid down the mud- and oil-splashed streets of Keifer," and that she had "danced bare-skinned on a tool shack roof as men tossed silver dollars at her feet."

She packed a pistol, wore silk stockings, and would try anything once, especially if she thought it would make her audiences applaud and scream. During one of her appearances at a stag party in an oil-camp mess hall, Ruby created such a stir that a lamp was knocked over and the building went up in flames. A member of her troupe called Ruby a "natural adventurer," and a popular saying of the day warned:

> If you've got a good man keep him home tonight,
> for Ruby Darby's in town and she's your daddy's delight.

During World War I she played the Army camps, but she liked the oil boom towns the best. She would roar into town in a big red flashy car, wearing only her fur coat and a smile. She'd always be stopped for speeding and pretty soon the news would be all over the streets: "Ruby's back and she's wearin' nothin' under that fur coat!" By show time the theater was invariably packed.

She liked playing Bartlesville, and during one of her engagements at the Oklah Theater, the audacious Ruby was hired by Jane Phillips to entertain guests—all summer bachelors—at John's birthday party. Ruby brought the roof down and almost all the walls as well.

In those oil-rush days in the Osage, whenever a gusher came in, the drilling crew was likely to yell: "It's a Ruby Darby!" Gradually the expression was shortened to: "It's a Darb!" Either way, everyone knew that a "Darb" was something extra special.

There were lots of "Darbs" worth bragging about in Bartlesville in the 1920s as Phillips Petroleum approached the end of its first decade. The rapid growth of the company took up all the available space in the First National Bank Building, and in 1925 Frank turned the first spadeful of earth for the new seven-story Phillips Building, which was completed in 1927. The new building—an impressive brick structure with elevators, boardrooms, and a mahogany-paneled office for Frank—was expected to be large enough to meet the company's needs for years to come. But growth continued at such a quick pace that an eighth floor and tower were soon added.

Phillips' assets had increased from $3 million in 1917 to more than $130 million, and 1,759 producing wells were turning out 30,000 barrels of oil each day. In 1925 it was known as one of the most progressive companies in the industry. Total payroll was more than $5,355,000. Frank was so proud of his company and his workers that he started a group life insurance plan to cover every single employee.

Much of the company's impressive growth during this time was due to the marketing of natural gas, the residue from the oil fields that at one time nobody seemed to want. Frank had decided to put natural gas to work, first as a feed stock in the manufacture of natural gasoline and then as basic fuel in the production of carbon black. Phillips pioneered development of the processes for extracting liquids from the natural gas vapors, and established a separate gas sales department with Carl Minning in charge. By 1925 the company had become the nation's largest supplier of natural gasoline, a position it was never to relinquish.

Later that year, however, Union Carbide & Carbon Company, itself a major producer of natural gasoline, filed suit against Phillips. The giant chemical firm contended that Phillips was violating Union Carbide's patent protecting fractionating, a method of processing natural gasoline to separate some of its constituent hydrocarbons. Frank was angry over the suit alleging patent infringement. He was determined to fight and win. After spending a few days with Dan and Doc Hammond out at his rustic log cabin above Rock Lake on Lot 185 in the Osage, Frank came up with a solution.

He'd create a research capability at Phillips to prove that Union Carbide's claims were unjust. Besides, thought Frank, oil industry technology was improving and it behooved Phillips Petroleum to keep up with the times. Research was the key to the future for Phillips. Frank immediately started his search for the right person to help fight the Union Carbide suit and lead his company into the modern era.

In no time, Frank found just the man he wanted right down the road in Tulsa. He hired the tall, towheaded former country schoolteacher named George Oberfell and told him his main mission was to help provide a defense for the lawsuit. Oberfell nodded, hung up his coat, and went straight to work.

A young man with a tremendous aptitude for science, Oberfell had studied chemistry at Ohio's Miami University. As an Army chemist during World War I, Oberfell helped develop gas masks, and at the same time he also hit on the idea of absorbing natural gasoline from natural gas. After the war, he became a consultant to oil companies in Tulsa and co-authored a book, *Natural Gasoline,* which instantly made him an authority in the field.

To help him prepare for the lawsuit, Oberfell hired Ted Legatski, a bright young chemical engineer, and Richard Alden, Oberfell's co-author of the book on natural gasoline. A short time later, he added Ross Thomas, Karl Hachmuth, Walter Podbielniak, and Frederick Frey. All of the team members had at least three things in common—they were extremely bright, they thrived on solving complicated problems, and they were all under thirty. Around Phillips Petroleum, this group of brainy young engineers and chemists came to be called "Oberfell and his Whiz Kids."

The team's first job was the Union Carbide allegation. They spent the greater part of a year preparing for the trial and it was worth every minute. In 1926, Oberfell and his boys demonstrated in court that the Union Carbide patent on the natural-gas fractionating process was invalid. They showed that the method of separating the various components of natural gasoline was not new, and that the ancient Egyptians had used a crude, but similar process in their stills to make a kind of Nile bourbon. Phillips won the suit. For Frank and his bunch of researchers, the Union Carbide victory was one hell of a "Darb."

Frank was not only gratified with the outcome of the legal action; he was pleased that his oil company now had its very own research department with a brilliant chemist at the helm. The oil industry was leaving the hocus-pocus ways of the old school behind. Men and women with scientific training—geologists, engineers, chemists—were coming on the scene. Frank knew they could help hitch his wagon to a new and bright star, and the name of that star was technology.

During this period of growth and optimism, Prentiss Moore, the diligent reporter who chronicled Frank Phillips over the years, caught up with the busy oil man again. This meeting, in the lobby of the Biltmore Hotel in New York, left Moore with a different impression of Frank. "This time he was in a hurry, moving quickly, with long-legged, decisive steps across the lobby, with three or four men trailing him. If he did not look back at them with some impatience, because they were not stepping up with him, I am mistaken," wrote Moore.

"What impressed me at the time was not triviality. Rather, it was something in his fast incisive steps that personified speed. I was seeing him in motion, where before I had seen him only at ease. There was a curious feeling that I had seen something powerful moving rapidly. That glimpse of the presiding genius of the company cleaving his way swiftly and strongly through the crowded lobby of the New York hotel was like seeing the spirit of the Phillips company in motion. Of all the times I have seen Mr. Phillips, that brief glimpse of him sticks strongest in my memory."

Although 1925 was a banner year for Phillips Petroleum, it paled in comparison to 1926, when the big Texas panhandle boom got underway. That year's annual report revealed company assets of $266 million, net earnings of $21.4 million, and nearly 2,300 producing wells averaging 55,000 barrels of oil a day.

The latest Phillips success was the oil field at Borger, about fifty miles north of Amarillo, in the bleak Texas panhandle. Like most boom towns, Borger was created from nothing after a big gusher came in. In less than ninety days, more than 30,000 people flocked to the oil oasis.

Borger was a carbon copy of the Burbank boom towns. All the roads were dirt and were filled with Model T's and trucks. The chuckholes were big enough for a small boy to hide in, and the ankle-deep dust of the dry season turned to thick mud when the rains came. Bachelor oil-field workers scrounged up meals at boardinghouses or else gobbled hamburgers for a nickel apiece at Chink-Link, the most popular sandwich stand in town, run by Pappy Oric.

There was little or no sanitation—there wasn't one flush commode in the entire town. As a result, there were so many rats running down the streets and darting across the sidewalks that in the summer of 1926 the Rig Theater started a bounty system and issued a ticket for every dozen rat tails turned in at the box office.

The town of Borger was made up of corrugated-sheet-iron buildings, tents, and wooden shacks. Some people lived in automobiles and trailers, others even found refuge in caves and dugouts carved from the barren landscape. As with other instant oil-patch settlements, the array of canvas tents and dilapidated shacks was called Ragtown.

Paul Endacott was sent to the panhandle in time to witness the creation of the town of Borger. He worked as a construction engineer and surveyor and laid out the boundaries for a nearby townsite that eventually was named Phillips.

In the bunkhouse, living at close quarters with the wetbacks, tramps, cowboys, and outlaws hired to create the new oil field, Endacott shared a steel cot with two other men. Each had the bunk for eight hours. That meant it was usually warm when one man got up to go to work and the next one slipped in beneath the sheets. When he got a toothache, Endacott found a fellow who looked like a roustabout but claimed to be a dentist. The man had a hand drill and two buckets of water—one for rinsing and the other for spitting. He sat Endacott down in a kitchen chair on the board sidewalk of the main street and put a silver inlay into the squirming engineer's mouth while trucks and cars drove by splashing water and mud on their legs.

Endacott adjusted to the hard life in Borger. He chewed the bland biscuits and he stood in line with the others to pay ten cents for a glass of warm water. He learned to avoid the patches of quicksand along the Canadian River, and he put up with the snow that blew through the cracks in the bunkhouse walls. He earned the respect of the other men in the camp. Before long he came to be known by the roustabouts as "the Mayor of Ragtown."

But life there was no day at the beach. Endacott found that Borger was as bad as, if not worse than, the Burbank boom towns when it came to crime and corruption. Fortunately, he was getting used to tough towns, though. Before arriving in the panhandle, Endacott had been in the Oklahoma boom town of Cromwell the Saturday night veteran marshal Bill Tilghman was murdered. Spawned by the Greater Seminole field, and surrounded by bobbing pumpers and oil-smeared storage tanks, in the blink of an eye Cromwell had become known as the "wickedest city in the world." Tilghman, the lawman who once ruled Dodge City and who sallied out to take on the likes of the Doolin, Dalton, and Jennings gangs back in Indian Territory days, was hired to clean up the town and restore law and order. He was seventy-one years old but could still put fear into the blackest heart. Tilghman had just started to make a dent in reducing the criminal population when he was gunned down in cold blood by a drunken prohibition officer. Because of his reputation and since he had been a state senator and an Oklahoma City police chief, the marshal's murder caused a large part of the criminal element to flee Cromwell. But they didn't disappear. They only went as far as the next oil field and boom town. Many undesirables ended up in West Texas and in Borger.

As violent a boom town as any, including Cromwell, Borger didn't disappoint anyone who was looking for trouble. There were enough dance halls, gambling dens, and brothels to satisfy a legion of roustabouts. Ladies of ill repute sat behind plate-glass windows advertising their wares, and dance halls were packed with girls who made good money waltzing with staggering drunks at a dime a dance. White lightning and homemade beer flowed freely in Borger and kept the local lawmen, or those who weren't on a bootlegger's payroll, busy. The jail was only a one-room building with tin walls. It could hold about ten prisoners. There was always an overflow. A string of drunks, gamblers, brawlers, and assorted felons were chained outside to a post anchored in a vacant lot or padlocked to a chain called the "trotline." At times it took a visit from the Texas Rangers to quell the disorders and keep the riffraff in line. Robberies and killings were everyday occurrences. Even the man the town was named for was murdered in the post office. Endacott had to laugh the evening he picked up the local newspaper and read the big headline across the front page that said: "Nobody Killed in Borger Today!"

But not every character Endacott encountered at Borger was unsavory. While surveying the site for the proposed Phillips gasoline plant, Endacott met Silent Carter, an oil-field eccentric who got his first name because he talked all the time.

"I was out near Borger surveying when I felt like someone was there watching me," recalled Endacott. "I could feel eyes on me. I turned around and there was this strange-looking fella standing there, holding up a copy of *Scientific American*. He was dressed like Boy Scouts dressed then and his face was pocked and scarred. He said, 'My name's Silent Carter and I'm a water engineer. This is my picture on the cover of this magazine standing in front of an oil mine.' "

Carter, who was so cold-natured he slept wrapped in an electric blanket even during the summer, went to the Tulsa Cafe in Borger every evening and sat down to a plank steak supper for two served by a tall waitress named Tree Top Mama. After every meal, Carter left Tree Top a new dress for a tip. They finally got married and lived in a shack on the edge of town and raised bulldogs.

Endacott, a no-nonsense engineer, was intrigued with the colorful Carter, who said he had perfected a secret method for locating water—a commodity as rare as diamonds in some parts and as precious as gold to an oil-field operation. Endacott offered him a job, and Carter accepted but only with the stipulation that he wouldn't share his secret for finding water. He also wanted $600 a month for his fee. When Endacott reminded him that was more money than the superintendent made, Carter politely reminded him that the superintendent couldn't find water. Carter was hired. Soon the company was blessed with more water than it could use.

Occasionally, late at night, when the snoring in the bunkhouse settled to a steady roar, Endacott, snug in his bunk, would think about the turn of events that brought him to Bartlesville and Burbank, and now to Borger. He'd think about Kansas and his family, and about playing basketball at the university and later for Phillips—he could still hear the fans cheering. He'd think about his work and about the first time he ever set sight on the Texas panhandle. It was in January 1926, just a few months before he moved to Borger. He had been out in New Mexico laying out a gasoline plant when he was summoned to Bartlesville. Mr. Phillips had a chartered railway car lined up to take his top executives to a big oil conference in Los Angeles. On the way out, he wanted to check on the lease potential in West Texas and perhaps make a few other stops. The party would need someone to handle the maps and serve as a kind of engineer-aide to Mr. Phillips. Endacott got the nod. The orders came from Uncle Frank himself. They left Bartlesville aboard Frank's private car, the *Manhattan,* bound for California. Endacott was understandably nervous. "I was just a lowly field engineer and here I was on a fancy train car with the big shots! Here were

Frank and L.E., John Kane, Clyde Alexander, and all the rest. I felt like a fish out of water."

The trip went smoothly and proved more eventful than anyone had imagined. In Los Angeles, the Phillips party headquartered at the Biltmore. One afternoon, during a convention lull, a man named Armais Arutunoff called on Clyde Alexander, the Phillips vice president and general manager. Arutunoff, one of eight children born to a soap manufacturer in a tiny village in the Caucasus Mountains in Russia, had been in the United States for three years. He was seeking citizenship and was also trying to find someone to back his invention—an electrical submergible motor and pump. Arutunoff claimed the pump could revolutionize the oil industry with the same dramatic effect as the Bolsheviks revolutionized his native land. Alexander wanted to hear what Arutunoff had to say and he was anxious to see his invention.

Like others in the industry, Alexander was frustrated when Phillips drillers had to abandon large amounts of oil because of intruding water in the hole which exceeded the pump's capacity. He was also weary of pumps that were inadequate when it came to recovering oil from great depths. When Alexander heard about the young Russian engineer and his invention, he decided it was worth a look.

Arutunoff escorted Alexander to an oil well near the city and demonstrated his submergible electric pump. Alexander was impressed. He returned to the Biltmore and told Frank about the Russian and his invention and what it could do for the petroleum industry. Frank quickly grasped the possibilities.

Arrangements were immediately made to launch tests with the pump on a Phillips well in the El Dorado field near Burns, Kansas. The new pump was a winner. Arutunoff was moved to Bartlesville, where, with the backing of Phillips Petroleum, he continued testing and development, and formed his own company. News of Arutunoff's successful invention, as predicted, caused a stir in the oil industry. Within a few years, the Reda Pump Company—Reda an acronym for the cable address of the Russian Electrical Dynamo of Arutunoff—would be a major manufacturer in Bartlesville.

But besides hooking up with the young genius Arutunoff and scouting the West Texas real estate which would become the Borger oil field, the Phillips party enjoyed a fine Western adventure during that whirlwind trip of 1926. It happened during one of those short detours Frank took on the way out to California. It was right after the chartered train stopped in the panhandle and then cut across New Mexico into Arizona. When the train approached the vermilion and pink mesas and cliffs, Frank gave orders to have his private Pullman switched off at Williams, and the band of oil executives from Bartlesville headed to the Grand Canyon. They stopped at the El Tovar Hotel, perched on the great canyon's rim, and spent the entire day exploring. Bundled in their overcoats, the oil men peered over the edge of the chasm. Endacott recorded their visit with his camera and hardly noticed when Mr. Phillips broke away from the others and walked to the hotel.

Later that afternoon, Endacott spotted Frank looking closely at the big log building and making notes as if he wanted to record his personal observations and remember the dimensions and just how the walls and windows and the roof looked. When Endacott went to the nearby Navajo trading post, he again encountered Mr. Phillips.

Frank was surrounded by Indians and he appeared to be buying as many blankets and rugs as he could. Endacott remembered Frank pointing out the ones he liked and saying, "Give me this one, give me that one . . ." When he finished, Frank pulled out a roll of money and peeled off $1,400—a great amount, especially for a young engineer like Endacott working for $190 a month.

At the time, only the money Frank spent at the trading post impressed Endacott. Nothing else, not even Frank's note taking, seemed very important. But years later, after he had paid his dues in the boom towns and bunkhouses and moved on to greener pastures, Endacott would think about the side trip to the Grand Canyon.

He'd picture Clyde Alexander and L.E. and the others standing there, gazing across the big gorge. He'd recall the enormous log walls of El Tovar, and Frank scribbling notes. Most of all, he'd remember the scene inside the trading post. He'd remember the pile of Navajo weavings, died indigo and red. And he'd remember Frank Phillips, the wildcat oil tycoon in the midst of those Indians, looking as happy as a farm boy at the circus.

CHAPTER NINETEEN

By the mid-1920s Frank Phillips was spreading himself thin. Too thin, he thought, for a man with a personal fortune close to $40 million.

The time had come to take stock of his situation and make some changes. Dashing back and forth between Bartlesville and New York was taking its toll. Yet, try as he might to change his pace, Frank was the first to admit that he not only needed to stay tuned in to Phillips operations in the oil patch; he also had to maintain a strong presence in the East, where he hobnobbed with other oil men, bankers, business executives, and politicians from all over the world.

In New York, Frank was royally entertained, wined and dined like a sultan. He attended Broadway shows, ate at the finest restaurants and chic clubs, and spent weekends at elegant Long Island estates or aboard lavish yachts. He made the rounds of cocktail parties, country clubs, and resorts.

Frank felt the need to reciprocate, but, try as he might, he couldn't find a suitable property to build an estate of his own. He and Fern scoured the countryside, but nothing suited him. The Ambassador and some of New York's private clubs were fine for intimate functions, and his town house in Bartlesville was adequate, but as his circle of friends and associates grew, Frank realized he'd have to find a large place in the country in order to entertain in style.

He also found that whenever he brought potential investors and members of his prestigious board of directors back to Oklahoma and Texas for tours of the oil field, they behaved like kids in a candy shop.

Sometimes as many as seventy-five directors and major stockholders would be included on one of the extravagant trips. Frank and Charles Musgrave, vice president of the Phillips transportation department, planned every detail. They'd start many of the tours at Grand Central Station, take a chartered train—stocked with the best liquor, food, and cigars—and head westward, with stops along the way to pick up more guests.

The train would take them to Texas and Oklahoma, with more stops in Fort Worth and Tulsa and at several of the boom towns in the patch. Frank and his top executives accompanied them every mile of the journey, and during the inspection of Phillips operations they always made sure that the visitors got to see at least one "gusher" come roaring in, a sight which seldom failed to impress them, even if it was staged by the flamboyant Uncle Frank.

The eastern bankers and businessmen were entranced with what they saw and experienced in the West. They especially liked the wide-open spaces of the Burbank and panhandle fields and the bluestem-grass country around Bartlesville. They relished Frank's stories of the frontier and they devoured the tales they were told of the outlaws and Indians and oil-field roustabouts. It seemed they couldn't get enough.

It was after he had several of the successful oil-field tours under his belt that Frank made a discovery—the place he had been searching for to entertain on a grand scale was under his nose all the time. The best way to return to his roots and at the same time establish a comfortable base for rest and entertainment was to go to the land he loved best. Frank turned to the Osage. The realization that the rugged landscape of the Osage would make the consummate retreat hit Frank like a blindside punch. After spending more than two decades in the banking and oil business, Frank, at long last, could actually visualize a place where he could relax and at the same time serve up ample portions of hospitality to friends and business associates.

Frank announced to family and friends his plans to develop a ranch and lodge on some land he already owned in the Osage Hills. He would also maintain a herd of cattle, some saddle horses, and as many wild animals as he could find and turn them loose on the acreage. The property Frank chose was near Lot 185, where years before he had built a gasoline plant and a camp for his workers and a schoolhouse for their children and where a string of gushing oil wells resulted in the formation of the Phillips Petroleum Company. For Frank, the creation of his ranch in the Osage was the fulfillment of a dream.

The Frank Phillips Ranch, a 3,600 acre spread, was fourteen miles southwest of Bartlesville but remote enough to allow Frank and his guests to feel like they had escaped the hustle and bustle of the business world. Amidst the rocky canyons and thick stands of blackjack oak, sumac, and persimmon, Frank's three-piece business suit and necktie were exchanged for a cowboy hat, open-neck shirt, and boots.

The area was rich in history. Ages before his ancestor Miles Standish had even stepped foot on North American soil, Indian warriors hunted on this land and camped near the freshwater springs scattered across the rolling hills. Chouteau Spring, named for the French fur-trapping family who came to the spring to barter with the Indians, was within pistol range of the spot Frank selected to erect the main entrance with its arched gateway bearing the inscription:

Frank Phillips Ranch
Home of Buffalo and Wild Game

There were other landmarks on the ranch—Outlaw Gulch, Daddy Miller Spring, Bison Lake, and Spencer Spring, named for Al Spencer, the desperado who teamed up with Frank Nash to rob the train at nearby Okesa in 1923.

The Spencer gang's train robbery was still a topic of conversation in the Osage, but all the old-timers, including Uncle Frank, knew it wasn't the first train job pulled at Okesa. In 1911, an outlaw named Elmer McCurdy carried off a daring robbery at about the same spot, but he was shot and killed a few days later. His body remained unburied in the corner of a Pawhuska funeral parlor for many years until a couple from California claimed it was a long-lost relative and hauled it away. Stories drifted through the Osage Hills that the mummified body of the outlaw was being exhibited in a carnival in Texas, and then the word spread that the body was in a wax museum in California. The mummy was being used as a prop during the filming of a motion picture when an arm was accidentally broken off and it was discovered that the body was not fake but had once been Elmer McCurdy, the Okesa train robber. In 1977, after sixty-six years of abuse, McCurdy was brought home to Oklahoma and laid to rest.

In the 1920s, though, he had not been forgotten. The story of Elmer McCurdy and his bold robbery was still being told around campfires and supper tables. It was just another one of the yarns that kept the past in the present and helped put fire in the bellies of the misfits who remained, like Al Spencer and his cohorts. The participants in the 1923 train robbery at Okesa made out much better than McCurdy. While Spencer and some of the other members of the gang looted the mail car of nearly $20,000 in Liberty Bonds, Nash stood guard over the passengers, telling jokes and kibitzing about the recent death of President Harding. Several of the gang were apprehended, but none of them ended up as mummies in a carnival sideshow.

Spencer and Nash and another of their pals, Henry Wells, the outlaw who helped plan the Okesa train robbery but had to drop out at the last minute because of a lame horse, were very familiar with the land that became the Frank Phillips Ranch. For many years the property had been a notorious refuge for several of the leading thieves, killers, and rustlers.

The notion of outlaws and thieves living in the area didn't bother Frank. He had never experienced any problems with bank robbers and he didn't plan to start now. Many desperate men had worked on a Phillips Petroleum oil lease at one time or another, and besides, the more colorful characters that were around, the more colorful stories were available for the eastern dudes to hear when they came out to Oklahoma for an annual meeting or a long weekend at the ranch. The wild animals, outlaw hide-outs, cowboys and Indians, roustabouts and oil derricks all exisiting on or near his land pleased Frank Phillips no end. He was in his glory.

Frank selected a site high above Rock Lake for his lodge. He had been fond of this particular spot since the days when oil activity was at its peak on Lot 185 and he used to come out on the rough road in his horse-drawn buckboard to visit the leases and occasionally spend the night in his crude log cabin overlooking the lake. This is where he wanted his southwestern showcase. It would be the place where family, friends, business associates, and others could let their hair down and taste a different culture—another style of life.

Frank not only admired the beauty of the surrounding area, including the small lake once used to supply water for the steam boilers and gasoline plant; he never forgot that this hunk of real estate produced some of his most important wildcat wells. He had rich memories of leaving the old office in Bartlesville, climbing the steep "44 Hill," and passing the Getty leases and the tall stands of bluestem grass. Early Phillips employees were often treated to an afternoon dip at Rock Lake. They changed into their bathing suits in some tents Uncle Frank had set up on the bank, and afterward they could usually persuade Daddy Miller into revealing the locations of the best fishing holes.

Arthur Gorman, a local architect and contractor, was selected to build Frank's lodge. He was told not to spare any expense. Construction of the first section—a dining room and kitchen—started in 1925. It would take the better part of two years to complete the lodge and other ranch buildings. Gorman used hand-hewn logs from the pine forests of Arkansas and others he found near Elk Springs, Missouri, to build the impressive structure. The fittings and furnishings were the best money could buy.

The dining room, built separate from the main lodge but later joined, was a replica of the one-room log cabin on the Nebraska prairie where Frank was born fifty-three years before. Frank got the idea from his mother, and he made sure she came down from Iowa and visited the work site so she could supply as many details as possible about the old homestead the Phillips family had to abandon after the grasshoppers swarmed over the place.

Frank insisted on keeping the exterior walls unchinked until, during an early inspection of the building, Jane discovered all the mice and ground squirrels which squeezed in to take up residence inside the lodge. In short order, the cracks between the big logs were filled with cement. Workmen also had to be on guard for rattlesnakes—big ones—that liked to nap in the sun on the rocks above the lake and weren't opposed to crawling into the lodge to hunt up a mouse lunch.

The main portion of the lodge, called the Great Hall or living room, measured twenty-eight feet east and west and fifty-two feet north and south with great native stone fireplaces at each end. The walls were finished in polished hewn pine logs, the posts and rafters were finished in logs with the bark still on, and all the closets were cedar-lined. There were large second-floor bedrooms for both Frank and Jane, guest quarters, billiard room, trophy room, fishing tackle and gun room, and a balcony that was ideal for all-night poker sessions. Dan Mitani, Frank's ever-present valet, was provided with a sleeping room off the balcony, just a closed door away from Uncle Frank.

Certain features of the lodge's rustic design were inspired by Frank's trip to the El Tovar Hotel on the rim of the Grand Canyon. Now Frank would have a place to hang the stack of Navajo blankets and rugs which Endacott watched him buy at the trading post.

Outside the lodge workers laid sandstone walks, cut from huge slabs found on the property. A spacious front porch was built along the east side of the lodge. From the porch—one of the coolest spots on the ranch—Frank and his cronies could sit and enjoy their cigars, shoot the bull, and look out over the lake.

To honor Clyde Alexander, his friend and vice president and the man who helped pioneer the early development of the ranch property, Frank ordered the name of the picturesque lake below his lodge changed from Rock Lake to Clyde Lake. The gesture moved Alexander, who in tribute to the Phillips named one of his sons Creston, after the Iowa town that always remained dear to both Frank and Jane.

As some of the work crews erected the lodge and the stable, also built of native sandstone, others stayed busy putting up fences around the border of the sprawling ranch. The fences were to serve a couple of purposes —keep Frank's herds of domestic and wild animals in and keep uninvited guests out. Still, it was easy enough for those who considered the ranch a choice hiding place, or who had a craving for fresh meat, to slip across the fence to butcher an animal or hole up for the night in Outlaw Gulch. Frank never had to put up with any large-scale rustling, and a trapper, hired to protect the herds of animals, kept the coyote and bobcat populations in check.

In January 1926—about three months before he officially unveiled his lodge—Frank's first major shipment of animals—a herd of wild buffalo— arrived at the ranch. He already had cattle grazing in the pastures, and a dozen buffalo were growing fat in a meadow, but Frank wanted more. The new buffalo Frank selected came from Pierre, South Dakota, and were part of the largest wild herd left in the country. A total of 183 buffalo were shipped—120 for Frank, 53 for the Miller boys at the 101 Ranch, and 10 for Waite Phillips' new ranch located in the mountains of northern New Mexico.

"I do not want to announce the exact date of the arrival of these animals," said Frank the day before the train carrying the shipment of buffalo chugged into Oklahoma. "I expect we will have trouble enough getting them from the station to the preserve."

To avoid having too many spectators, Frank and Clyde Alexander arranged for the train carrying the buffalo to arrive in the middle of the night at Okesa, not far from where Elmer McCurdy and later Spencer and Nash staged their train robberies. Frank made certain a bunch of Osage cowboys were on hand at the rail landing to meet the five express cars, and the Millers sent over their best buffalo man, who Frank said was so good at handling the big beasts that "he could drive them up on my front porch if I wanted them there."

As it turned out, Frank and Clyde's worries about wild buffalo stampeding off the train and trampling curious citizens were ill founded. No crowd showed up, and because the big animals were afraid of the lights around the railroad station, they were not unloaded until daybreak. The Osage cowboys had no problems herding the buffalo—docile as sheep—across the fields, down the snow-covered road, and through the main gate of the ranch.

That morning Frank could boast that he owned the second-largest herd of buffalo in captivity in the United States. His 132 buffalo put him ahead of Pawnee Bill, who kept a sizable herd of bison for his Wild West show. Only the Millers' herd of 200 buffalo was larger than the herd at the Frank Phillips Ranch.

In tribute to Buffalo Bill Cody, a childhood hero, and because of the importance of the buffalo in the development of the American West, Frank selected the big shaggy animal—the monarch of the plains—as the official symbol for his ranch. A buffalo-head illustration adorned the ranch stationery, and when the herd was thinned or an old animal died, their skulls were tacked on the lodge walls or in prominent places around the ranch.

But buffalo weren't the only animals Frank brought in to roam the range. The same month the large buffalo herd came from South Dakota, twenty-two Japanese sika deer were bought from a private preserve in Illinois. A short time later, forty elk from Montana were released, as well as herds of longhorn cattle, wild mustangs, and mule deer. Zebra were shipped from British East Africa. A flock of wild turkeys was released on the grounds and a colony of beavers was introduced to the ponds and lakes scattered over the ranch. Peacocks, with brilliant blue and green plumage and long tail feathers, strutted outside the lodge. Despite their loud cries and squawks, which reminded Jane of Frank's singing attempts, the handsome birds, with their iridescent plumes, were Aunt Jane's favorites and she had ranch hands save the feathers for servants to arrange in vases for the lodge and the town house.

It took little time before the inventory list of the animals on the Frank Phillips Ranch read like a Who's Who of the wild kingdom. The list included everything from water buffalo, camels, blue gnus, kangaroos, and llamas to Himalayan tahrs, yaks, macaws, ostriches, storks, monkeys, and Angora goats. Frank was able to brag to his business associates that more than five hundred wild and domestic animals made their home within the confines of his ranch.

The majority of the exotic animals, and most of the fowl, were purchased directly from the famed Hagenbeck Brothers, headquartered in Hamburg, Germany. John T. Benson served as the American representative for the Hagenbeck Circus, and shortly after the F.P. Ranch became operational, he arranged for a meeting to reacquaint Frank with John Ringling North, head of the Ringling Brothers Circus, the outfit Frank first encountered back in Chicago when he was handling booking arrangements for the Chicago Coliseum. Over the next several years, North, Benson, and even Heinrich Hagenbeck visited the ranch to consult with Frank about animal purchases. Later, Osa and Martin Johnson, the famous film-making adventurers; Frank "Bring 'Em Back Alive" Buck; and George Vierheller, father of the St. Louis Zoo, came to the ranch to offer their advice about animal care and the display of trophy heads on the walls of the lodge.

To make sure everything ran smoothly, Frank hired a former lawman named Grif Graham as his first ranch manager. Graham had served as a deputy sheriff in Washington County right after statehood and in 1914 he was elected sheriff. He was popular, served the citizens well, and was remembered for having shot a horse from under Henry Wells after a bank robbery.

Graham, and his wife, Orpha, lived in a tent while the lodge was under construction and eventually moved into the log gatehouse built next to the arched entrance. Ranch guests came to expect a greeting from Graham—a big howdy and a wave of his hat. As the official gatekeeper and guide, Grif wore a bright red shirt, a broad-brimmed cowboy hat, and high-heeled boots. He looked every inch a lawman. Just the sight of him, all tanned and fit, struck fear in the hearts of the lawless and even intimidated the haughtiest New York dandy.

In April 1926, Frank threw his first big party. He was ready to take the wraps off his Osage spread. He felt that even though his lodge was not quite finished, the time had arrived. It was a sneak preview and stag dinner—the biggest buffalo barbecue the Osage Hills had ever seen. The occasion was the annual meeting of the Phillips Petroleum Company stockholders.

"In whatever part of the world oil men or businessmen meet and talk of industry and affairs, the name of the Phillips Petroleum Company of Bartlesville, Oklahoma, has come to mean progress and prosperity," said the Bartlesville *Enterprise* in a front-page salute to Frank and his company. "Its rapid growth in an industry that is remarkable for rapidity of expansion and its basic, substantial strength combine to make it an institution of which Oklahoma can well afford to be proud."

The newspaper and all of Bartlesville had a lot to be pleased about in 1926. Rumors that Frank was seriously thinking about moving his company headquarters were laid to rest. There was not only a new Phillips

Petroleum office building going up on the city skyline, but Frank was bringing a bevy of distinguished visitors to his new ranch in the Osage.

Garbed in a big cowboy hat, high-top boots, and a leather coat, Frank welcomed prominent financiers and railroad executives from the East and introduced them to Oklahoma and Texas oil men and business tycoons.

It was an eclectic group—bankers, pioneer settlers, corporate presidents, Indian dancers, and cowboys. There was not one but three Du Ponts. Gordon "Hamp" Scudder, an old-time cowboy and rancher, showed up. Scudder worked on ranches near Hogshooter Creek and Chelsea, and while he was Will Rogers's bunk mate, the two old friends made a pact that whichever one died first, the other would provide mounted pall-bearers for the funeral. Both were still kicking—Hamp still a cowboy and Will, the newly appointed "Mayor of Beverly Hills."

Eugene Du Pont stood in the buffet line next to Amon Carter, the Fort Worth publishing czar, and listened to the colorful Texan explain the newspaper-war strategy he used against rival William Randolph Hearst. Grizzled cowboys rubbed elbows with bank presidents who didn't know a shooting iron from a drill bit. George Miller came over from the 101 Ranch to mix with the visitors, and another of Frank's good friends, Pawnee Bill, cut a dashing figure with his wide hat, gleaming boots, and stylish woolen overcoat.

Dinner was served in a tent set up in the natural amphitheater at the south end of Clyde Lake, below the high bluff where craftsmen were applying the finishing touches to Frank's lodge. Guests were handed plates heaping with barbecued venison and buffalo, and Grif Graham's dog, Smoke, made sure any tender bites that fell on the ground weren't wasted. After the big feed the guests were entertained by a troupe of buck-and-wing dancers, an all-black orchestra, and fancy Wild West stunts performed by some of the local cowpokes. As at most parties in that roaring era, the celebrating was feverish. And like everything that Frank Phillips touched, it was a booming success—only the first of hundreds of bashes Frank would host over the years at his ranch in the Osage.

That July, all the Phillips Petroleum employees and their families were also invited out to Uncle Frank and Aunt Jane's ranch for a picnic. There was plenty of food and drink, and the highlight was a bathing beauty contest featuring sixteen local lovelies, with bobbed hair and dimpled knees, clad in their stylish one-piece swimsuits. They gathered at the "Spring of Eternal Youth" to vie for the title "Miss Petroleum." The judges were all Phillips executives. Frank, his eye for the ladies sharp as ever, made himself the head judge. But his tenure on the panel was short-lived. He stuck around only long enough to personally record each contestant's vital statistics with the tape measure draped around his neck. Not wanting to be accused of favoritism and with what he considered the best part of the contest already out of the way, Frank "resigned" and left the rest of the judging up to Obie Wing and the others.

In September, Frank staged a Wild West show at the ranch, featuring many of his family and friends and some of his top executives. Jane, her close friend Winnie Clark, and the Phillipses' two foster daughters, Mary and Sara Jane, were given colorful Western names for the show. Winnie Clark was called "Cyclone Clark," Jane became "Lady Janett Phillips . . . descendant of Pocahontas," and daughter Mary was listed on the program as "the rifle queen." Everyone wore boots, chaps, and one of Orpha Graham's handmade shirts, each one so bright that they "competed with each other throughout the show for the most attention, so far as color was concerned."

The entire afternoon was tongue-in-cheek. "Fog Horn Clancy"—alias Stewart Dewar—was appointed master of ceremonies, and the only interruption came when two ostriches, recent arrivals at the ranch, broke loose and had to be rounded up like a pair of wayward steers. Several of Frank's best horses—Rex, Tony, Patches, Major, Copan, and Roundup—were saddled for the big parade. Sara Jane's pet goat, Billie, presented as a trick goat, and "the only goat in the world taking daily Turkish baths and beauty clay treatments," was a big hit. So was the Indian stomp dance performed by Jane and Paul McIntyre.

"If the buffalo which roam about on the big preserve had ever gotten sight of some of the loud shirts, fancy boots, and bright neckerchiefs, it is doubtful that the big eight-foot steel fence would have held them in that part of the country," reported the local newspaper. "Two of the baby deer were frightened out of a year's growth by the noise made by the prancing steeds."

The Wild West show was pure homespun entertainment spiced with a healthy dash of vaudeville. Frank loved every moment of it. And he loved spending as much time as possible out at the ranch, where he could ride horseback through the lush fields and across the rolling hills, or loll on the front porch of the comfortable lodge and enjoy a toddy prepared to perfection by his personal valet, Dan.

Frank had Fern Butler to thank for coming up with the official name for his Osage lodge on the hill above Clyde Lake. The naming of the lodge didn't happen at the ranch or even in Bartlesville. It occurred in New York. And there were two witnesses present who were the first to hear the name Fern and Frank conceived.

Frank was in the city, tending to business. Two teenage girls from Bartlesville—Elisabeth and Elise Seaton—with their mother, an ambitious woman and a classic stage mama who wanted nothing more than to see her daughters' names in lights, were also living in New York. The Seaton sisters were less than a year apart and looked enough alike to pass for twins. Their father worked for Phillips Petroleum, and they had lived out at Lot 185 for a couple of years before he went alone to work in the oil field at Borger. When he returned to Bartlesville, his wife had taken their two daughters to study drama and dancing in New York.

After making the rounds of talent agencies, they landed jobs with Howard Thurston, a magician, who called the two sisters "the Seaton Twins" and used them as dancers and assistants in his traveling magic act.

Their specialty was the disappearing trick. Thurston would place Elisabeth inside a cabinet, mumble a few magic words, and presto—the girl would vanish into thin air. A few seconds later, Thurston would snap his fingers and Elise, wearing a gown identical to her sister's, would walk from the back of the theater. The trick always worked and usually brought thunderous applause. Thurston and his attractive accomplices pulled their disappearing act on stages across the country.

Frank became acquainted with the Seaton sisters because their mother had paraded them in and out of the Broadway office, hoping some of Uncle Frank's New York contacts would help push the girls into the limelight. Both of them were pretty and pert, but Frank developed a special interest in the younger Seaton. Things never got out of hand. Elise was able to stay clear of Frank and made sure that whenever they visited him at the Ambassador, or even at his office, her sister went along. Elise may have been young, but she knew better than to be placed in an awkward position with a determined man like Frank Phillips.

One Saturday afternoon, both Seaton girls were invited to Fern's Lexington Avenue apartment to teach the svelte Miss Butler some of their dance numbers and the latest exercises of the day. When they arrived, Fern, looking slender and vivacious, greeted them in a pair of lounging pajamas. Seated in the living room was Uncle Frank, enjoying a cigar and the prospect of watching three good-looking females go through their exercise routine.

As she followed the girls' lead during the exercise session, Fern asked

Frank about his new ranch outside Bartlesville. She wanted to know everything. He told her about the buffalo and the cattle and the string of fine horses he rode every afternoon he was in town. He told her about the bathhouses at the edge of Clyde Lake and the picnic grounds with a pavilion and barbecue pits. He described the dairy barns and stable, the aviary, and the pastures filled with animals from around the world.

Frank told Fern that the ranch was turning out to be just what he had always dreamed about when they searched Long Island and Connecticut for an estate. But, he explained, he still wanted to come up with a name for the lodge itself. He said the big log building was much more than a ranch house. It deserved its own name. Fern agreed.

Still gyrating with the Seaton girls, Fern started repeating out loud words that came to her. Frank did the same. They bounced names off one another and considered anything that came to mind but just couldn't come up with a name they liked. They were after a name that would capture the essence of the place—a special name that everyone would remember.

Then, in the midst of the exercises, Fern stopped and grabbed paper and pencil. She wrote some words and then scratched them out and tried again. She paused and looked at Frank.

"Woo-la-roc," said Fern.

"What was that?" asked Frank.

"Woolaroc," she repeated. "Woods, lakes, rocks."

That was it. A name coined from the three natural elements which surrounded the lodge—woods, lakes, and rocks. Put them together and it became Woolaroc. It was the perfect name.

By the autumn of 1926, word was getting around about the Frank Phillips Ranch and the oil man's splendid Woolaroc lodge. That November, just before Frank's fifty-third birthday, the Kansas City *Star* assigned a correspondent to go to Bartlesville and spend a day with Frank Phillips both in his office and out at the ranch.

The reporter was surprised by the openness he found. Frank allowed him to sit in his office while he talked by private long-distance line to New York, signed scores of letters dictated to Marjorie Loos, and presided at the regular 11 A.M. conference of Phillips executives. Yet when the questions became too personal, Frank quickly changed the subject. His candor vanished.

The reporter noted:

"It is difficult to get this super-leader of the oil industry to discuss his career. He balks at any account of his life, his spectacular and daring success in the oil business which made him one of the wealthiest men of the nation, one who sits in the counsels of the mighty on Wall Street. A leading question can be spotted afar by Mr. Phillips. He would rather talk about the rolling hills out yonder in the Osage country, about his plans for the ranch.

A little later, the reporter recorded more of his impressions of Frank:

Mr. Phillips will be fifty-three years old November 28. He is tall, well proportioned, nervous with energy. He wears glasses. Expressive gestures are made with the hands in his conversations. His voice is deep and he knows how to emphasize his words. Mr. Phillips' own life has been dramatic because it has been crowned with success achieved entirely by his courageous attacks upon the citadels of fate; by his crashing will and acumen in business. He goes at high tension early and late. From conference to conference, from one task to another as the head and guiding genius of the Phillips Petroleum Company.

Just before departing for the ranch with his guest, Frank called in one of his officers for a final bit of business. "That matter we were discussing can go over until later," said Frank, and he named a date after glancing at his desk calendar. "Handle it then, will you?" The reporter noticed Frank acted as though they were discussing the cancellation of a golf date and not a transaction which involved more than two million dollars.

Once they were in the waiting limousine, Frank picked through a pile of new clothes stacked on the floor until he found some cowboy shirts with pearl buttons. "I'm taking these out to the ranch," said Frank, pulling at his necktie and making a face. "I don't believe I'll ever get used to wearing a white stiff collar. Pinches my neck."

When they passed the stately Phillips town house, Frank pointed out the window. "Here's a little place I built some time ago," he said. The big shiny limo roared on, and as they reached the city limits, Frank began to relax.

The reporter observed:

The ride to the ranch along a wandering graded road, which
soon is to be under construction for paving, was delightful.
The scenery slipped past like flashes of a movie view. It is all
cattle country. No cotton in this section. At the edge of
Bartlesville are some veteran oil wells, among the first sunk
in the Bartlesville field. The wells are still pumping. Mr.
Phillips points to the first field opened by the Phillips company. "There used to be a time when I handled all the details of my company, but I've got away from that now. I go
to the office in the morning—that is, when I'm in Bartlesville—and handle questions of policy in the business and
turn the details of the company's management over to my
executives. That gives me a chance to get out on the ranch
and roam around for a couple of hours every day. I've decided that I'll never retire from active business, but that I'll
not be a slave to my desk.

When the limousine arrived at the entrance to the ranch, Frank made
sure the reporter knew that "this is the only opening in eleven miles of
fence." They passed below the arched entrance made of brown native
stone, waved at Grif Graham in his bright red shirt, and motored down
the winding road which led to the Woolaroc lodge. Frank explained that
before he established the ranch he checked to see if there was oil in the
vicinity of the lodge. He was relieved to find there wasn't a drop. "I made
sure there would never be any oil on the ranch," said Frank.

Following a quick tour of the lodge, Frank took his guest outside for a
walk. They descended the stone steps leading from the front of the lodge to
Clyde Lake, and paused to examine a crypt in the rock cliff. Inside was a
pile of bones and a human skull. "That's old Chief Woolaroc in there,"
Frank told the wide-eyed reporter. "That stone there used to fit into the
opening of the tomb." He pointed at a huge rock with the toe of his boot.
Frank then showed the reporter some Indian "hieroglyphics"—crude pictures showing trees, a dragonfly, and rocks—chiseled in the stone. He
explained that, according to local legend, the row of trees stood for the
woods, the dragonfly represented the lakes, and the rocks were self-explan-
atory. It was from the first few letters of each of these words that the old
chief got his name—Woo-la-roc. And it was from Chief Woolaroc that
Frank came up with the name for his lodge. Frank could hardly keep from

breaking into a belly laugh as the reporter scribbled down the tale of Woolaroc and the origin of the name.

Frank figured it was as good a story as any he had told. Besides, the last thing Frank was about to tell a reporter from the Kansas City *Star* was the true story of how the lodge got its name. He could just see the headlines if the reporter ever found out that Frank was sprawled in a chair smoking a cigar when his lover, clad in her silk pajamas, came up with the name Woolaroc in a cozy Manhattan apartment while a pair of comely magician's assistants did their exercises. The tale about an old Indian chief would do just fine.

When they were back in the lodge after their walk, Dan fetched the twosome steaming hot coffee served in enameled mugs just like the ones the cowboys used. Once they were settled, Frank talked for a while about his purpose in founding the ranch. "This isn't all a dream about something —a place where I can fritter away time and spend money," said Frank. "The great difficulty with the American people today is that they are getting too far away from the fundamental things in life. They are getting away from the soil and from the basic things that made home life pure and happy. Today the home is not keeping pace with the growth in population, particularly among those homeowners who have no intention of making the home a real place of enjoyment. I have gone back to my hometown several times, and things seemed wrong to me. Something the matter? I decided the trouble all was with me and that the folks in the little hometown were the ones who were remaining steadfast, while I was getting away from those fundamentals. That's what I'm going to do here on the ranch. Build an ideal of the midwestern life, log house, rugged place for a cabin, a wild animal and game preserve. All part of the Frank Phillips 'back to nature' plan."

The reporter had a final question. What would Frank ultimately do with the ranch? "After the project is completed, I expect to endow it and, after my lifetime, turn the whole ranch over to some organization, probably the state of Oklahoma, to operate it as a wild game preserve," said Frank.

Satisfied with his day spent tracking Uncle Frank, the reporter returned to Kansas City and filed his report. It was a dandy. Frank got just the headline he was looking for: "Uses Riches to Preserve the Spirit of the West." The subhead was even better: "Frank Phillips Is Giving Expression to His Ideals on His Million Dollar Ranch in Oklahoma." Frank couldn't have paid the reporter enough to write the kind of story which appeared.

Woolaroc lodge is crested upon a hill top, mirrored in a crystal lake which sparkles like a great diamond about one hundred feet below the level of the pine log cabin. It is a splendid time to visit the ranch and Woolaroc lodge when the distant hills are splashed with autumnal pigments. Gossamer veils of purple cling to the trees and eminences that thrust into the sky at the horizon.

The Frank Phillips Ranch is the expression of an ideal in the life and career of the owner—a desire to build something monumental to the spirit of the West, particularly the West of the cattle, the Indians and the wild birds and animals that used to roam this country. It is more than the playground of a rich man. It typifies America's pioneer life.

The story, illustrated with a portrait of Frank and photographs of the ranch entrance, part of the buffalo herd, and a view of Woolaroc lodge, also made a stab at measuring the man behind the ranch:

Frank Phillips is an American, a western American, robust and scintillating, filled with the glory of achievement, of doing things himself, of making his own destiny. He has crammed a lot of living into his life. He prefers to idealize the Oklahoma he knew twenty-two years ago when he settled here. The vigorous men he met then—Indians and whites—are the ones he cherishes on the long list of friends. He found the leather boots and the wide hats and the leather coats the apparel of the men who were striking out for fortune in those first days of adventure. And Frank Phillips will continue to wear them on his ranch where he has made a pledge to himself that the ideals of western life will be preserved.

CHAPTER TWENTY

Frank wasn't the only Phillips interested in preserving the ideals of the West and building an empire based on the kind of revenues that black gold could bring. Down in Tulsa, his kid brother Waite was doing all right for himself. Oil had also been good to Waite. Damn good.

During the years since he severed business connections with his brothers in Bartlesville, Waite had become a millionaire many times over. His properties in Okmulgee County were regarded as the cream of oil production. In the early 1920s he purchased a refinery, remodeled it, and laid out pipelines to his oil wells so that the plant could get its crude at the least expense. Next, he completed the cycle by entering the marketing business. His first service station was in Tulsa, the second at Okmulgee. One newspaper reporter went so far as to say that "Waite Phillips' filling stations are things of beauty, wherever located."

But when federal and state taxes began taking big bites out of his income, Waite decided that even beautiful gas stations weren't joys forever. It was time to change his method of doing business. Waite switched from an individual ownership status and organized all of his oil properties into a single company, just as Frank and L.E. had done years before up in Bartlesville. In 1922, the Waite Phillips Company was incorporated.

Waite worked six days a week and had executive conferences on Sunday mornings to talk over which leases to buy next. He worked as hard as a roustabout and, like his mentor in Bartlesville, Waite wouldn't stand still if he thought he was paying someone who wasn't earning his keep. A parade of vice presidents walked in and out of Waite's office. Nobody crossed him or disagreed with him and stayed on. He had the true oil man's spirit. Waite believed in buying leases every place he even thought there might be oil. If he struck it rich—as he so often did—Waite made sure the wealth was shared.

"There is greater honor in being the best ditchdigger in a gang than in being a mediocre president of a company, because the first man has done something outstanding by means of his own efforts, while the latter is content to let the dignity of his position bear him along," Waite said in a memorandum to his employees. "Every job is important and should be done in a thorough manner if we wish to accomplish that perfection that sets us apart as experts in any line."

He controlled 75 percent of his company's stock and those who were close to him in its administration were given the other shares in lieu of bonuses for good service. A few got both.

Waite outgrew his suite of offices in the First National Bank Building in Tulsa, and he eventually outgrew his desire to preside over the Waite Phillips Company. In the autumn of 1925, Waite sold his firm—lock, stock, and oil barrel—for $25 million cash to Blair & Company, prominent New York investment bankers, who soon merged with Barnsdall Oil Company. Of the huge sum he earned from the transaction, Waite rewarded his ten key employees by doling out a total of $1.5 million in bonuses. They received their checks that year just in time for Christmas.

When he sold out, Waite was forty-four years of age and had built up a personal bankroll of close to $40 million as a successful oil producer. Easy street was in plain view, but Waite wasn't about to retire. Living the life of the idle rich wasn't in any of the Phillipses' cards, including those dealt to Waite. Shortly after the sale of his company, Waite put some of his enormous surplus of cash from the transaction to good use. Upon finding out that his wife Genevieve's father was in a bind, and needed to be bailed out, Waite immediately reacted.

J. B. Elliott, once the wealthiest banker in Knoxville, Iowa, and a man who had been opposed to his daughter's marrying a poor kid off the farm, was in serious trouble. In 1925, his little bank was about to close its doors. Whether from poor management, an overabundance of Iowa crops, or a lending policy based more on relationships than realism, the First National Bank of Knoxville was holding a lot of bad paper on farmers. They couldn't pay their mortgages or even the interest. Word got out and a run on the bank started. Waite didn't hesitate for an instant. Having just sold his company, he could well afford to be generous. And, as always, he was just that. It meant handing over $800,000, but he saved his father-in-law's bank. Waite took no credit. He said he didn't deserve any.

In 1926, J. J. McGraw, president of a venerable old Tulsa banking institution, evaluated Waite this way: "Waite Phillips says his success was due to luck. It wasn't. It was due to good judgment in the choice of men for his organization and a keen appreciation of real oil values. He isn't a gambler. He is a shrewd investor. I don't want to say he's the best oil man on earth—there are lots of good oil men. But I have never seen a better."

With the substantial cash he now had on hand, Waite invested heavily in stocks and bonds, oil and gas royalties, and real estate. Following the example set by Frank, he also invested in several banks, serving as the chairman of the board of the National Bank and Trust Company of Tulsa for several years. His land investments spread into Colorado, New Mexico, and Arizona. He soon owned more than a half million acres of ranchland, including a major cattle operation south of Denver and another prime ranch in the Apache country of Arizona.

His most famous landholding was a 127,000-acre ranch near Cimarron, in the mountains of northeastern New Mexico—an area rich in history. Waite's ranch had once been part of the vast Maxwell Land Grant, one of the largest pieces of land ever owned by one person in the history of the nation. The property had a colorful past. Prehistoric cave dwellers, Ute and Apache warriors, Spanish conquistadores, Santa Fe traders, fur trappers, ranchers, miners, and railroaders had all left their marks there.

Waite bought the huge spread purely as a business investment for diversified farming and livestock operations, but after he fished the canyon streams, rode horseback through the stands of spruce, pine, and aspen that covered the mountains, and learned more about the local history, Waite changed his mind.

He first planned to call the ranch the Hawkeye, in honor of his native state of Iowa, but before the name became official, the rugged beauty of the land inspired Waite. He dubbed his new ranch Philmont, a combination of his family name and the word "mountain." Besides being the first syllable of his family name, "Phil" also meant "love of." It was the best name Waite could have given his ranch. He brought his family there for summer vacations, and he found it a restful retreat from business pressures in Tulsa. It was Waite's favorite place.

In Tulsa, Waite and Genevieve, their daughter, Helen Jane, and son, Elliott Waite, lived in a spacious residence at the corner of Seventeenth Street and South Owasso, just a few blocks from Bill Skelly's place, in a neighborhood of elm-lined streets and wide lawns shaded by mimosas.

In the evenings, Waite changed into comfortable clothes and relaxed in a log cabin he had built in the backyard. It was filled with Old West furnishings and he'd stay there for hours, listening to the mockingbird's night song, watching the lightning bugs blink outside the windows, and imagining himself being nineteen again, roaming the great American West with his twin brother, Wiate. The log cabin became a link with his memories of those action-filled days when the twins traveled the West, but, like Frank and his first cabin in the Osage, Waite wanted something more. He admired Frank's town house in Bartlesville and also took notice when his brother started the development of his ranch in the Osage and the rustic Woolaroc lodge. As usual, Frank served as an inspiring example. Not to be outdone, Waite started building his own version of the great American dream.

Years before, just after he moved to Tulsa, Waite had purchased twenty-three acres of farmland about two miles south of town from two Indians. It was an undeveloped piece of land transected north to south by Crow Creek, a brook which would inspire the estate's name—Philbrook.

As he was making plans to develop his new property on Rockford Road, Waite consulted an old friend, J. C. Nichols, owner of the Country Club Plaza shopping center in Kansas City, and a prominent businessman whose wife was the sister of John Kane, the Bartlesville attorney who served as Frank's legal adviser.

Nichols had induced Edward Buehler Delk, a Dutch-born architect living in Pennsylvania, to come to Kansas City to create the master-plan design for Nichols' landmark Spanish-style shopping center with its plaza of towers, fountains, colorful tiles, and balconies. After hearing Waite's plans for his Tulsa property, Nichols strongly recommended that if Waite wanted to build a mansion worthy of his name and reputation, Delk was the man to hire. "You can't go wrong," Nichols told Waite. He was right. Waite did not go wrong.

He commissioned Delk to come up with plans for not just one but two large residences. Waite wanted a palace for his mountain ranch in New Mexico as well as one for the property in Tulsa. In 1926 ground was broken for Villa Philmonte, which would serve as his summer hideaway, and Villa Philbrook, the grand permanent home of the Waite Phillips family in Tulsa.

The design for the two residences was to be similar and some of the furnishings for both had been collected by the Phillipses during a 1925 Mediterranean cruise.

Villa Philmonte, or Philmont as it came to be known, was two stories and was built in southern Mediterranean style, with huge bedroom suites, guest rooms, private quarters for the servants and children, a swimming pool, and patios. Mounted wild game trophies and animal skins hung on the walls, and plenty of Indian and western artifacts, including buckskin drapes and Navajo rugs, filled the cavernous rooms. Besides the huge ranch villa, Waite also built a hunting lodge, remodeled the corrals to accommodate his polo ponies, and ordered the ranch hands to cut a series of trails throughout the property so he and his guests could enjoy long horseback rides in the mountains. Streams were stocked with trout, and the wild animal population—deer, elk, and bear—increased. Waite saw to it that bighorn sheep were imported and released on the ranch, as well as some buffalo, including ten head that were part of the bunch Frank had shipped from South Dakota.

When Waite's son, Elliott—a feisty youngster known as "Chope" by his many cousins and friends—visited his Uncle Frank's ranch outside of Bartlesville and spied a group of camels feeding in a pasture, he mentioned that he wouldn't mind owning one himself. In a New York second, Frank had his boys crate one of the big humped beasts and ship it to Philmont. The camel adapted to the mountains and had a long life, convinced that it was really a cow pony, even though Waite's herd of horses always shied away.

In time, Philmont became a southwestern showplace. Waite began to invite many of his friends and business associates to the ranch to relax and enjoy its natural beauty. Well-known politicians, authors, and business executives came out to visit. They watched the wild horses race across Buck Creek, caught rainbow and cutthroat trout, slept on hair mattresses in guest quarters flooded with icy mountain air, and sat like barons at a dining table that could accommodate sixteen to feast on gourmet fare prepared by Waite's best chef.

As magnificent as Philmont was, Waite's other architectural pride and joy—Philbrook—had become his haven from the daily grind of big business and high finance. The magnificent seventy-two-room mansion Delk designed on the outskirts of town was the talk of Tulsa.

Completed in 1927, the palatial villa—built in a sixteenth-century Italian Renaissance style—and the surrounding gardens, landscaping, and furnishings cost Waite a cool $1,191,000. It was the most elegant oil mansion in Oklahoma, its only rival being the fancy house Marland built for himself over in Ponca City.

Crowned by a roof of oversized Italian tiles, Philbrook was built with a steel framework and reinforced concrete walls and floors. The exterior glittered in the afternoon sun because of the ground white marble mixed in the stucco. Marble was used for floors, fireplaces, and fountains, and the general flooring was walnut, teak, and oak. The draperies, wall coverings, and curtains were fashioned of the finest silken fabrics. Other features included exquisite stained glass and ironwork, a sunken pink marble tub, and, beneath the white shag rug of the sunroom, a glass dance floor with alternating colored lights, modeled after one seen in a Paris nightclub.

From the organ in the Great Hall, music could be piped into several rooms throughout the house. There was an Italian formal dining room, a walnut-paneled library, a drawing room, and large bedrooms with private baths, dressing rooms, and sleeping porches. The mural in the first-floor music room was straight out of the 1920s. It depicted bobbed-hair flappers wearing Grecian frocks dancing in a mythical setting. Only the family's closest friends knew that one of the frolicking nymphs was modeled after the lady of the manor herself, Genevieve Phillips.

At ground level were several club rooms furnished in western and Indian styles. As a finishing touch for what he called the Santa Fe room, Waite commissioned Oscar Berninghaus, the famed Taos artist, to paint the Philmont Ranch across an entire wall.

Outside were formal gardens, large open lawns, wooded areas, an orchard and vegetable garden, and a swimming pool bordered by evergreens, willows, and poplars. Just outside the sunroom, Waite planted a magnolia tree so he could pick the blossoms for his sweet Genevieve. As a reminder of how Philbrook came to be, remnants of oil and gas wells on the property were used on occasion as flares.

And there were many occasions worthy of celebration at Philbrook, starting with the housewarming in 1927 after the Phillipses and their ten year-old son and sixteen-year-old daughter moved into the villa.

It was a party that Tulsa would not soon forget. The guest list numbered more than five hundred—the elite of society—including General Patrick J. Hurley, a prominent Oklahoman who would become Herbert Hoover's Secretary of War; a member of the Hohenzollern dynasty; and E. W. Marland, who had to admit that Waite's new place wasn't too shabby. Truckloads of roses and chrysanthemums from Mrs. DeHaven's Flower Shop began arriving at dawn the day of the party. Food for the formal dinner took nearly three weeks to prepare at a cost of $25,000.

Chauffeur-driven limousines started lumbering through the villa gates about seven o'clock. It was a perfect evening. The temperature was mild,

the women were all dressed in the most fashionable gowns, and the gentlemen in white ties and tails. The waiters all wore tuxedos and served dinner
at round tables covered with blue, yellow, and white linen and set with the
best crystal and china money could buy. Imported champagne and French
wines flowed freely and the guests danced until nearly dawn.

The most memorable moment in the evening came much earlier.
Around eight-thirty a hush fell over the crowd gathered in the Great Hall,
where they were sipping bubbly and gushing over each other. In perfect
unison, the guests looked up at the first stairway landing. There was Genevieve Phillips, descending the stairs. She had paused for just a moment to
look at her party below before going on and she had a faint smile on her
face. She carried flowers and wore a fine gown, and as she came down the
stairs to join the party, the crowd broke into applause. By the time she was
in their midst, they were cheering. Upon reflection, it was said to have
been the most dramatic moment in Tulsa social history.

There were grand days in store for Waite and Genevieve and their
magnificent Philbrook. When Will Rogers walked into the Great Hall, he
took one look, broke into his trademark grin, and proclaimed: "Well, I've
been in Buckingham Palace, but it hasn't anything on Waite Phillips'
house."

Rogers was dead right. In the 1920s, the Phillipses of Oklahoma—led
by Frank and Waite and, to a lesser extent, L.E.—were about as close to
royalty as anybody could get in the oil patch. But it was down-home
royalty—a combination of the Protestant work ethic and Wild West frivol-
ity with a touch of class. They were a purely American family with ties to
Pilgrim fathers as well as roots which grew deep in the nation's heartland.
They had risen above their humble beginnings and become an enormously
wealthy and powerful force. They'd evolved into a sort of Oklahoma-style
Medici family.

The Phillips clan, except for a few, gathered from time to time at
Woolaroc or Philmont to catch up on family matters. Everyone—especially Frank and Waite—tried to be civil when talk turned to the oil business, as it usually did. They had made the decision long before that they
couldn't afford to allow business to get in the way of the family. The
brothers knew that just because they couldn't work together didn't mean
they couldn't still be friends. There was a time and a place for competition.
It was, as Waite continued to point out, all a question of proportion.

At one of the family reunions at Woolaroc, Lucinda Phillips, an aging
farm widow, came from Gravity and sat on the front porch of the lodge.

Still alert despite her age, Josie had only one wish—for Lew Phillips to be around to see his family and Frank's big log home in the Osage. She wrapped herself in an Indian blanket and proudly posed for photographs with the whole family and then for a portrait with just the five Phillips boys. One of her daughters, Lura, married to Tulsa businessman Johnson Hill, came with her brood and there were grandchildren and great-grand-children everywhere.

The Phillips family was growing.

L.E. and Node's two sons—L.E., Jr. and Phil—spent their summers working on Uncle Waite's Philmont Ranch. There were also long trips to Hawaii or Europe for the entire family, including little sister Martha Jane, who, at her high school graduation, was presented by her Aunt Jane with a brand-new cream-colored Cadillac roadster, with wire wheels and red up-holstery.

In the late 1920s, Phil, with a degree from the University of Kansas, joined the production department of Phillips Petroleum. Lee, Jr. and Martha Jane were finishing college. Both of them were also in love—Lee with a wealthy young lady from Wichita he met at the University of Kansas, and Martha Jane with Wilbur "Twink" Starr, a University of Kansas football hero she met at a fraternity dance while she was back from the Erskine School in Boston visiting her brother. Both couples would eventually marry.

Never the force that Frank became in the oil industry, L.E. continued to be bothered by health problems, ranging from hemorrhoids to heart trouble. He and Node traveled as much as possible and spent time on their farm outside of Bartlesville. Their large brick residence on Cherokee Avenue, next door to Frank's town house, was comfortable but never became a showplace.

"Hard work and loyalty and use of common sense have put Phillips in the strategic position that it maintains," L.E. told the San Francisco *Chronicle* after returning from one of his countless trips to Hawaii.

There were few who could contradict that statement, but many of the Phillips employees saw L.E. from a different point of view than he saw himself. They didn't consider him a strategic warrior, as they did Uncle Frank. Instead they knew L.E. as a pompous nitpicker, prone to scolding workers for not using pencils until they were worn to a nub, or for missing the wastebasket with wads of paper.

After the company was ensconced in the new office building, there were tales about L.E. going on daily patrols of the nearby drugstores and coffee shops, with notebook and stub of pencil in hand, hoping to catch some secretary or clerk who extended his or her coffee-break time. It was not unusual to see people ducking behind counters or running out back doors when they saw L.E. coming. On one occasion in the coffee shop of the Maire Hotel, L.E., his pencil at the ready, demanded to know one woman's name and her department. "I don't work for Phillips Petroleum, so get the hell away from me!" she screamed. L.E. did a quick about-face and moved on in search of new quarry.

Frank heard all the criticism about his brother, but he continued to value L.E.'s presence with the company. After all, L.E. had stuck it out with Frank from the beginning, when they were just learning about oil. He was there when the Anna Anderson blew in and when the wildcat wells started to hit out at Lot 185. Others came and went—including Waite— but despite his poor health and personality quirks, L.E. was still on board. Besides, Frank had more serious family problems to contend with than dealing with the tug of egos between himself and Waite or answering the gossip about L.E.'s pomposity.

John Gibson Phillips—Frank's only son—remained a constant headache and a major source of concern to both of his parents.

Through the years, Frank tried very hard to get John more involved in the operations of Phillips Petroleum Company. His efforts did little good. John simply wasn't interested. The plain truth was that John had been protected and indulged all his life by his mother and, except for a few business matters, largely ignored by his father, who was forever involved with the management of his successful oil operations. Some said that when John was a young man he helped deliver a baby aboard a train and was inspired to become a doctor. But nothing came of it. Nothing much came of any of his plans. John didn't finish school; instead he married young and then, like a trained puppy, followed the course of action laid out by his parents.

John's time spent with Phillips Petroleum was a waste. Jane prodded Frank into keeping John involved in company matters, but it just didn't work. He was a total disappointment to his father. Some concluded that John Phillips was the only person that Frank Phillips never really understood. The communication gap between father and son, severe depression, and a desire to be someone he wasn't contributed to John's alcoholism. Early in life, John had developed an almost unquenchable thirst for liquor.

That thirst never diminished. His marriage to Mildred and the birth of daughter, Betty, didn't slow John's drinking a bit. He could be a likable chap, pleasant and mannerly, but one whiff of an open bottle and he'd be off on a tear. Sometimes, when things really got out of hand and John went on a long drinking binge, Frank would banish his son to the Broadmoor Hotel in Colorado Springs in order to get him out of town and from underfoot.

John and Mildred, a young woman with a pleasant disposition who continually tried to keep her husband sober, were at the Broadmoor in May 1925 when their second child, and Frank's first grandson, was born.

John vowed to Mildred that he'd not take a drink until after she had the baby, and for once, he kept his word. John was sitting in the waiting room, as sober as a judge, in the wee hours of the morning when his son arrived squalling like a renegade Indian. They named him John Gibson Phillips, Jr., after his father and his great-grandfather. Frank and Jane arrived by train two days later to see their grandson, whom they immediately nicknamed Johnny.

When the family returned to Bartlesville, the newest Phillips grandchild was moved into the nursery in a new residence on Cherokee. The house was a gift to John and Mildred from Frank and Jane, conveniently located just across the street from the Phillips town house. It was a striking brick structure with a mahogany staircase, an interior fountain, servants' quarters, and a room for Mildred's mother, Betty Beattie, widowed when her banker husband died of cancer in Colorado Springs, about a month before Johnny was born. A hired nurse, Zibba Zee, slept in the nursery with baby John and little Betty while John and Mildred served bootleg liquor to guests in the entertainment room down in the basement.

Shortly after John and his family moved into their new home, his grandfather, and Frank's old adviser and mentor, John Gibson, and his wife, Bertha, arrived in Bartlesville from Chicago. They occupied the residence on Delaware where John and Mildred used to live. Betty and her playmates and neighborhood children quickly discovered the kindly old man and gathered on the Gibsons' front porch to hear him recite his poetry.

For Jane's forty-eighth birthday, Gibson, then seventy-five, presented his daughter with a handwritten manuscript containing the scores of poems and prose passages he had memorized over the years. Gibson attached a note:

My beloved, My Precious Daughter:

I can never forget the ecstasy of joy which came into my life forty-eight years ago today, when a German midwife laid in my arms your good self.

Your mother wanted you to bear the names of her mother and father. I interposed no objection and you were so named, but it did not satisfy my desire so I bestowed upon you a name to suit myself. The name given you by me was Betsie. I am not selfish in the least degree; it pleases me to hear your kind, noble, generous husband occasionally in his letters and conversations, call you Betsie. I feel that we have in common, personal ownership in one of the most charming and beautiful personalities this world affords.

I wanted on this anniversary day to give you something of my handiwork which would endure while you live. I have prepared this Anthology. It may not be free from errors. If such there be I think you will overlook them because it was conceived and accomplished as a labor of love by

> Your loving father,
> John Gibson

In 1929, when Gibson was in his eightieth year, Jane arranged for the collection of memory work, written in long hand, to be published as a tribute to her father. In the foreword of the handsomely bound Gibson anthology, Jane spoke of her father's uncanny memory and how he entertained strangers aboard ship and friends in his home with "these lovely gems which years of love and constant usage have almost made his own." She said people listened to the poetry for hours at a time and were never bored.

Hardly a soul was bored during the Roaring Twenties in Bartlesville, Oklahoma—not with two powerhouses like Frank Phillips and H. V. Foster doing business in the same town. Added to the Phillips and Foster one-two punch, was Henry L. Doherty, founder of Cities Service Company, and some of the dynamic executives associated with that firm—Burdette Blue, W. Alton Jones, Herbert R. Straight.

The men and women attracted to Bartlesville by Phillips, ITIO, Empire, and Cities Service were bright, energetic, and ambitious. They were a blend of well-educated professionals from affluent backgrounds who were used to sophisticated behavior and rough-and-ready oil-field types who had to learn which fork to use first at a fancy dinner party. It made for a curious but potent combination.

During this decade a select group of young Bartians emerged on the scene and climbed straight to the top of the social ladder. Many of them were Frank's best young execs—Stewart Dewar, Paul McIntyre, Howard Sherman, Obie Wing—and their wives.

John and Mildred Phillips were part of this crowd, as was their good friend and neighbor Dana Reynolds. Hal Price, a young man who turned some borrowed money and a notion about electric welding into a prosperous pipeline construction firm, ran with them. So did Jake May, youngest of the five May brothers and the son of H. M. May, founder of a popular clothing store located where the old Right Way Hotel once stood. There were several others—couples and singles. They were carefree and cavalier young men and women, eager to celebrate their good fortune. They called themselves "the Halcyons."

The name for their group came from the mythical magical bird named the halcyon, once fabled to have had the power to calm the wind and waves. Halcyon came to mean a time of golden prosperity, a time of elegance and peace. The term "halcyon days" stood for a number of days occurring about the winter solstice, when the weather was calm. They were days of tranquillity.

While there was definitely a flair for elegance associated with the Halcyons of Bartlesville, there was nothing especially calming or tranquil about any of them. The Halcyons' parties were memorable. They'd throw costume balls, stage elaborate dinners, and hire orchestras at the drop of a top hat. At one impromptu bash, Halcyons were invited to come dressed the way they felt at that exact moment. One young man, already enjoying a crippling hangover, arrived prone on a hospital stretcher. Another guest wore his dress shirt, tie, and coat but no trousers. John Phillips said he felt stylish that day but his feet were killing him, so he wore his tuxedo and went barefoot.

In the hot months, several of the married Halcyons became "summer bachelors" when they sent their wives and children to Colorado Springs or another cool clime. Immediately, many of the young husbands became rakes again and spent their evenings gambling, drinking, and pursuing every stray young lady in sight. John Phillips was usually in the center of this pack.

One summer evening, when John didn't come stumbling back to his home, Frank began calling around town in an attempt to track down his wayward son. He finally located some of John's pals playing bridge and he ordered them to go find him—immediately. The card game was over before the telephone receiver was hung back up.

They searched all over Bartlesville for John, but at every bootleg joint it seemed like they just missed him. Finally, on the other side of the tracks, in front of a scuzzy dive filled with tough guys and whores, they spied John's lettuce-green Packard. Inside they learned that John had just been there, almost too drunk to stagger, and had pitched through the back door muttering about jumping a train. At that moment they heard a loud commotion coming from the railroad tracks out back. There was John inside the caboose, holding a half-empty whiskey bottle in one hand and a wad of money in the other. He had awakened a railroad worker and was trying to hire the train to take him to Tulsa. Once they calmed down John and parted him and his bottle, the search party called Uncle Frank. His new orders were direct and to the point—take John straight to the hospital, get his stomach pumped, and bring him home.

Frank was growing more and more weary of his son's antics. But there were more to come. A dapper gambler with the ominous name of "Titanic" Thompson came to Bartlesville and put another big nail in John's credibility coffin. Thin as a reed and always impeccably dressed, Thompson could play pool, poker, blackjack, craps, and golf like nobody else. He was a master hustler and would bet on anything.

Once, when he was hustling a country club just outside Chicago, Titanic came up with a scheme so he could make money by just driving to and from the golf course. He dug up a road sign near the club, which said it was thirty-five miles to Chicago, and he moved it several miles closer to the city limits. Titanic was careful to note—to the second—how much time it took to make the drive. Then, while shuttling some old duffers he'd just skunked in a locker-room card game, Thompson pointed to the sign—in its new location—and casually mentioned that he would bet he could make it to the city limits in an incredibly short amount of time. Seeing a chance to recover some of their losses, his passengers promptly made hefty wagers that Thompson was wrong. When his big car sped past the city limits sign at precisely the time Thompson predicted, his stunned patsies could do nothing but reach for their wallets once again. He had cleaned them completely out.

When Thompson and the handsome Ky Lafoon, his trusty sidekick on the links and at the gaming tables, breezed into Bartlesville like an ill wind, they took the place—mostly some gullible Halcyons—by storm. John Phillips was near the top of Titanic's sucker list.

One night, John and Titanic tangled in a poker game at the Maire Hotel. As usual, John had enjoyed more than his share of liquid refreshment before the cards were even out of the wrapper. When the first few hands were played, he was looped. It was no match. All Titanic had to do was shuffle the deck, deal the cards, and rake in pot after pot. Easy pickings in an oil town, playing seven-card stud with a multimillionaire's son who was so drunk he couldn't tell a queen from a joker.

It only took a little time for Titanic to clean out all of John's cash. Then the card playing got serious—serious as a heart attack. His wits blurred with ninety-proof bourbon, John upped the ante one more time. He was going for everything and he didn't stand a chance. He put up his house—the nice big brick residence on Cherokee, the one his father built for him and his family.

Titanic barely smiled as he dealt the last hand. He glanced down and saw nothing but royal faces and wild cards. The smile broadened. John had difficulty focusing on his hand of cards. It didn't really matter—they were all losers. He threw them down in disgust. Titanic won. He won it all. John's house with the mahogany railings, the fancy fountain, little Betty's brick dollhouse in the backyard, and the party room with Frank Phillips's name carved in the wall—it all belonged to Titanic Thompson.

But Titanic never got a chance to move into the house on Cherokee. He owned it less than twenty-four hours.

In the sober light of the next day, John, as he always did when he was in trouble, went to his parents' town house. Jane comforted her son and helped him explain the predicament to his father. Frank was livid. He was so angry that at first he couldn't speak. When he recovered his voice he cursed and stormed for a long time. There was nothing wrong with gambling, but Frank couldn't understand a man gambling dead drunk and then risking his own home. Most of all, he couldn't abide a man who seemed to be good at nothing but losing.

That afternoon, Frank remedied the problem. He bought back John's house from Titanic Thompson and also made it clear to the high-rolling gambler that in the future his son was no longer fair game. Then, to ensure that nothing like that would ever happen again, Frank had the title to the residence placed in the name of Johnny Phillips, his six-month-old grandson.

John continued to drink like a fish and he'd always be a loser at cards, but never again would he bet the roof over his head. That was one lesson he learned.

Titanic Thompson in the meantime continued to have good luck in Bartlesville. He and young Ky Lafoon spent a lot of time out at the Hillcrest Country Club and worked the course like it was a violin they'd played all their lives. The country club was just south of town and too close for Titanic to try his sign scam. Nobody would have gone for that one anyway. But playing golf by day and cards and dice at night was good enough.

Titanic and Ky hustled the members and they hustled the pro. They teased and tantalized their victims into thinking they were playing a couple of bumpkins and then they'd lower the boom. Titanic would set them up and Ky would shoot the dimples off the ball. Hillcrest was bubbling with money and the pair of gamblers came back for three seasons to reap their rewards. They got along just fine by observing a couple of simple rules—never play cards with John Phillips and steer clear of Uncle Frank.

Actually, there was little danger they'd ever run into Frank at the country club, unless they were in the dining room.

Designed by Delk, the architect who dreamed up Waite Phillips' Philmont and Philbrook residences, Hillcrest was completed in 1926 and took the place of the Oak Hill Country Club. Both H. V. Foster and Frank helped steamroll the plans for the new club, and Paul Endacott surveyed the site for the eighteen-hole golf course, which overlooked the Caney River and surrounded the old Silver Lake Delaware Cemetery.

Although he helped plan and promote the country club and enjoyed social outings there for many years, Frank was never much of a golfer. He tried to master the game, but his lack of patience usually got the best of the situation.

Frank would stick with horseback riding. That was something he could do well. Hillcrest and the other fancy country clubs would be fine for banquets and dinner parties, but instead of putting on knickers and chasing a ball around, Frank found that the feel of a great stallion beneath him could not be beat. It was the best way he knew to relax and clear his mind.

Most late afternoons when he was in Bartlesville, as Phillips Petroleum prepared to celebrate its first decade of doing business, Frank went to his ranch and rode Buster or Skiatook or one of his other spirited horses.

He'd leave his problems with the company back in the office. He'd put any concerns about John and his family out of his mind. Then for an hour, maybe two, he wouldn't be Uncle Frank or Mr. Phillips or anybody's father, husband, brother, or lover. He'd just be Frank.

He'd head through the brush, cut across the fields, and ride next to the long rows of fences. The horse would snort and prance, and as they made their way back to the stable, Frank could look off in the distance, across Clyde Lake, and see the buffalo, silent as ghosts, feeding on the hillside. And all the way back all he would hear was his big horse panting and the sound of the empty wind.

PART FOUR

"66"
1927–1939

"Oh, give me a home where the buffalo roam,
Where the deer and the antelope play,
Where seldom is heard a discouraging word
And the skies are not cloudy all day."

—"Home on the Range"

CHAPTER TWENTY-ONE

In 1927, the year Phillips Petroleum turned ten years old, the world was beginning to flex its wings. Something exciting was astir. The heavens—still the domain of daredevils and romantics who chased the wind—were about to be conquered. That spring all eyes looked skyward while a gangly airmail pilot named Charles Augustus Lindbergh captured everyone's imagination. In the early hours of May 20, a slender twenty-five-year-old Lindbergh climbed into a single-engine airplane, *The Spirit of St. Louis,* and took off in a dawn drizzle from the muddy sod of Roosevelt Field on Long Island.

On May 21—thirty-three and a half hours later—Lindbergh landed at Le Bourget airstrip outside Paris. The crowd there was delirious. Within hours so was the entire world. Lindbergh was the first person to fly non-stop and solo across the Atlantic Ocean.

Lindbergh instantly became an international hero. He won the $25,000 prize money offered by a wealthy hotel owner for the first transatlantic flight and among the multitude of gifts and awards bestowed on the modest pilot were the Congressional Medal of Honor and the first-ever Distinguished Flying Cross. He also won the admiration of millions. During a New York ticker-tape parade a frenzied crowd showered "Slim" Lindbergh with tons of paper and all of their love. He couldn't appear on a street without being mobbed. At a picnic in St. Louis, a swarm of well-dressed women battled for possession of a corncob the young aviator left on his plate.

When the news of Lindbergh's success spread around the world, Frank was in New York. He read all the accounts of the flight in the *Times* and for weeks he talked about almost nothing but the dramatic event.

Aviation and the future of commercial flight became the main topic of discussion during lunches at the Bankers Club with Kane and Wing. In the evenings, while relaxing at his Turkish bath, Frank pondered the impact of Lindbergh's historic flight, and the subject came up again when Frank and Kane and some other business associates motored to the races at Belmont.

For some time, Frank Phillips had seen the clouds of change for the entire aviation industry on the horizon. The praise being showered on Lindbergh didn't surprise him in the least. It was well deserved and, even more important, this singular feat meant the beginning of travel by means of motorized wings.

Those motorized wings, as Frank saw it, would require fuel—lots of gasoline—in order to get off the ground and remain aloft.

In the early days of Phillips Petroleum, as World War I was being fought, the great demand for oil and gasoline needed for the military vehicles and aircraft had given the company a healthy boost. Since the war, Phillips researchers, led by Oberfell, continued to develop gasoline for aircraft, and by 1927 the company was already marketing a new, lighter product called Nu-Aviation fuel.

Besides supplying fuel for the fledgling aircraft business, Frank saw another reason for Phillips Petroleum to become involved in the world of flight. Frank realized that even though his company had achieved great commercial success in only ten years, the company's name was not as well known by the general public as it could be. All the Phillips production of crude oil and natural gasoline was sold to other refiners, who converted what they bought into finished products for final sale under their own brand names. For a decade, the motorists of the nation, who were growing in record numbers, had not been exposed to the Phillips Petroleum name. By 1927 there were more than 22 million automobiles on America's roads. All of them needed gasoline and oil. But most consumers were unaware that what was being pumped into their tank was actually a blend of Phillips gas sold under somebody else's trade name.

That Phillips Petroleum was the leading manufacturer of natural gasoline in the world was of little consequence to the public as long as the company remained in the wholesale end of the business and allowed others to do all the refining and marketing. Frank sensed the need for his firm to acquire a refinery and get into the retail business. He could no longer depend on others to market his company's spiraling production. The time had come for Phillips Petroleum to reach the ultimate consumer. By taking the company's name aloft, Frank believed he could get the public's attention.

But before these critical changes could be accomplished, there was also widespread speculation about what would happen to independent producers such as Phillips. Rumors surfaced that the company was in the market to merge with one of the larger, full-service oil firms. Frank flat out denied such talk.

"If there are any mergers involving Phillips Petroleum, you can bet that Phillips Petroleum will not end up being submerged," said Frank in a public statement. "We'll always come out on top and retain our name and identity." Privately, Frank was forced to admit that the possibility of a merger with a larger company that had refining and marketing capabilities had to be considered.

Kenneth Beall, Frank's traveling secretary and assistant for a period in the 1920s, recalled that although Frank had no wish to relinquish control over a company that he and L.E. had worked so hard to create, he was willing to consider a merger if it meant keeping the firm in sound financial shape.

About the only truly serious threat of a Phillips merger with a larger company was with the Texas Company—the organization, which would later be called Texaco, that was headed by the shrewd Amos L. Beaty. Negotiations between Phillips and the Texas Company went on for the better part of two years, but in the end no deal was made. "There were probably several reasons it didn't go through," said Beall. "But the one Mr. Phillips gave me was that he damned well wouldn't leave Bartlesville or be party to any deal making his employees leave town."

Despite the quieting of the merger rumors, it was also being said that Frank, in his mid-fifties, was simply tired of running a big oil company and was ready to quit and go to his ranch. One observer predicted: "Before the year is out, there will be no more Phillips Petroleum Company." Frank didn't even dignify that statement with a response. But his actions spoke loud and clear.

Long before Lindbergh flew his monoplane into the annals of aviation history, Frank and his crew of business and research experts launched plans to extend their operations to include refining and marketing. Two days after Lindbergh completed his flight, Frank enhanced those ambitious plans by buying Oklahoma's largest gas firm, the Oklahoma Natural Gas Company (ONG), for $25 million. Financial problems plagued ONG for some time and Frank saw a chance to turn the company around. At the same time it gave Phillips a major market for its natural gas.

Five months later, Frank turned right around and sold ONG for $40 million to the American Natural Gas Corporation. Besides the respectable profit he gleaned from the transaction, Frank also negotiated in the sales contract a stipulation that Phillips would continue to provide ONG's natural gas through a proposed pipeline connecting the Texas panhandle and Tulsa.

Frank was on a roll and he didn't want to stop.

Always attuned to the thinking of the masses, he also understood the necessity of acquainting motorists with the company name. More and more automobiles were traveling America's highways and soon aircraft would fill the skies. As Frank saw it, aviation would enable his company to grow in the heavens and on earth. It was innovative planning tempered with one of Frank Phillips' "old best dreams."

The Atlantic may have already been conquered by Lindbergh, but Frank knew the Pacific still remained unchallenged. He looked westward like many others.

Shortly after Lindbergh's transatlantic journey, James Dole, founder of the Hawaiian company which bore his name and a major pineapple tycoon, decided to capitalize on all the publicity about Lindbergh's achievement. Dole wanted to encourage commercial flights between the mainland and the islands. He believed there would be no better way than by sponsoring another air race.

Dole put up $25,000 for the first commercial pilot who could fly from California to Hawaii. The pineapple czar immediately invited Lindbergh to enter the competition, but the Lone Eagle turned him down. He felt it was too dangerous. Hawaii was a considerably smaller target than Europe. The slightest error in navigation and an airplane could easily miss the tiny string of islands and fly out over the empty ocean until there was no more gasoline. But Dole was undaunted. Even without Lindbergh, the race would go on. And he soon found there were plenty of others either brave or foolish enough to risk their necks.

When Frank heard about the proposed Dole race between San Francisco and Honolulu, he saw an opportunity. Not only could he show that his new product, Phillips Nu-Aviation, had the advantage of "less weight per gallon, greater power because of more complete combustion, and superior efficiency in many respects than average aviation gasoline"; he also had a perfect opportunity to get public exposure for Phillips Petroleum Company.

In Bartlesville, an honest-to-goodness aviation department was developing at Phillips under the watchful eye of W. D. "Billy" Parker, a native of Oklahoma City whose parents made the big land run of 1889 and donated the land on which the state capitol building was erected.

Billy Parker was a dashing pilot who had wanted to do nothing but fly since he was ten years old. His mother took him to the 1908 World's Fair in Toronto, where he saw Lincoln Beachey go up in a balloon powered by a motorcycle engine. Manned flight was still in its infancy when Parker was a boy watching Beachey's balloon rise into the heavens. It had been only five years since some brothers named Wright, who ran a bicycle shop in Dayton, Ohio, built their odd-looking flying machine and journeyed to the sand dunes of North Carolina and a spot on the map named Kitty Hawk.

At fourteen, Parker's dream about flying came true. His family had moved to a farm near Fort Collins, Colorado, and the boy spent every waking moment building a biplane in his father's barn. With the help of a friend, Parker pushed the plane into a pasture, fired up the engine, and flew a few feet through the air before settling back on the ground. It wasn't an extraordinary flight, but young Billy had gotten his wish.

Within a few years he got even more. He joined the British Royal Flying Corps and flew with distinction as a captain during the early days of World War I. When Uncle Sam got into the fray, Parker transferred to the U.S. armed forces and reported to Dewey, Oklahoma, one of the sites where the federal government had established an airplane factory and flying school. Parker was to serve as manager, chief engineer, test pilot, and instructor.

Parker was presented with the forty-fourth pilot's license issued in the United States, and on Christmas Day 1917, just six months before the Phillips brothers founded their company in nearby Bartlesville, Parker took the controls for the test flight of the state's first commercially built airplane. On New Year's Day, Parker piloted the *Dewey* in its first public flight, with Joe Bartles, son of the founder of Dewey, as his first passenger. Bartles had been the main promoter of the factory and flying school and donated the land north of town for the facility.

The factory, with thirty employees, turned out ten airplanes before the war ended and production ceased. Even though the production plant folded, Parker continued to run the flying school until 1919, when he took off in his biplane operating as a charter pilot and barnstorming his way around the country.

In 1925, Billy Parker was back in Bartlesville, and the next year he went on Uncle Frank's payroll as Phillips Petroleum's first chief pilot. His job included flying firefighters and nitroglycerin to well fires and hauling Phillips executives on scouting trips to new oil fields. Parker also assisted with the research and development of the company's new aviation gasoline.

When news broke about the contest sponsored by Dole, Parker was busy touring the country demonstrating the performance of Nu-Aviation gasoline. Frank quickly tracked him down and summoned him to his office to plan strategy for the Dole race.

After a lengthy conference, it was decided that Billy could best serve the company by directing operations from the ground. Bennett H. Griffin, a crackerjack pilot from Oklahoma City, and navigator W. A. Henley would represent Phillips Petroleum, flying a Travel Air plane named the *Oklahoma*. Both Benny Griffin and Al Henley were considered to be first-rate and had Parker's blessing. Their monoplane was powered by a Wright Whirlwind, just like the engine in *The Spirit of St. Louis*.

"Henley and I will load her up with all the gasoline she'll hold and see how long she'll run," said Griffin. "Sure, we're going to do it!" The Phillips flight team began training for the big race, scheduled for August 16, 1927.

At the same time in California, a Hollywood stunt pilot named Arthur C. Goebel heard about Dole's race and decided to enter. Goebel was thirty-one and had flown a variety of airplanes during his career. He made his living flying loop-the-loops and other hair-raising stunts over early movie lots for National Pictures, Inc., and had a real flair for taking risks; he was considered to be a skilled pilot with nerves of steel.

After studying the proposed course, Goebel decided he wanted to fly a Travel Air monoplane in the Dole race. He visited the Travel Air plant at Wichita, Kansas, in June. After five days of meeting with the young pilot, Travel Air's management was convinced he was competent and a good risk and they agreed to provide a custom-made aircraft. Goebel gave them a $5,000 deposit and left with assurances he could take possession of his new airplane in early August.

Upon returning to California and continuing to prepare for the race, Goebel had to beat the bushes looking for additional financial support. Many of his friends came through with contributions to the cause and

Goebel invested every dime he had saved, but when the time came to make the final payment on his plane, Goebel found he still needed $15,000. That's when someone mentioned Frank Phillips, the Oklahoma oil man whose firm had just developed a new aviation fuel and already had an entry for the Dole race. Maybe, Goebel thought, Frank Phillips would be willing to sponsor two airplanes. It was worth asking.

Goebel approached Frank and explained his predicament. Frank listened to Goebel and he also listened to Billy Parker, who endorsed the idea of the company hedging its bets by entering not one, but two contestants using Phillips gasoline. Goebel's grit and good looks also appealed to Parker and Frank. They figured he'd make a fine showing for the company. After briefly deliberating on Goebel's request, Frank agreed to sponsor the pilot, pay for the aircraft, and provide all the fuel. There was just one stipulation. The airplane would have to carry the name *Woolaroc,* in honor of Frank's lodge in the Osage. Goebel, concerned not about his plane's name but about accomplishing the mission, instantly concurred.

Because a solo flight would permit him to carry more fuel and not have to share the prize money, Goebel wanted to fly the race alone. "I believe I'm going to get along without a navigator," said Goebel. "I sure don't want one, and I'm practically sure, now that I've tried it, that I can handle the instruments myself." But Goebel's solo plan was quickly shot down. The race rules made it clear that a navigator aboard each airplane was an absolute requirement. The *Woolaroc,* like all the other entries, would have a crew of two.

One of Goebel's friends knew a young Navy officer, Lieutenant William V. Davis, Jr., a skilled navigator who was willing to guide the *Woolaroc* across the Pacific. Davis, a graduate of Annapolis, had trained at the naval flight station in Pensacola, Florida. He knew how to use radio equipment and he was well versed in both marine and celestial navigation techniques. But even more important, the naval officer had served aboard the USS *Langley,* a converted coaling ship that became America's first aircraft carrier. If Davis could help bring an airplane to rest on a pitching deck that looked the size of a postage stamp in the ocean, Goebel figured he could help him find Hawaii.

Davis was granted temporary leave from his duties at the naval air station at San Diego and hooked up with Goebel in early August, just in time to pick up their craft at the Travel Air plant in Wichita. Griffin and Henley were also in Wichita to take possession of the *Oklahoma.* From there, they all had orders to fly directly to Bartlesville so Frank could give the two Phillips planes a pre-race inspection.

Frank was getting more and more excited about the "Dole Derby." The month before, he and the rest of his brothers spent a long Fourth of July holiday at Philmont, where Waite held a big rodeo in their honor. For a change, most of the talk on that trip wasn't about new oil fields or production figures but centered on Frank's airplanes and the chances they had—if any—of winning the highly publicized race.

The day before the *Oklahoma* arrived in Bartlesville for Frank's inspection, Fern Butler, in town for a visit from New York, accompanied Frank to the ranch for an afternoon away from company matters. All the way out to Woolaroc, during their long horseback ride, and all the way back to town, Frank told Fern about his two airplanes. He said he believed they would both do the company proud in the race, but Fern secretly hoped that the airplane which bore the name she helped create would be the winner. In either case, Fern was proud of Frank and intrigued with his sudden passion for flying.

The next day, the *Oklahoma* landed at the small airport on the edge of town and Frank and several guests turned out for a photo session. A couple of days later, Goebel brought down his plane—an orange-and-blue beauty with the name *Woolaroc* and below it *FPR,* for Frank Phillips Ranch, painted on both sides of the fuselage. Once they'd received Uncle Frank's blessings, the two airplanes left for California and final preparations for the August 16 race.

In San Diego, Goebel and Davis installed an earth inductor compass, a drift sight, and other important navigation equipment. The mammoth nine-cylinder Wright Whirlwind engine purred like a cat and appeared to be in tip-top condition, but that wasn't good enough for Goebel. He checked and double-checked everything. Goebel firmly believed the race would be won on preparation alone and he went over every inch of the monoplane with his own hands. When the work in San Diego was completed, Goebel flew to Oakland and rechecked all systems one more time. For the next few days, the crews made final adjustments and each contestant was given several brief flight tests to prove his basic competency.

The morning of the race, both the *Woolaroc* and the *Oklahoma* were perfectly tuned and ready to fly. There were 417 gallons of Nu-Aviation fuel in the tanks of the *Woolaroc* alone and seventeen gallons of oil aboard. All the extras were in place—collapsible life rafts and paddles, signal flares, and navigation equipment. Goebel's final check of the *Woolaroc* tallied 5,520 pounds of plane, equipment, fuel, and passengers to travel the 2,437 miles separating Oakland and Honolulu.

Originally, there was talk of twenty-five entries, but as the date drew nearer that number shrank. Shortly before the big event fifteen entrants were eligible for the grueling race. By the day of the race, some had withdrawn; three had wrecked their planes and one was disqualified. The eight remaining pilots drew lots for takeoff positions. Griffin and the *Oklahoma* would take the number one slot. The *Woolaroc* was in the seventh position.

Dawn broke on August 16 and a dense coastal fog shrouded Oakland and covered the airport. By late morning the fog burned off, revealing the eight planes lined up on the sandy runway. A crowd had gathered, most of them curious to see if all the planes would get off safely, let alone reach Hawaii.

Griffin and his navigator, Henley, were in the *Oklahoma*, anxiously awaiting the signal for the race to begin. At last the word came down from the officials and the Travel Air was pushed to the starting line at high noon. In seconds the monoplane began its taxi, and after 3,000 feet the *Oklahoma* was airborne. The Dole race was underway.

As the plane started its slow westward climb, every other pilot below watched the ascent and waited for his turn. The next plane, *El Encanto*, piloted by Norman Goddard, crashed on takeoff. Goddard lost control and his plane lay on its left side with the right wing pointed upward in a pathetic salute to those planes which would continue the race.

The other planes moved into position and one by one motored down the runway. At last, the *Woolaroc* stood on the starting line, waiting only for the order to go from the official starter. Behind were grueling days of sacrifice and hard work. Ahead lay Hawaii.

The signal came at 12:36 P.M., and they were on their way. The eighth —and last—plane followed.

In less than an hour, there were problems for several of the craft. *Miss Doran* came chugging back to Oakland with engine trouble. *Dallas Spirit* returned next, trailing a hunk of its lower fuselage. A little later, spotters reported that one of the Travel Air planes was returning. Binoculars focused on a small dot as it came closer to the airstrip. It was the *Oklahoma*. After landing, Griffin said his engine had been running hot. It was whispered that perhaps the Phillips fuel was to blame.

Billy Parker quickly squelched the gossip. He told the race officials and others in the crowd that the Phillips fuel had been thoroughly tested and had already been proven all across the nation. He also made it clear that one of the primary problems with the *Oklahoma* had to be blamed on its pilot, who had been known to run planes at full throttle on the ground for long periods of time, without regard for what this might do to the engine.

It really didn't matter. The crew at least returned safely. What did matter was that Phillips Petroleum was down to one airplane and two men flying somewhere over the Pacific.

Aboard the *Woolaroc* everything was under control. Goebel could look out his tiny window on the side of the fuselage but was completely separated from Davis by the huge tank holding hundreds of gallons of precious fuel. The two men exchanged messages scribbled on paper and attached by clothespins to a wire on pulleys.

Davis took readings through a roof window, listened for dots and dashes of the Morse code to give him a bearing as they flew over the endless sea, and dropped smoke bombs into the ocean to calculate the effect of wind drift on course headings.

"Above us was the glaring sun for a few hours, then the darkness of a long night and finally sunlight again," Goebel later wrote. "Below us was nothing but the vast ocean. There was plenty of time to think. Lieutenant Davis and I were completely separated as if we had been at extreme ends of the earth. We could not see or talk to one another. Not ten feet apart, we were strangers together in the air over mid-ocean.

"The sun went down. It had been a good friend, and the weather had been fine. Now night was coming on, and with it a loneliness that cannot be described and which has not been experienced by any except those who, like us, have headed a plane into darkness over water. Regularly the radio signals came in. The instruments on board were working perfectly. Everything was going as anticipated days and weeks before. Air perfection cruising toward a little spot in the vast Pacific."

Each man had a thermos of coffee and some sandwiches, but they weren't hungry. They could only stare into the darkness and the fog and watch the mountains of night clouds rushing past their ship. They listened to the mighty engine's constant song of power while the plane flew deeper into the Pacific night.

Goebel stuck to the routes given to him by Davis, who laid them out one at a time as the *Woolaroc* flew westward, level at 4,000 feet above the waves. The navigator took sightings on Polaris—the North Star—a star of the second magnitude. He tuned the crackling radio and transmitted position reports by Morse code. At midnight, the *Woolaroc* climbed to 6,000

On March 28, 1931, Frank Phillips was adopted into the Osage Indian tribe in a ceremony held at Woolaroc. Following the ceremony, Frank—Eagle Chief —was dressed by the Osage chiefs in an official costume and was presented with a split buffalo hide by Zack Miller of the 101 Ranch. The adoption resolution was etched in English and Osage on the hide. *(Frank Phillips Foundation, Inc.)*

ABOVE:

Indian guests gathered with host Frank Phillips at the annual Cowthieves and Outlaws Reunion held at the F.P. Ranch (Woolaroc), September 27, 1930. Seated, left to right: Zack Miller, Major G. W. Lillie (Pawnee Bill), Frank Phillips (holding Chief Lookout's baby grandson), Osage chief Fred Lookout, and F. Revard. *(Frank Phillips Foundation, Inc.)*

LOWER LEFT:

Chief Baconrind, Chief Fred Lookout, and Eagle Chief (Frank Phillips) are pictured, left to right, with the Osage Tribal Council at the F.P. Ranch (Woolaroc), March 1931. *(Frank Phillips Foundation, Inc.)*

Following the 1930 merger of Independent Oil & Gas Company, headed by Waite Phillips, and Phillips Petroleum, the five Phillips brothers gathered at Woolaroc for a picnic and barbecue. Left to right: Fred, Waite, Ed, L.E., and Frank. *(Frank Phillips Foundation, Inc.)*

With the financial backing of Phillips Petroleum Company, Colonel Arthur C. Goebel (seated on the plane) flew *Woolaroc,* a single-engine monoplane, nonstop from Oakland, California, to Honolulu to win the Dole race on August 17, 1927. Goebel, pictured here with Frank Phillips (right), is getting ready to leave Bartlesville on the first leg of his victorious flight. *(Phillips Petroleum Company)*

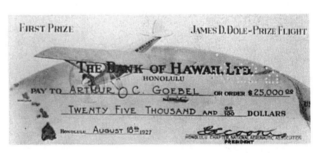

ABOVE:

Goebel divided the Dole race prize of $25,000 with Davis, his indispensable flight navigator. *(Frank Phillips Foundation, Inc.)*

LOWER LEFT:

Virginia Wilson of Wichita, Kansas—site of the first Phillips 66 gas station—was one of several aviatrix models in the Phillips "Phill-Up and Fly" gasoline advertising campaign, which capitalized on aviation, the national craze of the day. *(Phillips Petroleum Company)*

Woolaroc, Arthur Goebel's plane, was named by Frank Phillips when he agreed to sponsor the Hollywood stunt flier in the Oakland to Honolulu Dole race. Following a victory in the race, Goebel flew *Woolaroc* around the country promoting Phillips aviation fuel. *(Frank Phillips Foundation, Inc.)*

Woolaroc II, a Ford trimotor with streamlined "pants" on its landing wheels, provided high-speed transportation for executives in the company's midwestern headquarters and served as a flying laboratory before it was replaced by a more rapid model. *(Phillips Petroleum Company)*

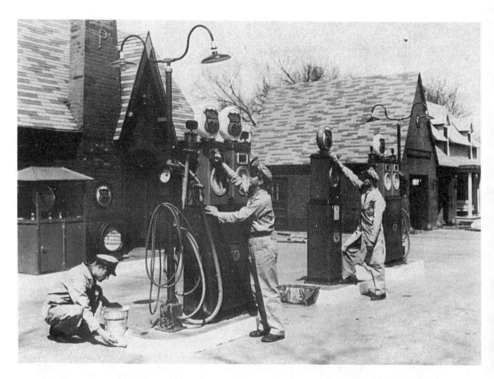

Phillips service stations were designed to look like cottages so that they would blend into residential neighborhoods where they were located. Dark green with orange and blue accents, a departure from popular colors of the time, were chosen for the stations, known for helpful attendants and frequent free-gasoline bonuses. *(Phillips Petroleum Company)*

1. Phillips 66 logo, 1927   2. Phillips 66 logo, 1928   3. Phillips 66 logo, 1930   4. Phillips 66 logo, 1959

1. "66" gasoline banners were paired with disk-shaped shields bearing the Phillips name at the first service stations, opened in 1927. *(Phillips Petroleum Company)*

2. The following year, "Phillips" and "66" were combined on a disk-shaped sign. *(Phillips Petroleum Company)*

3. In 1930 the familiar six-pointed "66" shield appeared in orange and black. The emblems closely resembled the national highway signs. *(Phillips Petroleum Company)*

4. The exaggerated curves of the shield were diminished in 1959 and the colors changed to red and white. *(Phillips Petroleum Company)*

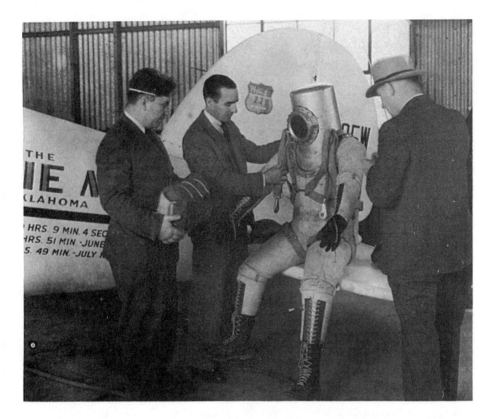

ABOVE:

Wiley Post (left) made his record flight into the stratosphere—the near border of space—in 1934. For the flight, which was sponsored by Frank Phillips, Post designed and used a pressurized suit that served as a prototype for suits used later by aviators and astronauts. The suit and plane are now part of the Smithsonian Institute collection. *(Phillips Petroleum Company)*

LOWER RIGHT:

Art Goebel scrawled "Phillips 66" in mile-high letters of smoke around the country, including high above St. Louis, where one of Phillips' busiest stations was situated at Big Bend and Clayton Road, as well as at the air races at the 1933 Chicago World's Fair. *(Phillips Petroleum Company)*

While Goebel wrote "Phillips 66" across the skies, an associate kept in radio contact and touted company products on the ground. *(Phillips Petroleum Company)*

A wide array of guests visited Woolaroc during the 1930s and '40s, including concert violinist Dave Rubinoff, pictured here with Frank Phillips, and future company presidents Paul Endacott (far left) and Boots Adams (far right). *(Frank Phillips Foundation, Inc.)*

On April 26, 1937, upon the recommendation of Frank Phillips, chairman of the board, the Phillips Petroleum Company board of directors unanimously approved the appointment of Boots Adams as president of the company. Adams and Phillips are pictured at a luncheon held for visiting directors at the annual stockholders' meeting in Bartlesville. *(Phillips Petroleum Company)*

Will Rogers, pictured at Woolaroc in 1930, was a frequent guest and close friend of Frank and Jane Phillips. *(Frank Phillips Foundation, Inc.)*

L.E. Phillips (left) and J.S. Leach, a local newspaper publisher, pay close attention to Will Rogers as he entertains guests at Woolaroc. *(Frank Phillips Foundation, Inc.)*

feet, just above a layer of clouds. Goebel remembered the scene: "Slowly the night was passing. Every little bit I turned to look back east out of the rear of my compartment to catch the first glimpse of daylight—welcome daylight! And it was long in coming. The fog was still below us at daybreak—nineteen and a half hours of it altogether on this trip."

Morning approached. Davis prepared to drop more smoke bombs for a drift check. The plane might have been blown off course during the long night. The northeast winds appeared that Davis and Goebel had hoped for, and the ground speed of the plane reached nearly one hundred miles per hour. Precious fuel had been saved thanks to the warm sea winds. When more drift checks showed the wind shifting to the east and southeast, Davis attached a course correction to the wire and sent it to Goebel up front, telling him to turn to the south. Goebel disagreed, but he did as he was told.

"The morning passed—the weather cleared, and below us again was the rolling blue Pacific," wrote Goebel. "Noon came by our watches, with still the steadily humming motor, the distinctly audible beacons. One o'clock . . . two o'clock. Minutes ticking slowly. A speck on the horizon that might be a cloud or an island or a mountain!" The "speck" was not a cloud. It was Maui. When he saw the Hawaiian Islands below, Goebel let out a loud whoop. His steady hand and the faultless navigation of Davis had paid off.

Goebel steered his plane toward Honolulu. Just after they flew past Diamond Head, a speedy Army pursuit plane from Wheeler Field appeared and tucked in close to the *Woolaroc*. Both Goebel and Davis looked out and saw the smiling pilot holding up one finger. At that moment they knew for the first time that they had won the race. They turned inland and touched down twenty-six hours, seventeen minutes, and thirty-three seconds after taking off from Oakland.

Two hours later, the only other surviving entrant arrived. Martin Jensen was its pilot, Paul Schluter the navigator. Two other planes were lost, never to be seen again. Searchers who went out to look for the lost planes were lost themselves. In all, thirteen had died in the "Dole Derby."

When Goebel and Davis climbed out of the *Woolaroc* a crowd of more than 20,000 surrounded them. "Honest to goodness, do you mean that I am really the first one here?" asked Goebel as his feet touched the runway. Many of the well-wishers had waited all through the night and the morning, scanning the bright skies for a sight of the planes. James Dole was there to shake their hands and present them their check for $25,000. The governor of the islands and all sorts of Army and Navy top brass also came out to cheer the winners. Native dancers and musicians performed, and Hawaiian girls, with hibiscus in their hair, placed fragrant orchid leis around the weary fliers' necks.

Reporters ran up to the intrepid Goebel, who flashed his best Hollywood grin and told them: "We had the right kind of gas . . . a new aviation gas made by the man who bought this ship and financed our flight —Frank Phillips of Bartlesville, Oklahoma."

Goebel later wrote about how he and Davis felt after landing. "We weren't hungry. We weren't tired. We weren't sleepy. We wanted but three things—to cable home, then a shave, and a dip in the beach at Waikiki—a swim—and that from two men who had seen only water for twenty-six hours."

But first things first. As soon as the generous welcome ceremony was over, Goebel and Davis were driven to the telegraph office so they could cable Uncle Frank at Bartlesville. The wire was delivered late that afternoon. It was the kind of message Frank liked—short, sweet, and positive:

WE'RE HERE PHILLIPS AVIATION GASOLINE
TURNED THE TRICK HAVE RECEIVED THE
WINNING MONEY FROM DOLE

GOEBEL

Frank handed out cigars like a proud papa. "We have tremendous faith in the future of aviation," Frank said afterward. "It is because of a desire to do our part in its progress that the Phillips organization has been working so diligently on the airplane fuel problem." L.E., also bursting with pride, immediately sent his old friend Dole a wire suggesting he was going to send "a couple boys over" to Hawaii to collect the prize money. In a more serious vein, L.E. added that "Goebel's indomitable courage, Dole's pineapple money, and Phillips' gas turned the trick."

When he heard about the triumph of the *Woolaroc,* Lindbergh praised Goebel and Davis for their ability to fly to a speck of land in the middle of the Pacific. "The flight from California to the Hawaiian Islands was the greatest air feat in history," said Lindbergh.

As crews disassembled the *Woolaroc* and loaded it aboard the steamship *Monoa* for its return voyage to the mainland, the Phillips name appeared in headlines from coast to coast. Goebel's winning the Dole Prize brought Nu-Aviation fuel to the public's attention in a most dramatic fashion. Soon after Goebel's victorious flight, Boeing Air Transport selected Phillips' "high compression" gasoline for all its mail service flights between Chicago, Omaha, and Salt Lake City.

Not only did Phillips' new aviation gasoline get some timely publicity from the Dole race, but people all over the country—from school kids to grandparents—were now taking notice and were aware of Phillips Petroleum Company. Frank's strategy was working. His company's aviation fuel had established a reputation for performance in the skies. The next step was to bring that reputation down to earth. That was easier said than done.

The Midwest was still the territory of Standard Oil of Indiana, an industry giant, and there was ruthless competition among the gasoline retailers. It would require a very bold move. Phillips would have to take on not only Standard of Indiana but also Standard of New Jersey and the Texas Company and some other major firms. For Phillips Petroleum, an independent oil producer—with no refineries, terminals, bulk stations, or marketing outlets—to even imagine getting involved in the retail field was, according to many, plain foolish. But not to Frank Phillips. He had taken risks all his life and he wasn't about to stop now.

"Boldness was a strong trait of Frank Phillips," said Kenneth Beall, Frank's assistant. But all of Frank's bold moves, including this one, were preceded by careful investigation.

"It seemed almost insurmountable to find a market for natural gasoline when the big refineries and marketing companies did not take the products of our plants with any degree of constancy or with any assurance of continuation," Frank told a writer for *American Business Magazine*. "I had built up a large production of crude oil and natural gasoline which was at the mercy of a market adamant to salesmanship. There was but one course for me." Frank took the big step. He called upon his ace research team and the top men in the Phillips' gasoline division. By the time the *Woolaroc* won the Dole Prize, the Phillips researchers had already come up with a new gasoline that they believed would solve their problems.

The company's first aviation fuel came from the Burbank plant, after F. E. Rice, head of the gasoline division, installed fractionating units to remove the lighter hydrocarbons of butane and propane from the natural gasoline. The natural gasoline which remained was a stable motor fuel. Working out in the Burbank field, in a crude galvanized-iron building they called a "laboratory," Oberfell's whiz kids next aimed their sights on a new target—automobile fuel.

The research department went to competitors' filling stations and bought gasoline to bring back to the lab for detailed analysis. Next, various blends were tried in road tests in different makes of autos—Fords, Dodges, Buicks, Lincolns, Chevrolets. As a result of the testing, the researchers developed, through selective blending of Phillips' natural gasoline and naphtha, a fuel with a high gravity that made automobiles easier to start in cold weather and seemed to have more pep. With this simple scheme and foolproof selling point Frank was now ready to go into the refining business.

Phillips Petroleum bought a skimming plant near Borger from the Alamo Refining Company to supply the blending naphtha for the new fuel. Naphtha, a colorless flammable liquid obtained from crude petroleum, could be used as a solvent and as a raw material for gasoline, and was essential to the process. Phillips' new high-quality automobile motor fuel was a radical departure from other gasolines. It was an original blend of Casinghead gasoline extracted from natural gas and lower volatility gasoline refined from crude oil.

The whiz kids had developed a gasoline that was, in Frank's words, "radically different" and "superior to any motor fuel on the market." It only needed a name. A committee was formed to select the name and Frank was made chairman.

In those early days of retail fuel marketing, the term "high-gravity" implied quality because gravity rather than octane rating was the main measure. Since the new fuel was in the gravity range of 66, an especially high gravity mark, someone suggested calling the gasoline just that— "66." A few others agreed and thought it was as good a name as any. They liked the sound and thought it looked good. But the research department had just conceived of the rather revolutionary scheme of varying gravity and controlling volatility so that the Phillips gasoline could fit all seasons and different locales. With that in mind, the notion of tying the trade name to one gravity quickly faded.

Other names, trademarks, and catchy titles were tried, but none of them seemed to work. Frank was anxious to open his first filling station, but he wanted a suitable name for the gasoline he'd be selling.

The marketing department leaned toward a current corporate fad of having a trade name which combined a numeral with a word or two, such as Heinz 57. The words were no problem, but what about a number? Time was running out. Frank called a special executive committee meeting just to settle the question of the trademark.

On the eve of that meeting John Kane, one of Phillips Petroleum's top executives, was returning to Bartlesville after tending to some company business in Oklahoma City. A Phillips driver, Salty Sawtell, was at the wheel of the company automobile being used to road-test the new Phillips gasoline.

The auto was speeding up Route 66, a federal highway christened only the year before, which spanned two-thirds of the continent. The two-lane highway started at the corner of Jackson Boulevard and Michigan Avenue in Chicago and eventually reached across 2,238 miles and three time zones and wound through eight states before it dead-ended in the Pacific breakers at Santa Monica Boulevard and Ocean Avenue in Los Angeles.

It was a highway steeped in the history of the land it crossed. In Illinois, the road traversed the gently rolling farmlands where the soil was the color of licorice. In Missouri, Route 66 closely tracked the old Osage Indian trail that started at the mouth of the Missouri River and cut across the state in a southwesterly direction. The highway briefly caressed the Kansas prairie before crossing the oil fields and ranchlands of Oklahoma. It continued through the Texas panhandle, not far from Borger, and then climbed the steep plateaus of New Mexico and Arizona, crossed the Mojave Desert and the San Bernardino Mountains in southern California, and found its way to the coast. It became known as "the main street of America," and linked the industrial states of the East with the golden West.

Nowhere was Route 66 more at home than in Oklahoma, where the West and East collided and the pavement followed the contours of the land as though it had always been there. Along the highway, Oklahomans watched a new culture emerge—complete with hitchhikers, Burma Shave signs, and flashing neon lights. Route 66 also cut directly through the prime territory targeted by the Phillips marketing department. Coincidentally, almost every major midwestern city on the Phillips Petroleum marketing list happened to be located on Route 66.

The company auto, with Sawtell at the wheel and the new gas in the tank, was making good time. Very good time, thought Kane, whose mind had been on company business until he finally noticed how fast the automobile was hurtling down the highway.

"This car goes like sixty with our new gas," Kane said.

"Sixty nothing," answered Sawtell, "We're going sixty-six!"

The two men looked at each other and grinned. Going sixty-six miles per hour on Route 66. There it was again. That same name kept coming back—sixty-six. Maybe it was some kind of a sign.

The next morning at the conference, Kane shared his story about the fast trip from Oklahoma City thanks to the high-performance gasoline. Frank later wrote about the reaction:

"Someone inquired where the incident took place. 'On Highway 66, near Tulsa.' Already we had envisioned Highway 66 as the backbone of our future marketing area . . . Sixty-six kept creeping into our discussions. We began to like to say it. It sounded catchy. Someone suggested it might fit our needs even though it wasn't descriptive of the product. A quick decision was imperative, so we just decided then and there to use 'Phillips 66,' and we never have regretted our hasty decision."

For years after the Phillips 66 trade name first appeared, stories surfaced about the meaning behind the name. Several of the name theories sounded believable and a few had a touch of truth, but most were pure fable. Some said that Frank and L.E. were down to their last $66 between them when the Anna Anderson hit. Another story claimed the number in the name meant 66 octane, even though the method for determining octane wasn't developed until five years after the name was chosen. Another popular version held that Frank was sixty-six years old when he founded Phillips Petroleum. Had that been the determining factor, the gasoline would have been named Phillips 43. Down Borger way, the word was that a Phillips executive won the company's first refinery in the panhandle when the original owner rolled "double sixes" in a dice game and the directors liked those "boxcars" so much they named the refinery's product Phillips 66. An ecclesiastical story that the 66 was arrived at because there were sixty-six books in the Bible became so widespread that an elderly lady sent Frank a letter protesting his commercial poor taste. Some said the name was adopted because Frank's ranch covered sixty-six sections of land, and others believed the name stood for the year Frank's parents were married back in Iowa.

Even though there were so many theories and stories circulating about the Phillips 66 name that the company finally published a booklet explaining the origin, Frank wasn't especially upset by the confusion. All the talk about the name brought more attention to the company and its line of products. Let people theorize and gossip all they want, Frank told the executive committee, as long as they buy Phillips gasoline when their cars get thirsty.

On November 1, 1927, Phillips was operating its first refinery—a small operation, dubbed a "teakettle," on the outskirts of Borger that turned out about 1,500 barrels a day. By the following year, production leaped to 7,500 barrels a day, and in 1929 the Borger refinery was churning out 15,000 barrels every day of the week.

The company now boasted its first refinery and had a gasoline product ready to sell to the public. Frank was painfully aware of one thing however: the rival brands were already entrenched across America. Many parts of the country were practically overrun with filling stations. The gasoline business promised fierce and intense competition. But Frank didn't flinch. He summoned the executive committee instead.

When the committee members dutifully reported to Frank's office, Marjorie Loos made sure everyone seated at the long conference table had a pad of paper and a pencil—brand-new ones, much to L.E.'s chagrin. She turned on all the indirect lights and lamps so the room, paneled in blond mahogany, was as bright as noon, just the way Frank liked it. Instead of starting the meeting with the usual handshakes, Frank, wearing a three-piece suit and bow tie and clutching a cigar, got right down to business. "Gentlemen," said Frank, his eyes quickly moving around the table so each executive present got his share of Uncle Frank's stare, "I move that we build ourselves the damndest string of filling stations that anybody ever saw." He let his words sink in for only a few seconds before he added: "If it's all right with you, I move that we make the vote unanimous."

The motion carried.

Now that his firm was prepared to launch marketing efforts, Frank ordered the sales department to select the toughest market they could find to locate the first Phillips Petroleum retail site. Led by Oscar Cordell, the company's sales manager and a former partner in the refinery Frank bought near Borger, the department carried out an exhaustive search. After several weeks they targeted the most competitive sales point in the entire Midwest—Wichita, Kansas.

The theory behind this strategy was to take the competition head-on and build from there. If Frank could sell his new Phillips 66 in the Wichita market, he could sell it anywhere. It was a difficult challenge, even for the old Mountain Sage peddler.

A prototype of what would eventually become the standard Phillips filling station was erected at 805 East Central Avenue, near a railway line and within sight of Wichita's downtown business district. There was a dairy across the street and a commercial bakery next door. With the addition of the Phillips station, all the important products the public could ever desire were being sold in this single stretch of Central—milk, bread, and, soon, gasoline.

The actual filling station was built to resemble an English cottage. The idea was to build the stations near neighborhoods and make them look like bungalows so the consumers would feel comfortable. The brick building was painted white with black trim, and the first Phillips Petroleum shield, a disk-shaped logo, was placed on the sign out front and on the glass globes mounted atop the tall, cylindrical hand pumps. A giant letter "P" was mounted on the false chimney.

The first "Phillips 66" to appear on the shield sign would not greet motorists until 1928, when the color of the bricks changed from white to dark emerald green with a bright orange roof. In 1930 the logo would be an orange-and-black six-pointed "66" shield, similar to the signs used to mark national highways. But in 1927 the only mention of "66" appeared on banners and signs at the station and in newspapers, the natural outlets for Phillips' first advertisements.

As the grand opening of the company's first retail outlet drew near, Phillips ads were placed in the Wichita *Eagle & Beacon*. The ad was an eye-catching marketing device featuring an illustration of a female aviator, in her flying helmet and boots, with a beguiling smile and her arms akimbo. Virginia Wilson, one of several local young women selected for the advertising campaign, served as the model. Behind the lady flier was a drawing of a Phillips filling station, where several cars were pulling up to the gasoline pumps. Directly above, its propeller spinning away, was the *Woolaroc*.

If the aviation-oriented illustrations weren't enough to convince motorists that Phillips' new gasoline was fit for both autos and planes, the copy did the trick. "Phill-up and Fly," read one of the headlines, a not too subtle reminder of Phillips' recent triumph in the Dole race. "Free 5 Gallons of Phillips 66," read the other headline, ensuring that everyone couldn't help but read on.

"The new Phillips highly volatile gasoline which was used by Col. Art Goebel on his successful flight to Honolulu in his *Woolaroc* ship, and used extensively by aviators throughout the country, is now available in a grade adaptable to automobile use," stated the ad.

"The new winter gasoline—gravity 66—will be introduced to Wichita motorists at the formal opening of our new service station . . . We want to meet every automobile owner in Wichita and vicinity . . . and to every motorist who fills his tank at our station on the opening days, we will give a coupon good for five gallons of Phillips '66' gasoline. Just drive in and say 'Phill-er-up' and the free five gallon coupon will be yours."

Anxious for his first service station to be a success but confident that his tremendous stock of crude oil and gas was in no danger of depletion, Frank had given Oscar Cordell a flip response when the sales chief sought his approval for the free gasoline offer.

"Sure, go ahead," said Frank. "It isn't worth as much as water anyway. Give 'em all you want to."

On November 18, the day before his first service station officially opened in Kansas, Frank returned to New York after spending almost a month in France and Spain. Accompanied by Beall and William N. Davis, his vice president supervising the gasoline sales department, Frank had journeyed to Paris at the invitation of the French government to speak to the Chamber of Deputies about petroleum conservation.

On the crisp October afternoon they had departed New York, there had been the usual gala bon voyage at the docks. Frank's stateroom on the ocean liner was filled with presents and flowers—red roses and forget-me-nots from Jane; a basket of fruit from Fern; chrysanthemums from L.E.; sweetheart roses from Mary and Sara Jane; plus caviar, cigars, dates, nuts, and books. "The parlor of his suite looked like a greenhouse," remarked Beall.

In Paris, Frank was met at the train station by several dignitaries, including Dr. H. B. Baruch, brother of Bernard Baruch, the financier and economic adviser to Presidents. Baruch was a gracious host and treated his fellow Americans to a dinner which lasted until after midnight. Afterward the dinner party took in a bit of the nightlife, except for Beall, whom Frank deemed too young to taste the pleasures of Paris.

During his formal address before the petroleum committee of the Chamber of Deputies several days later, Frank—recalling some of the crises that had confronted his company and conscious of marketing difficulties in the United States—urged the French to avoid wasting their natural resources.

In his remarks to the assembly, later published in the New York *Times,* Frank said: "In my country, excessive competition has led to over-expansion of refining and such unreasonable duplication of filling stations and other distributing facilities as to multiply several times the cost of doing business, which must eventually be paid for by the consumer. It would seem that at this stage of the petroleum industry in France some steps might be taken, in harmony with your national policies of government supervision of this industry, to prevent such economic waste."

Frank's address was brief and to the point, but it so impressed his French hosts that they made him a Chevalier of the Legion of Honor. For the rest of his life he proudly wore the tiny red ribbon symbolizing his rank in the lapel of his coat.

When Frank returned to the United States, he wasn't thinking about waste or the French or his red ribbon; his mind was on gasoline sales. He wanted to know if the service station in Wichita was ready to do business. The reports he received were excellent. Everything was in place. All that was needed was some customers. Despite the optimistic predictions, Frank was as nervous as a groom when he switched off the bedside lamp in his Ambassador suite and tried to go to sleep. Frank need not have worried.

The next day—November 19, 1927—the first Phillips 66 filling station opened for business without a hitch. Before the grand-opening festivities could even begin, the customers Frank was praying for turned out in droves. There was a line of automobiles stacked up and down Central Avenue in Wichita.

It was a memorable occasion. For that moment, when the first automobile pulled in and one of the attendants, wearing white coveralls, tipped his cap and started pumping gasoline, marked Phillips Petroleum's leap into the ranks of full-service oil firms.

The opening was better than anyone imagined. More than 24,000 gallons of Phillips 66 gasoline flowed through the pumps by the afternoon of November 20, when the station was forced to close until a new shipment of gasoline arrived. Vic Peters, one of the attendants at the opening, described the kickoff as "the talk of the town . . . and the competition." The new service station attracted so much attention that one irate competitor even filed suit against Phillips for taking all his customers away. After a couple of days of stewing, the suit was dropped.

Encouraged by the success of the Wichita location, before the year ended Phillips opened two more Kansas service stations in Salina and Topeka and one in Oklahoma near the railroad depot in Bartlesville. At Topeka, hundreds of motorists came to the station to buy gas at seventeen cents a gallon and pick up their coupon for free gasoline.

The marketing department stepped up its activities. Over the next few years, Phillips executives, including Uncle Frank himself, sped from state to state, by plane and car, buying out other marketing companies and outlets, building new filling stations, and leasing wholesale bulk plants.

Phillips was either building or acquiring new service stations at the extraordinary rate of more than fifty a month. From the handful of stations opened in late 1927, the number increased to more than 1,800 retail outlets in 1928, and by the beginning of 1930, Phillips 66 was being sold at nearly 7,000 filling stations in a dozen states.

"Never in the history of the oil industry was a complete system of sales outlets established with greater speed," said one company summary of the Phillips marketing effort.

During those early years of marketing, the record for opening-day gasoline sales was established at the first Phillips station in St. Louis. While attendants pumped more than 35,000 gallons of gasoline during the opening day, a Phillips technician, wearing what could have passed for a military uniform, lectured on the finer points of Phillips 66 and kept up a running banter with Art Goebel as he flew in great circles above the station.

Goebel's airplane twisted and turned among the billowy clouds and then he streaked for a clear stretch of blue. The crowd at the station—school kids playing sandlot ball, old folks nodding off in the shade, and drivers stuck in traffic—all looked up in time to see the airplane release a thick trail of white smoke. For a long time, Goebel maneuvered from left to right, writing in the sky.

And when he finished and the plane cut away, there in mile-high letters and numbers made of smoke, for all the world to see, was the name that had become synonymous with success—Phillips 66.

CHAPTER TWENTY-TWO

The 1920s were drawing to a close, and Frank Phillips, wildcatter extraordinaire, was on top of the world. All his dreams were coming true; his plans were falling into place. He was on a lucky streak that made Titanic Thompson look like a dupe.

Phillips Petroleum was no longer a simple producer of oil and natural gas. Nor was the company still dependent on others for a market. Phillips had become a fully integrated firm with refineries, pipelines, and service stations plus several promising offshoots.

The decision to add refining and marketing was significant. The company now had a much steadier outlet for its products, and the early test service stations exceeded all sales expectations.

Bulk plants were built or purchased in state after state, and Phillips filling stations appeared all along Route 66 and throughout the Midwest. Free-gasoline openings, hosted by pretty young women dressed in flight suits handing out coupons for five free gallons to those who wanted a "Phill-Up," brought motorcars by the thousands to each new station.

Through a selective blending process to match seasonal conditions, Phillips researchers had perfected motor fuels. The research department helped create a high-gravity gasoline that was more volatile, resulting in engines which could start easier and run smoother. To help educate the public, "controlled volatility" was stressed in all the Phillips advertising.

Specially constructed laboratories on wheels, called Volatility Trucks, traveled throughout the sales territory so technicians could lecture about the quality of Phillips high-test gasoline. The cottage-style filling stations, operated by friendly attendants pumping "the gasoline that laughs at cold weather," became known as *the* places to stop for great gas and free air.

News was spreading fast about the company's line of products. Radio stations broadcast ads pitching the superiority of Frank Phillips' new and different gasoline that was custom-tailored to match monthly changes in the weather. As a result, Phillips 77, a natural gasoline highly touted for its controlled volatility, was selling better than bootleg beer as a premium motor fuel. Phillips Benzo-Gas, promising "power, no-knock, extra miles," and going for three cents more per gallon than regular gasoline, was another popular seller.

But the gasoline that fueled the company's growth was Phillips 66. Ever since that blustery November day in Wichita, at the first station opening when customers bought the pumps dry, Phillips 66 was on a growth track. It was the company's standard of excellence.

"Phillips 66 is the result of many years' research for a superior motor fuel," said the blurb on the free road maps the company handed out in Missouri, Illinois, Iowa, and all across the marketing territory. "It is a product built to a Phillips ideal—not just to a price."

Inside the map, below the mileage table, there was more company prose designed to instill confidence: "Phillips products are backed by a tremendous organization which has every facility for the producing, refining and distributing of petroleum from the oil well to the ultimate consumer. No expense is spared to insure the public that in the purchase of Phillips products they will get only the highest quality. You are safe with Phillips."

Frank wanted all the Phillips messages to build confidence and win the public's trust. He also wanted the ads to help part the public from a few bucks at the gas pump. The advertising was doing just that. Everywhere Frank went, he heard his name or saw it in lights or print—Phillips 66. His only regret was that he didn't have any advertising copywriters when he was trying to sell Mountain Sage back in Iowa.

Frank saw to it that either the first syllable of his last name or "66" was ascribed to as many subsidiaries or new ventures as possible. In 1929, even the established company basketball squad picked up on the popularity of the hot-selling gasoline's name, and started calling themselves the Phillips 66 basketball team. Most players and fans shortened the name to the 66ers.

Lou Wilke, former athletic director at Phillips University, a private four-year institution in Enid, Oklahoma, which was not related in any way to Frank or his family, came to Bartlesville and coached the amateur team. In their first season using the flashy new name, the 66ers ran up thirty-two victories against only five defeats, including a post-season win over the Wichita Henrys, National AAU champions. For many years to come, crowds from Bartlesville to Madison Square Garden would be thrilled and delighted by the championship play of the Phillips Petroleum quintet known as the 66ers.

While the basketball squad was making news on the sports pages, stories about the company's successful refining and marketing operations were appearing in the business sections of newspapers across the land.

Phillips 66 had became synonymous with the man who founded the company. The gasoline his firm created and sold was so closely associated with Frank that he received mail bearing only the address "Mr. 66, Bartlesville, Oklahoma," or simply "66 USA." It got to the point that when he checked into a hotel, Frank would often sign only "66" in the guest register. "66 is almost a part of me," said Frank. And in the sky over hundreds of cities and towns, Art Goebel and his airplane belching white smoke made certain nobody forgot either Phillips or 66. The name and the gasoline became indelible in many minds.

But 66 wasn't the only product Phillips Petroleum was marketing during the late 1920s. The company had solved the problem of what to do with the propane and butane, the basic components which were fractionated out of natural gasoline. Up to this point they had been considered waste.

Along with a few of his boy geniuses, Frank immediately saw the potential of liquefied petroleum gas, or LP. He understood that the light gases that were extracted during the process of taming the "wild" natural gasoline were too volatile for use in automobile motors. But they could be used for heating, cooking, and industrial purposes, if enough time and energy were devoted to solving some major problems, including supply, transportation, and service.

As with any other new venture the company explored, the problems associated with LP gas were not going to be easy to solve. Frank thought it was worth the trouble, but several members of the executive committee didn't agree. They couldn't quite see the value of formulating a plan to enter yet another new business.

Frank wouldn't be swayed. So, just as he did when he called for unanimous support of his motion to go into the retail business, Frank again took control. He simply overrode the executive committee and made sure that Oberfell received the backing he needed to allow the LP-gas project to fly. "Let's give George and his kids some money and let 'em take a whack at it," Frank told the executive committee. There was no further argument. Frank had spoken.

The company's new LP-gas business was christened Philgas, and Oberfell picked his top administrator, Ross Thomas, to run the new enterprise. Oberfell had implicit trust in Thomas and the rest of the research team's abilities to solve the problems of processing and handling the natural gas liquids.

Karl "Hack" Hachmuth, one of the research department's top chemists, was assigned the task of finding a smell to put in the odorless LP gas so it could be safely detected by the consumer. Hachmuth searched high and low for just the right stench and finally settled on ethyl mercaptan, a pungent odor which would alert consumers using the product if there was a gas leak.

Protection against liability and safety precautions were key to the success of LP gas, but there were many other hurdles to cross if the LP-gas business was going to succeed.

A potential market for LP gas already existed in the rural and suburban areas beyond the reach of city gas mains. What Phillips had to do was systematize sales, supply, and service. It would take someone with an analytical mind and plenty of endurance. Paul Endacott, the former sharp-shooting basketball guard from Kansas who became the "Mayor of Ragtown" out at Phillips, Texas, got the nod.

Endacott was transferred from Phillips' engineering department and was dispatched to Detroit, where the Philgas general offices where located. His main task was to launch the LP-gas business and keep it growing at a steady pace and make the company a profit. His work was cut out for him. As usual, Endacott didn't complain about his new post. He went to work and looked for solutions.

Although he didn't know it at the time, Endacott would wind up staying in Detroit for seven years, coming back to Bartlesville in 1930 just long enough to marry his sweetheart, Lucille Easter, a former school-teacher and dietitian, who had worked at the Mayo Clinic. Lucille's father had served both as mayor of Bartlesville and in the state legislature, and when Lucille was a little girl, sitting with her family in a pew at the Methodist church, Frank Phillips playfully tugged on her long curls. Now she was married to one of Frank's brightest young engineers and found herself moving hundreds of miles from home with a husband who was trying to find ways for affluent suburban housewives to cook their meals with gas and for farmers to heat their houses and run their tractors with liquefied petroleum.

"We had so many problems you couldn't count them all," recalled Endacott. "Philgas was a wild idea." Rail tank cars had to be found to haul the liquid gas, and there were difficulties with corrosion and transferring the gas into the tanks located next to the customer's home. Even figuring out how to keep customers supplied regularly and billed properly were chores.

But, one by one, the problems were solved. F. E. Rice helped design a special pressurized tank which could keep the hydrocarbons in liquid form. Smaller, noncorrosive tanks were connected to gas lines at the customer's home, enabling suburban dwellers to heat their houses and enjoy meals cooked with gas just like the city folks.

"Each time we came up with a solution for one of the problems, we'd turn around and get a patent on it," said Endacott. "Our competitors were always trying to pry into every detail of our developments and we were just as determined to protect our business from them."

Eventually all the problems were solved. Philgas became another profit center for the company. During 1929 alone, Philgas customers increased from 800 to 8,000 and a wholly owned subsidiary, Philfuels Company, was organized to take over the Philgas business. Philfuels enjoyed substantial growth as a distributor for liquefied gas. Retail domestic sales for 1928 were 368,000 pounds, and by the next year had reached 2,957,000 pounds. Wholesale and special product sales jumped from 3,052,000 gallons in 1928 to 6,137,000 in 1929.

The wild idea worked. Philgas not only proved to be a boon to thousands of consumers but also played a role in a couple of historic undertakings. Admiral Richard Byrd's men used Philgas as cooking fuel during both the 1929 and 1934 expeditions to the South Pole, and Philgas was also used in the early days of liquid-fuel rocket development.

For their devotion and hard work, Frank remembered those he called on to help with Philgas. F. E. Rice was made company vice president in charge of all manufacturing, including carbon black, a product important in the production of tires, paints, and inks. Endacott became vice president and manager of the Philgas division.

While the new refining, marketing, and Philfuels segments of the firm were getting off to a flourishing start, Frank continued to spend much of his time in New York, meeting with Wall Street figures, members of the board, prominent stockholders, and others. In 1927 he logged eighty-one days there, and in 1928 he made eight trips to New York for a total of sixty-six days—the magic Phillips number.

In 1928 Frank also stole time from his hectic schedule for visits to Iowa to see his mother and occasional holiday visits with Jane and his two daughters. When not with his family or at one of his two offices, he still managed to tend to business in St. Louis, Kansas City, Chicago, and other cities where Phillips maintained either sales offices or strong financial contacts.

There were also other diversions. In June that year, he visited Gravity and Lincoln and then took a train to Baltimore to look after Fern, who underwent a hysterectomy at Johns Hopkins. Frank checked into the Belvedere Hotel in Baltimore but stayed at the hospital until he determined that Fern was okay and that there were no further complications. Then he departed for New York.

On August 7, while in Bartlesville to celebrate Jane's fifty-first birthday the following day, Frank took the first airplane ride of his life. Appropriately, his maiden flight was in the *Woolaroc,* piloted by Art Goebel. The flight took place the day after the first anniversary of Goebel's visit to Bartlesville with the airplane before the start of the "Dole Derby."

After months of listening to Goebel and Parker trying to convince him to "go for a spin," Frank finally consented. Goebel and Frank flew over Bartlesville for several minutes and then they headed southwest and circled over Woolaroc lodge, fulfilling Frank's wish of seeing the lodge from the air. Then Goebel took Frank to an altitude of six thousand feet and cruised back to the airstrip. When he climbed out of the plane Frank was all smiles and summed up his first flight with a single word: "Thrilling!"

It was the enthusiasm of Goebel—and Billy Parker—that helped make Frank a total aviation buff.

At first, Frank was reluctant to fly, except when Billy Parker was at the controls. But even when he was flying with Parker, Frank despised the least bit of turbulence, and unless it was absolutely necessary, he never flew after sundown.

"It was really hard to get people to fly at first," said Parker. "They would only go if it was an emergency or something, or else they'd take the train. Frank Phillips used to tell me he would fly with me if he could keep one foot on the ground."

After much cajoling, Parker was able to convince the Phillips executives, even Frank, that flying was the only way to travel. "I just told Frank

Phillips that the company employees needed to use the planes more," said Parker. "It's just good business. I told him if he were in the business of making bicycles, he would probably ride a bike. So, since he was in the business of selling aviation fuel, he should fly."

By 1930 the company began to build a commercial air fleet by buying a Ford trimotor and hiring Clarence Clark. "We ended up with the finest company aviation fleet anywhere," said Clark, who became Frank's personal pilot shortly after he joined the firm.

When Clark first started flying there was no radar, no sophisticated radio systems, no weather reports. Clark had learned to fly in 1923 at age seventeen and had five years' experience as a test pilot for Beechcraft in Wichita. Early pilots depended on roads, train tracks, and other landmarks to guide them to their destinations, but mostly they relied on instinct. "You never had any kind of extended weather forecast," said Clark. "You would just go up when the weather looked good, and if it got bad along the way, you landed and waited for it to change."

There were few paved runways, mostly grass strips. Night flying was especially tough because there were no runway lights. Often, when coming home to Bartlesville, the pilots would circle over the city alerting their wives or girlfriends, who would then drive to the runway and use their car lights to help guide the plane down to a safe landing.

During one early flight, Clark and copilot Fidel "Pancho" Vela were flying Uncle Frank from Columbus, Ohio, to St. Louis. It was winter and the ceiling was low, so they had to fly over the clouds. Then the rain started. It got heavier by the second. The pilots knew their fuel was low, but they didn't know how far they were from their destination. They decided to land.

"I was really scared because we could go through the clouds into the heavy rain and never even see the ground," said Vela. "As we got lower and lower it suddenly was clear right on top of the river and we landed in East St. Louis at Curtis Field. Mr. Phillips got out and didn't know a thing about it. When we started fueling the plane we realized we were practically out of gas. We would have never made it to the next airfield."

Years later, when the planes got bigger and the equipment improved, Frank flew directly to Newark or New York. But through those first few years of flying, Columbus—strategically located, as Goebel pointed out—served as a base for most of the flights back East. Frank would fly into the Ohio capital city and then take an overnight train to Grand Central Station, only blocks from the Ambassador.

In New York, with his valet, Dan, and sometimes Doc Hammond, the osteopath with good hands, Frank maintained a fine lifestyle at the hotel apartment and kept a high profile in the better clubs and restaurants, where he conducted business with local financiers, as well as visiting oil men and Phillips executives from Bartlesville.

He also continued his personal relationship with Fern, who continued to run the Phillips office at 120 Broadway with an efficiency and flair that made her one of Frank's most valued employees. But the work took its toll on Fern Butler. Despite the passion between them, during working hours Frank Phillips was a tough boss—demanding and often unreasonable. As a result of the tension and pressure, Fern suffered from colitis, an inflammation of the colon associated with nervousness and anxiety. To find relief, in 1928 Fern bought a summer home near Westport, Connecticut. It was the smartest move she ever made. It saved her relationship with Frank, and quite possibly her life.

Settled in 1645 and located on Long Island Sound, only forty-five miles from Grand Central Station, Westport was a community of distinction and a haven for artists and writers. Fern's home was in Owenoke Park, an area both rural and residential in character, situated on the Saugatuck River not far from Compo Beach, where British troops splashed ashore in 1777 and where Scott and Zelda Fitzgerald had lived in a comfortable beach house the summer of 1920, just after they were married.

Fern bought a four-story home with plenty of guest rooms, quarters for a maid, and a living room large enough to entertain a small army of bankers, brokers, and oil men. From her bedroom, Fern could see yachts and sailboats scooting up Long Island Sound.

Fern maintained her New York apartment, but every summer she locked it up and commuted daily from Owenoke Park to the city. She bought a bright red 1928 La Salle convertible and hired John Graham, a pleasant Irishman born in County Cork, as her chauffeur. Lempie Krank, a native of Finland, was Fern's live-in maid and cook at the summer house, and during the winter Lempie moved to her own place in New York but appeared at Fern's apartment every morning to get her dressed, fed, and off to work.

In the summers at Owenoke Park, Fern rose before dawn, and while she bathed, Lempie laid out her wardrobe, complete with matching hat and purse. After breakfast, Graham drove Fern to the train station, and during her one-hour-and-fifteen-minute commute to Grand Central, she read the New York *Times* and circled with a red pencil any items she

considered important enough to bring to Frank's attention. Once in the
city, Fern would proceed directly to Wall Street in time for the opening of
the stock exchange and then she'd go to the Phillips office for a full day of
conferences, telephone calls, and correspondence.

Although she was technically Frank's secretary, Fern actually func-
tioned as the office manager and bossed a sizable staff, including three of
her own secretaries. By 1930, Fern had become so valued, not only to
Frank but to the entire Phillips organization, that she became the first
female officer at Phillips Petroleum and was made an assistant secretary
under Obie Wing, the company's secretary-treasurer, rumored to be
Frank's choice as the next Phillips president. The promotion caused Fern
to work harder than ever and made the house at Owenoke Park even more
important as a place to unwind and escape the world of business.

After a long and hectic day in the city during the summer, Fern
would take the commuter train back to Westport. She wouldn't speak to
anyone until she got home, changed into her bathing suit, and took a
swim. If the weather was sour or the tide was in, she'd put on her English
riding outfit and go to the Longshore Country Club or the Fairfield Hunt
Club and ride one of the horses she kept there. With a long swim or a
horseback ride behind her, Fern was refreshed and ready for dinner—
always with candlelight, even if she ate alone, unless Frank was there
wanting the lights turned up as bright as a boardroom.

There were always lots of guests at Fern's home during the summer
months, and while the ladies went upstairs to change for dinner, the men
and Fern would head for the living room to sip some of the fine liquor
always in stock and discuss business and the latest news from Wall Street.

Lempie kept the house sparkling, and Fern insisted on serving only
the best. She budgeted at least $200 every month just for food for her
guests. Overnight visitors were provided with thick towels and slept under
silk sheets, all monogrammed with the initials "SFB," for Sydney Fern
Butler.

Like Frank, Fern was a staunch Republican, and could talk politics,
stocks and bonds, or the price of oil. She was an excellent chess player and
adored going to the theater. She went as often as possible with Frank or
one of her friends and caught the late train which served theatergoers
living outside the city.

A woman with definite opinions—again like Frank—Fern knew who
and what she liked and didn't like. She liked diamonds and expensive
clothes. She liked Billy Parker and was amused with the flirtatious Art

Goebel. She met J. Paul Getty and didn't like him at all. She was fond of
Dan Mitani, the quiet Japanese so devoted to Uncle Frank. She couldn't
abide a messy desk or room. She enjoyed Obie Wing and his wife, Oral.
She rooted for the New York Yankees. She loved horses and dogs—especially Welsh Highland terriers and Belgium sheep hounds.

She also loved Frank Phillips, but still had no desire to break up his
marriage. In fact, Fern didn't limit her social activities with the opposite
sex to just Frank Phillips. Occasionally she went out to dinner and a show
with a banker or stockbroker, and during her busy career in New York she
was pursued by many eligible and attractive men. Her family, back in
Oklahoma, encouraged her to consider marriage. She told them that she
wasn't interested in getting married or in settling down. Fern's career was
quite enough. And although there were times when she was angry with
Frank over something he said or did, usually in the heat of business battle,
she still looked forward to the times they could be together.

Frank liked spending weekends in Connecticut, and even after he
established his ranch in the Osage and built Woolaroc, he continued to
look at properties in the Westport area, including an estate he almost
purchased very near Fern's waterfront home.

In the late 1920s and during the 1930s, Westport was a comfortable
retreat for anyone with a zest for life and money, including a busy and
bothered oil tycoon. Nearby Compo Beach was equipped with bathing
houses and dining and dancing pavilions. Canoes and Japanese umbrellas
were the order of the day. There were usually polo matches underway at
one of the chichi clubs and any number of celebrities either lived or vaca-
tioned in the area, including Charles Lindbergh and his family, Edna Ferber, the Gish sisters, George Gershwin, Oscar Levant, and William S.
Hart.

On occasion, Frank and Fern took in one of the local night spots. The
Penguin Hotel, originally known as the Miramar, was a popular club and
during Prohibition it was a speakeasy where tuxedos were required and
not one but two orchestras serenaded the patrons. But Fern preferred the
Longshore Country Club, a 168-acre pleasure complex opened in August
1929, which became a hangout for the "show biz" crowd and well-known
personalities such as Noël Coward, Wisconsin senator Robert La Follette,
Bill Tilden, Fay Wray, Helen Hayes, and many others. While patrons
spent the afternoon in the beauty parlor or barbershop or at an exercise
class, their automobiles were serviced and cleaned.

In the summers Fern's sisters came from Oklahoma for visits. Zee had quit teaching and become a secretary at a Tulsa bank. After she met and married J. C. Halliburton, a prominent businessman, they moved to Oklahoma City and he opened a large department store. Fern's other sister, Monta Locke, and her husband, Cleve, also left Bartlesville and moved to Muskogee. In 1928, Monta gave birth to a daughter, Nancy, and every summer for many years—starting when she was only nine months old—the little girl traveled to Owenoke Park to spend time with her Aunt Fern.

The Phillips family came for visits too. Fern doted on Frank's foster daughters and she kept photographs of both Mary and Sara Jane at her home. Fern also maintained a cordial relationship with Jane Phillips, and at times in the late 1920s and during the early 1930s, Jane and various friends came out to Owenoke Park to spend time in the sun and water. On one occasion, Jane brought her grandson, Johnny, a good-looking boy, who horrified Fern's niece, Nancy, by snaring butterflies with his net and sticking pins through their bodies.

The harmony between Jane and Fern would not last forever. But for many years after Fern bought her home in Connecticut, they were friendly with each other. They traveled together, attended some of the same social functions, and even exchanged gifts, including flowers and perfumes, expensive clothing, and a piano bench with a needlepoint cover made by Jane.

There were other gifts. Starting in 1928 with a large buffalo rug, and over the course of several years, Frank and Jane shipped a variety of animal-hide rugs to Fern for both her New York apartment and the Owenoke Park house. Jane found it great fun to send an exotic skin, such as a zebu or eland, and have Fern guess what animal it came from. The rugs were striking additions to both residences, except for a deer-skin rug which had to be thrown out after it was attacked by Pirate, one of Fern's beloved sheep hounds.

The hides came from the variety of beasts roaming the Frank Phillips Ranch. Many of Frank's wild animals from Africa and Asia had a difficult time adjusting to their new surroundings and ended up as trophy heads for the lodge walls or were made into rugs.

The loss of the expensive creatures troubled Frank, but he didn't give up on his idea of creating a wild animal preserve. He imported other varieties and turned them loose to see if they'd fare any better. Some did. A water buffalo Frank's grandchildren named Ed got along fine for many years living near one of the ranch lakes. The huge animal with great, crescent-shaped horns was a familiar sight to guests as they drove to the lodge. Many of the imported hooved animals, including the sika and fallow deer, also adjusted and grazed in the ranch pastures.

The white-tailed deer, elk, and buffalo—all native to North America —were self-sustaining. Hardy animals that needed little attention, they foraged on native bluestem grasses and acorns and occasionally waged battles in their ranks over territory.

During the mating season the bugling of the male elk, some weighing half a ton, with antler spreads of more than five feet, could be heard in the night and the early morning hours echoing over the Osage Hills. It was a strange, hair-raising sound, but it never disturbed Frank if he was out at the ranch for a weekend of rest. He'd be deep in sleep, snug in his upstairs bedroom at the lodge.

Frank's room was comfortable, with plain hickory furniture, an easy chair, and a stand-up ashtray made of animal bones. Bear and cougar rugs —their glass eyes open wide and their mouths frozen in eternal growls— were stretched side by side across the floor. Frank kept small framed photographs of family members, including one of his father when Lew was a seventeen-year-old soldier ready to take on the Confederacy. Lithographed portraits of Indian chiefs hung on the walls. The faint smell of cigars and bay rum lingered in the air. Frank's bed was just a single cowboy bunk. The brands of his favorite ranches—the Miller brothers' 101, the Cross Bell, J. H. Bartles, the Johnstone and Keeler, the Frank Phillips, and others—were burned into the wooden headboard and footboard. On a stand next to the bed rested Frank's eyeglasses. Pairs of tall leather cowboy boots stood against the wall and a gun belt and lariat were draped over the bedposts. It was more like a room for a ranch foreman and not the president of a big oil company.

CHAPTER TWENTY-THREE

Frank Phillips could be as predictable as Christmas. But there were times when he was a total enigma and puzzling as a cat's cradle. He was like that all of his life—a stream flowing deep and straight but with sudden twists and turns and rapids. When Frank was a young barbershop owner he was bald, yet he sold tonic to make hair grow. The barber-turned-banker-turned-oil-tycoon built an empire that included a huge network of gas stations and refineries, all of which had a profound impact on the automobile, yet Frank never learned to drive.

Frank was a self-proclaimed tough guy and shrewd corporate captain who some competitors believed had ice water in his veins, yet he was often a compassionate man who would worry over injured employees or sick animals out at the ranch, and not rest until he knew they were out of danger or on the mend. He was a stern disciplinarian, with a temper hot as a branding iron, who would fire a worker on the spot for the slightest infraction, yet a forgiving employer apt to hire the same worker back by the end of the day. He was an expert at delivering off-the-cuff speeches and a masterful communicator able to win over just about any audience he faced, yet a father who never really learned how to express himself with his only son.

Frank Phillips was a riddle waiting to be solved. A complete contradiction.

The main incongruity in his life became evident out at the ranch. That's where it became public. A few family members and associates picked up on this personal quirk earlier in Frank's career, soon after he and L.E. opened their first bank. Others noticed it when Frank went out into the oil patch. It had to do with outlaws. Real, honest-to-goodness desperadoes—the kind who packed shooting irons and wore masks and made their living robbing banks and trains and stealing horses and cattle. The Henry Wells variety.

When it came to outlaws—especially the ones like Wells who frequented the Osage and found refuge in the nooks and crannies near the F.P. Ranch—the ramrod-stiff banker and conservative businessman side of Frank Phillips melted like a cube of sugar in a cup of hot jamoke. He may have been the descendant of a Pilgrim father and counted many preachers and God-fearing souls in his family tree, but Frank would just as soon hunker down and swap lies with a pack of rascals from Okesa as endure a meeting in some fancy conference room with a bunch of spruced-up and slicked-down executives. To Frank, outlaws were like the wild mustangs, the longhorns, and the buffalo that lived on his ranch—all were symbols of the Old West.

Outlaws absolutely fascinated Frank. He loved their look and swagger, enjoyed their company, listened to their tales, banked their money and loaned them some if they asked for it.

After his ranch operations were under control and the lodge was built, Frank even devised a way to salute the old-time outlaws as well as the cowhands and the Indians of the Osage. For several years, starting in 1927, Frank hosted an annual party for them at his ranch. Some called it a picnic, others a barbecue. Frank gave the affair a more flamboyant name and touch. He organized the hard-core regulars—genuine cowpokes and tough guys from the area—into an association and summoned them to the F.P. Ranch each year. He called these annual blowouts the "Cow Thieves and Outlaws Reunion." Nothing before or since ever quite compared.

Frank soon found there were plenty of experts around willing to offer advice about how he should stage his colorful gatherings. Only a few were worth listening to and Frank did just that.

Gordon W. Lillie, better known as Pawnee Bill, was a close friend and was willing to share his experiences from Indian Territory days, as well as pointers he picked up on the road when he was touring with his Wild West circus. Lillie had combined his show with Buffalo Bill's, and was an expert at putting on unrivaled western-style spectaculars. Lillie's wife, a former Philadelphia socialite who could ride like a wild Indian and became the "Champion Lady Horseback Rifle Shot of the World," also befriended Frank and Jane, and after the Lillies retired in 1913 the two couples visited at each other's homes. Frank brought them out to Woolaroc after it opened in the mid-1920s, and Pawnee Bill hosted the Phillipses at his mansion, which he built atop Blue Hawk Peak on his buffalo ranch near the town of Pawnee. Frank learned firsthand from Pawnee Bill, one of the masters of cowboy and Indian pageantry and a man who had actually been Buffalo Bill's partner, how to throw a genuine Wild West shindig. It was invaluable counsel.

Another big influence on Frank's outlaw wingdings were the Miller brothers, who sponsored rip-roaring rodeos at their 101 Ranch and, starting in 1924, also came up with a new style of racing that developed into an annual event during every Labor Day roundup.

It was an altogether different type of racing and it didn't involve sleek cow ponies. Instead, they were tongue-in-cheek contests, before audiences of hooting and hollering cowboys and ranch visitors, in what the Millers called the "Terrapin Derby"—or, for the unwashed, turtle races.

For only a two-dollar entry fee a guest at the 101 could go to the terrapin pit and make a selection from the hundreds of turtles rounded up on the range by the Millers' cowboys. All the money went for prizes, and "Foghorn" Clancy, known from the Osage Hills to the Rio Grande for his loud voice, announced each race.

By 1925 the second derby drew 1,679 entries, and the next year there were 2,373, including one owned by Frank, which didn't win, place, or show. The entry numbers were painted on the turtle's shells and some of the better-known contestants were even named. There was "Jenny Lind," "Bridesmaid," "Marie Antoinette," "Easter Bells," and "Star of the Night."

Old cowboys who made up a group known as the Cherokee Strip Cowpunchers Association met for reunions at the 101 about the same time as the annual roundup and the derby. The old-timers liked to gather on a bluff on the south side of the Salt Fork River known as "Cowboy Hill" and they'd sing old trail songs and recall the stories of cattle drives, Indian fights, and stringing up rustlers.

After the derby, Frank and the other guests mixed with the old cowboys and ate suppers of barbecued meat, cider, and biscuits, served out under the stars. The days and nights at the 101 Ranch left an impression on Frank, and he remembered his experiences there when he began planning for his outlaw bashes at the F.P. Ranch.

Grif Graham, the former sheriff who served as Frank's first ranch manager, was the driving force behind the Cow Thieves and Outlaws Reunions—annual affairs intended to salute those survivors from the territorial days "when men were men and women were respected."

Although, as the name implied, the idea was to get together as many cow thieves and outlaws as possible, there were usually a good many law-abiding citizens in attendance, including ex-sheriffs, ranchers, and others.

When the first reunion was held at the ranch in 1927, only about 100 guests showed up. Within a couple of years there were more than 500 in the Cow Thieves and Outlaws Association and, counting members' families and other guests, the attendance swelled to more than 1,200 when the fourth annual reunion took place.

From the start, the reunions were colorful affairs attended by a variety of locals—old trail riders, horse traders, Indians, U.S. marshals—and usually a sprinkling of Frank's personal guests, including a few of the Phillips directors and others he was trying to impress.

"Cowboys, real cowboys, old riders of the plains in the days when the cattle business was the only business and every man had to ride his stuff and throw a wicked rope to hold his job, gathered at the Phillips ranch Thursday for an all-day barbecue and picnic as guests of Frank Phillips," said a newspaper account of the first reunion.

Some prominent ranchers and old-time cowboys showed up, including George Miller; Mert Keifer, who rode for Jake Bartles; and Jimmy Rider and Bright Drake, both of whom rode with Will Rogers.

A kangaroo court was in session throughout the day and a mock trial was held, with one of the guests playing the role of "Hanging Judge" Parker of Fort Smith, the famed jurist who had tried to tame Indian Territory during the tail end of the nineteenth century. Several guests were fined on a variety of trumped-up "charges" in order to collect fifty-five dollars to buy Frank a fancy Stetson hat with an inscription in the band: "With the compliments of Cow Thieves and Outlaws Reunion. F.P. Ranch, June 2, 1927." During subsequent reunions, Frank received other gifts, including fancy boots and a pair of leather chaps with his name and the initials "FPR," for Frank Phillips Ranch, emblazoned down each leg.

When the second annual outlaw party rolled around in October 1928, Frank and Grif had every detail organized. This time they even prepared an official invitation. From New York to Pawhuska, the invites to the F.P. Ranch became coveted items.

INVITIN'

yu an yer wimmin folks to

Second Annual

COW THIEVES AND OUTLAWS
REUNION
at F.P. Ranch
Saturday, Oct. 6, 1928

Aims to throw chuck about noon, if the
Boss can borry a side of meat and some flour.

No guns er store cloze is purmitted
er no golf breeches.

Show this here invite to the Brand
Inspector feller at the big gate, cause
he wont pass yu thru without none.

Hopin' to meet yu all
at the F.P., we begs to remain

THE COMMITTY

P.S. The Boss wants yu all to be at the wagon
at 10 o'clock forenoon.

2nd P.S. Leve yer nives an guns with the Boss at the gate,
cause we aint allowin' no shootin'.

The comical invitation, which included an illustration of Frank dressed in western clothes, was sent to hundreds of cattlemen, peace officers, known outlaws, Indians, bankers, and business executives.

Not everyone could make the reunion, but if they couldn't come they were quick to send word so they wouldn't get bumped from the invitation list for the following year.

J.L. Johnston, one of Frank's most valued board members, sent a short wire after he received the invitation to the 1928 reunion: SORRY I CANT BE AT THE RANCH PARTY BUT AFTER ALL WHAT CHANCE WOULD A WALL STREET CROOK HAVE AGAINST COW THIEVES AND OUTLAWS

Amon Carter also sent his regrets for that year's reunion: SHO WUSH I COULD BE WITH YOU COW THIEVES TODAY AS I NATURALLY HAVE A HANKERING FOR EITHER STEER OR BUFFALO MEAT   WHETHER YOU

STOLE EM OR NOT DONT MAKE NO DIFFERENCE WITH MY APPETITE I SHO THINK YOUR PICTURE ON THE INVITE IS DARN GOOD AND I HOPE YOU AND GRIF HAVE NEW CLEAN RED SHIRTS AND PLENTY OF EXTRACTS AS NOTHING HELPS TO TENDER TOUGH MEAT MORE THAN THE KERECT KIND OF SASE PROPERLY DISTRIBUTED BEFORE AND AFTER CHUCK   THANKS A HEAP FOR THE INVITE

Frank's ground rules for the celebrations, held in the picnic area near Clyde Lake, were simple. Any wanted desperadoes would be granted a day of grace for the reunion. If there was an outstanding warrant, the law officers in attendance would have to wait for another time and place to make their moves. Frank wanted to be sure all of his guests had ample time to sleep off their hangovers and get a few miles' headstart. All guns and grudges had to be checked at the main gate.

There were some who swore the story was true that Frank arranged for a few of the outlaws serving time to be released from jail for the day so they could come to the reunion, and that Frank posted his own personal bond guaranteeing their return. More than likely that was one of the many tales cooked up and perpetuated by R. C. Jopling, called "Jop," Frank's ace public relations man. Besides insisting that all lawbreakers and law enforcers abide by his one-day moratorium, Frank left instructions at the main gate to "admit any cowboy with a saddle horse, admit any American Legion boys in uniform and all Spanish-American War veterans in uniform; also any full-blood Indians in costume."

It was never really clear just how many bona fide outlaws actually showed up at the ranch, but a good number of questionable characters always appeared, such as Henry Wells, who had served five years and one day of a prison sentence for bank robbery before winning his release in the early 1920s. Within a week of getting out of prison, Wells robbed another bank just to see if he had lost any of his criminal prowess while he cooled his heels in the "cross bar hotel."

Frank liked Wells, and from time to time the two men played poker at Okesa with a few of the outlaw's shady friends. Wells added an air of authenticity to the reunions and other ranch functions. Frank put the outlaw on the ranch payroll whenever there were guests—especially eastern guests—coming out for a visit, and Henry, an imposing six-footer who stood straight as an arrow, called on Frank at his office in downtown Bartlesville to confer about ranch matters.

On one occasion, Frank arranged for Henry and a few of his boys to have some fun at the expense of some unsuspecting Easterners invited to

spend the day at Woolaroc. Frank's guests wanted to experience the Old
West, so he had them brought out to the ranch in an authentic stagecoach
built in 1869. Shortly after the half dozen horses pulling the stage made
the steep climb up "44 Hill," a gang of outlaws, wearing bandannas over
their faces and with six-shooters drawn, rode out of nowhere and halted
the stage. In a flash, the startled passengers were relieved of their wallets
and jewelry and the stage was sent on its way. Several of the victims
figured the holdup was Frank's doing, but then again they had also heard
his stories about outlaws operating in the Osage. There was always the
chance that the bandits were for real. When the stage pulled up in front of
Woolaroc and the excited guests were greeted at the lodge, all of their
belongings—down to every penny and watch fob—were laid out waiting to
be claimed on a table inside the door. Out back, Henry Wells and his
cronies were laughing in their beer and barbecue, their pockets filled with
pay from Uncle Frank.

As gutsy as a bull buffalo in rut, Wells was proud of the fact that
many bankers closed their doors and declared a holiday if they learned he
was in the vicinity—something, Henry was quick to point out, that bankers usually did only for George Washington or Abe Lincoln.

Word around the Osage was that Wells was actually the bankers' best
customer, and that many times they exaggerated their losses after a holdup
and actually made more money from the robbery than Wells did. Some
bankers thought so kindly of the Osage bandit that they reportedly sent
Wells Christmas cards each year without fail.

At one gathering of bankers at the F.P. Ranch, the wily outlaw sidled
up to a cluster of distinguished gentlemen sipping highballs and puffing on
Frank's best cigars.

"Henry robbed me once," bragged one of the bankers.

"Say, I've been meaning to ask you something," Henry shot back.
"What did you ever do with that $22,000? You told the bank examiners I
got $25,000, but I just counted $3,000. Where's the rest?"

The banker choked on his drink and dropped his cigar as Henry
walked away laughing without waiting for an answer.

To make certain his guests—especially the renegades—were content
and didn't take it upon themselves to swipe one of his prized animals,
Frank had his Woolaroc staff lay out an enormous feast. There was always
enough to feed an army, but it was far from normal picnic fixings. For
example, at the 1929 reunion it took two grown buffalo alone to make a
dent in the crowd's appetite. The menu included slabs of barbecued buf-

falo, beef, and pork; platters of pickled buffalo; mounds of boiled potatoes, baked beans, and cabbage slaw; hundreds of hard-boiled eggs and ice-cream cones; and plenty of bourbon and draft beer.

To the delight of the guests, more than three hundred old-time range riders came from Oklahoma, Texas, New Mexico, Colorado, Kansas, and California. Local wranglers showed up from Bartlesville, Okesa, Dewey, Ramona, Pawhuska, Nowata, Copan, and Ochelata. Hamp Scudder and Joe Bartles came. Jop Jopling mugged for photographers with Doc Hammond and Henry Wells. The kangaroo court did a brisk business.

Grif Graham, wearing a new flaming-red shirt, served as the master of ceremonies and supervised the trick-riding and roping exhibitions. One of the judges was Pawnee Bill, who came, as usual, with a full complement of Pawnee Indian dancers. Chuck wagons and pack outfits competed for prizes, and there were fiddlers and square dancing and Indian stomp dancers all dressed up in feathers and paint.

At one point, Frank, wearing his cowboy best—big hat and high-top boots with fancy tooled designs—walked into the midst of the throng and bawled out: "I can outrun any man or woman in this crowd half my age!"

Quick as a flash, Grif shouted: "That's seventy-three, folks!" The crowd roared and so did Frank.

But sometimes incidents occurred at the ranch that were not laughing matters. In 1927, just a month before he hosted the first annual Cow Thieves and Outlaw Reunion, Frank put on a big barbecue feed to honor Chief Baconrind, the Osage tribal leader who had been friendly with Frank ever since the early days when the Phillips boys came to town and opened their first bank. Through the years, Frank had great success drilling for oil on Osage tribal land and as a result both the Osage and Frank Phillips prospered. It was important for Frank to maintain goodwill with the Osage Nation and he wanted the dinner at his ranch to go off without a hitch.

Boots Adams, the aggressive young Kansan who had played basketball for a few years for Phillips Petroleum while quickly working his way through the corporate ranks, was assigned the important task of checking names off a master invitation list as people arrived at the front gate of the ranch.

"At one point," Adams recalled years later, "I was in the middle of a mass of people at the entrance, busy with a hundred details, when I noticed an elegant touring car draw up with a number of persons in it. Because there were so many cars already at the gate, this new arrival had to wait."

But after idling for several minutes, with only a glance from the young man at the gate, the big automobile maneuvered out of line, turned, and sped away. Boots was sorry he hadn't been able to get to the car to check off the occupants' names, but there was nothing he could do. He forgot the entire episode until about an hour later when Frank Phillips appeared.

Frank had seen young Adams playing basketball and he was aware of his steady climb through the various departments at Phillips, but that meeting at the ranch gate was the first time that Boots Adams and Frank Phillips ever spoke to one another. It was a conversation neither man would soon forget.

Frank, his face creased with frowns, said he was worried because the party was ready to start at the lodge and the guests of honor had not yet arrived. He asked Boots if he had seen anything of Chief Baconrind and his entourage.

"No, I haven't sir," answered Boots.

"Didn't you see a group of Osage in a large car?" Frank asked.

As Boots tells it: " 'Oh, those,' I said, with a sudden sinking feeling. 'There was a Pierce-Arrow with a chauffeur and a group of Indians in it. They drove up, but they had to wait a couple of minutes because there were other cars ahead of them, so they drove away.' Mr. Phillips looked at me for a moment as if he were about to take my head off. Then he growled, 'Those were my guests of honor—you idiot!' "

Frank was so mad he couldn't even fire Boots—at least for the moment. Instead he jumped in his waiting car and ordered the driver to go straight to Pawhuska so he could personally apologize to Chief Baconrind, who was sulking at his home, angry at "the young upstart" who wouldn't let him in to his own party. Frank rode back to Woolaroc with the chief, more apologies were offered, and the barbecue continued with no further mishaps.

The next morning Frank summoned young Mr. Adams to his office and proceeded to give him a ten-minute ass chewing. Boots was also ordered to write a detailed explanation of the incident and a formal apology to the chief.

"That was the first time I had ever met Frank Phillips, the first time I came to his personal attention—and I made some impression," said Boots.

The timing of his faux pas couldn't have been worse. Only two weeks after he snubbed the Osage chief at the ranch gate, Boots was the subject of a memo drafted by his immediate boss—O. K. Wing, the secretary-treasurer of Phillips Petroleum. Handwritten on the company's official blue memorandum paper, the single paragraph from Obie Wing was a recommendation about Boots for Frank Phillips to consider.

> K. S. Adams has been with the company about eight years. He is now my chief assistant looking after credits, etc. He has had a fairly good education and has good potential possibilities. Has been receiving $350 per month since 1926 (Sept.) and I now recommend that his salary be increased to $400 and that he be elected to the office of Assistant Secretary and Treasurer.

Although his temper had cooled by the time Wing's memo reached his desk, Frank had not yet forgotten Boots's blunder at the ranch. He stewed over the recommendation for a while and wondered how a young man who wasn't able to recognize a guest of honor when he saw one could possibly handle the responsibilities of an executive position. Fortunately for Adams, the respect and confidence Frank had in Obie Wing prevailed. The memo was returned to Wing with "O.K., Ex. Comm., F.P.," and a few changes. The words "and Treasurer" were crossed out and the suggested salary of $400 was cut to $375.

Less than a year later, Adams was also given the assistant treasurer title and another raise. Only a few months after that promotion he was again called to Frank's attention. This time the results were better for Boots. In an effort to build up a loyal following at the Phillips 66 filling stations, Boots had started selling coupon books, good for discount purchases of gas and oil at the Phillips stations, to various business people he encountered during the course of his workday. The campaign was successful and many of those Boots reached remained Phillips customers long after the coupons were gone. When he heard about Boots going out and selling up a storm, it struck a chord, and reminded Frank of himself when he was a young barber in Iowa trying to become a master salesman.

Frank was touched by Boots's effort and he sat down and wrote Adams a note of thanks:

> I am taking this opportunity to express to you my appreciation of the spirit and initiative that has led you to go beyond your regular duties to further the interests of our company. Such a spirit throughout our organization could not help

but insure for us a tremendous success in any field we
choose to enter.

Boots saved the note from Frank Phillips and kept it tucked away. It
took the sting out of the reprimand he received the year before for his
blunder at the ranch. Boots felt that he had finally been redeemed, but
whenever he drove under the big arched gate at the F.P. Ranch, he still
thought about the night he was in charge of the guest list and the first time
he came face to face with Uncle Frank. Now the memories would be
sweeter and Woolaroc would always be special for Boots Adams.

Even before he built the lodge, Frank invited thousands of Phillips
employees out to his place in the Osage to swim and picnic, and the grassy
grounds and the dance pavilion next to Clyde Lake became a favorite site
for company outings for many years. Although he believed in hard work—
and lots of it—Frank also saw the value in good, hard play.

In July 1928, when he perceived the summer doldrums were beginning to set in, Frank drafted what at first glance appeared to be a stern
message for his employees. It was circulated on Frank's letterhead for all
hands to read.

BULLETIN

Very seldom do I have an opportunity to take an hour off
and visit the various offices. A few days ago, however, I
visited most of the floors and my impression was that the
oldtime pep which dominated this organization in the past
was not in evidence. Spirits seemed to be lagging and in
many rooms it occurred to me that there was about fifty per
cent efficiency, with fifty per cent of the employees doing
most of the work. The conditions which I met up with disturbed me very much. An office is a place for work only; if
you cannot find something to do probably we do not need
you.

Perhaps I am to blame, or maybe I do not understand and
all of you are already overworked and need an outing. In
any event, let's all go out to my ranch next Saturday afternoon and jump in the lake. I have appointed J. S. Dewar
chairman of a committee on arrangements. A later bulletin

will announce to you plans for next Saturday afternoon and evening.

More than seven hundred took Frank up on his offer. They not only jumped in the lake, but they ate, drank, and danced until long after the moon rose over the ranch. The following Monday, Frank noticed a decided improvement in everyone's spirits.

But Frank used his ranch for much more than employee outings and outlaw reunions. Early on in the life of the F.P. Ranch, Frank hosted a variety of groups, organizations, and individuals, ranging from Scout troops, classes of schoolchildren, and ladies' clubs to delegations of Catholic bishops and the directors of major railroad lines and well-established banks and corporations. If there were no conflicts with family or company activities, Frank was usually amenable to allowing outside groups use of the ranch. He'd arrange for the lodge to be open at a designated hour so the visitors could look at the animal heads mounted on the walls, and he even provided a Victrola they could use for dancing in the pavilion next to Clyde Lake.

Fern Butler, acting on Frank's orders, went to Abercrombie & Fitch —"the Greatest Sporting Goods Store in the World"—on Madison Avenue and ordered a diving board for the lake and dozens of bows and arrows and all the accessories, including an instructional book about archery, for ranch guests to use.

In 1928, Frank was elected as a director of the Chatham Phenix Bank in New York, one of the largest banking institutions in the nation. He also served on the board of the First National Bank of St. Louis, along with Charles Lemp, a Phillips director. Officers and fellow board members from both of these banks were frequent guests at Woolaroc, and for one of the dinners he held for the First National board at the lodge, Frank arranged for Jack "Backlash" Lamb, the champion angler from Fort Worth, to give a fly- and bait-casting demonstration.

Lamb had caught more than twelve thousand bass during his career and could fill a washtub with bass in less than three hours. Frank's guests, all dressed in suits and ties, stood with their mouths agape as Lamb quietly cast a fly one hundred feet across Clyde Lake with pinpoint accuracy.

When he entertained fifty of the nation's top railroad executives at Woolaroc, Frank seated them at a huge U-shaped table set up around an electric train representing the crack Overland Limited. The train blew its whistle and puffed tiny clouds of smoke as it raced around the guests, who were enjoying a breakfast of oatmeal, eggs, sausage, and buffalo and elk steaks. Afterward, Grif Graham initiated everyone into the "Woolaroc Klan" and sent them on their way.

A group of veterinarians were given a more animal-oriented tour of the ranch when Frank invited thirty-nine carloads of vets and their families to spend two days as his guests at the ranch. There was the usual barbecue and dance, but Frank also took the animal doctors on a close-up inspection of his huge buffalo herd and the other wild critters. Later, two of Frank's cowboys gave an exhibition in riding bucking horses and mules and several guests took camel rides. For all his trouble, Frank received plenty of free advice and tips about animal care, something he was constantly seeking.

Not only large groups visited the ranch, such as the bankers, or the veterinarians, or the eight hundred members of the Izaak Walton League who came for a barbecue in the spring of 1927. Many well-known individuals also came—actors and actresses from New York and Hollywood, authors, politicians, religious leaders, and other celebrities.

One weekend in May 1928, Frank and Jane hosted not one, but two Pulitzer Prize-winning writers—Edna Ferber and William Allen White.

White, the popular editor and author from Kansas who was known as the "Sage of Emporia," brought his wife and son to the ranch. The spokesman for grass-roots and small-town America and an influential figure in Republican politics, White had won the Pulitzer for editorial writing in 1923 and later ran unsuccessfully for governor of Kansas on an anti-Ku Klux Klan plank. Frank and his guest had more in common than their political party affiliation. Both men were blunt but humane and had risen from virtual obscurity to national prominence. All afternoon while they toured the ranch and later at the lodge, they sized up one another. After dinner, while their wives and the other guests visited, Frank and the gregarious White found comfortable chairs on the front porch and talked long into the night about world affairs, White's campaign to discredit the Klan, and the merits of Herbert Hoover, only a month away from winning the Republican nomination for President.

Edna Ferber, unfortunately, did not find her stay at Woolaroc nearly as stimulating or pleasant. The popular novelist and playwright's bestseller, *So Big,* won the Pulitzer Prize in 1925, and her romantic *Show Boat,* published in 1926, had just been transformed by Jerome Kern and Oscar Hammerstein into the perennially appealing operetta. Ferber was interested in developing yet another romantic novel, this one with a western theme, which would focus on life in territorial Oklahoma through state-hood. It would be called *Cimmaron*. For picking up the flavor of the land and the people, Ferber couldn't have picked more inspiring subjects than Frank Phillips and his Osage ranch.

She made it through the first day and was ready for bed when the trouble started. The writer's peace of mind and train of thought were shattered. It wasn't all the political talk between White and Frank that drove her off. It was another sort of racket and it came from Jane's peacocks. The birds' plumage was handsome but their voices were loud. Very loud. The first night at the lodge, the big birds perched near the windows of the guest room where Ferber was trying to sleep. Their shriek-ing was constant and not even two feather pillows over her head could give Ferber any peace. Either everyone else was used to peacocks or they were sound sleepers, because the next morning only Ferber complained.

The second night Ferber could hardly keep her face from falling into her plate of buffalo steak and potatoes, she was so weary. As soon as she ate her last bite of pie, she excused herself and retired to her room. It was as quiet as a graveyard outside. She slipped into bed and was on the edge of slumber when the high-pitched screams of the peacocks started again. Their cries were louder than ever and twice as shrill. Before daybreak, Ferber had her bags packed, and as soon as breakfast was over she bid hasty goodbyes to the Phillipses, told them what she thought of their noisy peacocks, and left in search of some peace and quiet. Edna Ferber never returned to Woolaroc, which was just fine as far as Jane was concerned. Anyone who insulted her peacocks, as Ferber had done, wasn't welcome.

But Frank and Jane hosted other guests who didn't mind the peacocks and weren't afraid of the rattlesnakes sunning on the rocks leading down to Clyde Lake or the wild animals sniffing around in the pastures and pens.

In the summer of 1929, Tom Mix, now a big movie star and on tour with the Sells-Floto Circus, came back to Oklahoma after a fifteen-year absence and was the guest of honor at a dinner party at Woolaroc hosted by Jane. With Frank in New York, John Phillips was quickly recruited to stand in for his father. Jane found the reckless Mix attractive and was entertained by his stories about his wonder horse, Tony, and life in Holly-wood. John remained reasonably sober and Grif Graham initiated Mix into the "Woolaroc Klan." Mix sent Jane an autographed photo of himself as a token of his esteem. He had seen many ranches, lodges, and resorts, but Mix was especially impressed with Woolaroc and the collection of wild and domestic animals roaming the ranch.

Will Rogers, another guest, made his first visit to Woolaroc just a few weeks after Mix. "Thought I'd just drop in and be neighborly while I'm back home," said Rogers.

After he toured the ranch, inspected Frank's horses and wild beasts, and gawked at the growing collection of paintings, Indian blankets, and animal trophies in the lodge, Rogers turned to the rest of the party and declared: "Well, boys, she's a success—there ain't no doubt about it." A few years later, in his autobiography, Rogers had more to say about the F.P. Ranch: "When you are visiting the beauty spots of this country, don't overlook Frank Phillips' ranch and game preserve at Bartlesville, Oklahoma. It's the most unique place in this country." Strong praise from a country boy who had seen more than his share of the world.

As the years went by, the compliments about the beauty of the lodge and the lakes and the land continued and the flow of guests to the F.P. Ranch never slackened.

Jane hosted summer parties at Woolaroc for her circle of friends— Winnie Clark, Noretta Low, and others from Bartlesville. Minnie Hall, her old chum from Creston, and close friends from out of town also came for visits. Jane's daughters, Mary, or Mary Frank as she preferred, and Sara Jane, or Jane as she preferred, began spending more time at the ranch.

The two teenagers were only a grade apart when they went to Garfield School in Bartlesville, just a couple of blocks from the town house on Cherokee Avenue. But in the late 1920s, both girls were sent to the Ogontz School outside Philadelphia, and by attending a few summer sessions, the sisters managed to end up in the same class. Ogontz was a girls' school founded in 1850. The students there wore military-style uniforms, drilled with wooden rifles, and were required to spend long hours with their noses in books in order to obtain a classical, and expensive, education.

At Thanksgiving and most other holidays, Mary Frank and Jane went to New York and joined their parents at the family suite at the Ambassador. But at Christmas break or after classes ended in the summer, they headed straight for Bartlesville and the ranch. Out at the ranch, everyone, including Frank, seemed to let down their guard when only family and close friends were out for the weekend or a holiday visit.

"Our parents were very strict with us," recalled Jane. "They weren't when John was growing up, but they made up for it with us. Especially Father. He was very strict." Jane would stay up late chatting with her daughters or friends. They'd sit in her bedroom, with the great animal-skin rugs and noble dog, Fidac, to keep watch, and talk the night away. Jane, with her deep voice—the voice that was full of money—would sit on her bed and take long, almost regal, drags from her cigarette holder as she told stories to her girls.

Frank, usually in bed hours before the rest, would rise early and go for a horseback ride. He'd ride again in the afternoons before cocktails and dinner. "I loved to go riding with Father, but it was always the same—he didn't even know I was along," said Jane. "His mind was on something else and he didn't talk. Finally, I'd just give up and go back to the barn and he wouldn't even miss me. He'd be thinking about business. But I kept on riding with him. He loved it so much. We rode every day we were out at the ranch."

Jane arranged for the two Seaton sisters—Elisabeth and Elise, the magician's twins—to give Mary Frank and Jane dancing lessons during the summer vacations. A driver picked up the Seatons at their home in Bartlesville and brought them out to Woolaroc for the Phillips girls' daily lessons.

Elise was still aware of Uncle Frank's fondness for her and was always on her guard. One night, after a party at the ranch, Frank caught a ride back to town with the Seatons, who had driven out to Woolaroc in their father's Buick. Frank insisted that Elise sit in the back seat with him while Elisabeth drove and chatted with a Tulsa banker up front. It was a harrowing ride for Elise. She had to wrestle Frank off all the way from the ranch until they dropped him off at the town house.

Although he always paid close attention to the ladies—especially pretty ones—Frank still didn't hold a candle to the antics of his son, whose womanizing increased in direct proportion to his drinking.

"There were always stories about John Phillips and the ladies," said Tom Sears, a caddy at Hillcrest Country Club from 1928 through 1933. "The caddies would hear all the locker-room talk. The members would get in there and start drinking and we'd wait around to collect our tips and we'd hear all these wild stories. A lot of them were about the Phillipses. Frank had a real reputation for liking women, and there was even talk about Aunt Jane. But I think much of that was just jealous talk. Now, John Phillips was another question. We knew he had some real problems. Most of the time I saw him he was dazed. He'd strut around the club, looking like he was in a trance."

The alcoholic antics—skirt chasing, drinking sessions, gambling— were all taking their toll on John's marriage to Mildred. Frank was also growing more disgusted with his son's behavior. By 1929, John's alcohol problem was severe. He found plenty of company for his bouts with some of the other Halcyons. Art Goebel, under contract with Frank to barnstorm the Phillips marketing territory with his skywriting plane, also served as a drinking companion for John.

One of their more infamous escapades occurred when John was flying back to Bartlesville from Dallas with Goebel in a two-seat, open-cockpit airplane. As they flew over the Red River, Goebel and his passenger developed a thirst and broke out a fresh bottle of whiskey to help wash some of the dust from their throats. They passed the bottle back and forth, and by the time they approached Tulsa, neither of them was feeling any pain. Just on the other side of Tulsa, they spied a herd of goats grazing in a field, and in a flash, the plane dove from the clouds and landed in the pasture. Goebel leaped from the plane, ran into the midst of the bewildered goats, grabbed one, and threw it into John's arms in the rear cockpit. Then they were off again in a cloud of smoke and dust, John grasping their reluctant cargo with all the strength he could muster and still hold the bottle.

Thinking some of the whiskey might calm the goat, they poured generous swigs of the stuff down the animal's throat, and by the time Goebel made his rather wobbly landing at Bartlesville, all three of them were thoroughly soused. The sun was setting when John, Goebel, and the pickled goat piled into an automobile left waiting for them at the airport. They drove directly to the residence of O. K. Wing, where John remembered Obie and his wife, Oral, were hosting a fancy dinner party. The auto screeched to a halt at the curb just as the Wings began seating their twenty guests at the dining-room table. Included among those invited was Mildred Phillips.

The china and silver sparkled in the light of the blazing candelabra, and Obie, his glass raised high, was about to deliver a toast, when the doors swung open and in burst Goebel and John bearing the bleating, bellowing, burping billy goat, which they proceeded to dump square in the center of the dining-room table.

Dinner was over before it could even be served. Platters of food, broken glass, and shrieking guests went every which way. Mildred was so embarrassed she felt like plunging into the Caney River. She was also angry, very angry, but as she had done for years, Mildred eventually gave in and forgave her wayward husband and the reckless, but repentant, Goebel.

Goebel injured his knee in an aircraft accident in Kansas City a few weeks later, and when he was able to hobble from the hospital, he promptly took a taxi to the airport and chartered a plane to Bartlesville to be close to his friends. Goebel spent six weeks convalescing in the children's nursery at the John Phillips residence on Cherokee.

Mildred tended to the crippled pilot while his leg mended, and Goebel, who had a schoolboy crush on Mildred, whiled away the hours entertaining young Betty and Johnny with stories of his flying exploits. As soon as his injuries were healed, Goebel departed. Mildred, who was pregnant, had plans to put the nursery to better use than housing a handsome rascal whose wings were temporarily clipped.

On December 23, 1929—a cold Sunday morning—Mildred gave birth to twin sons at St. John Hospital in Tulsa. She and John decided to name the boys for their grandfathers—the late Robert Beattie and Frank Phillips. The babies weighed only between four and five pounds each at birth, so the physicians attending Mildred told her and John that the boys would have to spend at least a month in a hospital incubator. When news of the twins' arrival reached Bartlesville, the family was delighted. Especially Frank. There were twins in the clan once more—Phillips twins, and one of them bore his name.

Frank's joy lasted but a couple of days. On Christmas Eve, his namesake—Frank Phillips II—tiny and weak, died. Frank went to Tulsa to see his daughter-in-law and returned to Bartlesville that evening with his dead grandson. Jane, John, Bertha Gibson, and Mildred's mother stayed with Mildred and the surviving twin.

Frank hosted a dinner on Christmas Day at the town house for part of L.E.'s family, Betty and Johnny, and the two foster daughters. It was a quiet meal, and that afternoon, instead of singing around the piano or calling on friends and neighbors, Frank faced a solemn chore at the White Rose Cemetery. Back in New York, Fern faithfully recorded Frank's movements in his daily business diary: "Mr. Phillips took baby Frank II corpse to cemetery, where simple service was held in the mausoleum." At White Rose, a chill wind stirred the cedars and the naked branches of the catalpa trees. After the brief service, Frank, wrapped in his overcoat and grief, read the inscriptions chiseled on the front of the mausoleum:

> "There is no death! What seems so is transition. This life of mortal breath is but a suburb of the life elysian whose portal we call death."
>
> —Longfellow

"Some evening when the sky is gold I'll follow day into the west nor pause, nor heed, till I behold the happy, happy hills of rest."

—Paine

Frank returned to Tulsa the next morning because baby Robert, or Bobby as the family called him, appeared to be weakening. By noon, when he was assured his grandson was out of danger, Frank came back to Bartlesville and met for several hours with his security chief and two Tulsa police officers. Frank was very conscious of his family's high profile and didn't want to take any chances. His namesake was gone, but Frank was determined that no kidnapper looking for a fortune in ransom money would snatch Bobby from the hospital nursery.

Between Christmas and New Year's, Frank stayed busy with duties at the office, executive committee meetings, and year-end strategy sessions with L.E., Kane, Wing, Alexander, Parker, and others. Work was always the best therapy for Frank, and now more than ever it helped take away the sting of the loss of his grandson.

On New Year's Eve, Frank reached the office by midmorning, in time for several conferences and a final executive committee meeting—the last of the year and the decade. By four o'clock he had concluded his business day and left for the ranch.

Out at Woolaroc, surrounded by family and friends, Frank nursed a strong Scotch and water carefully prepared by Dan. A chill winter wind swept across Clyde Lake, and from the windows of the lodge Frank watched the waters ruffle like a field of prairie grass in spring. Flames crackled and danced in every fireplace, and there was a tall fir dressed in bright lights and garlands. Bunches of mistletoe with fat waxy berries the color of ivory that ranch hands cut from trees near the lodge were tacked above the doors and mixed with the boughs of holly spread across the mantles.

Frank called for another whiskey and reviewed in his mind events and situations from the past decade. It was a time for reflection—a bittersweet moment for Frank Phillips and all the world. The Roaring Twenties were about to vanish. There was much to consider and recall. A new decade, a new age was about to dawn.

Dan brought a fresh drink and Frank took a deep sip and saved the rest for a final toast. The whiskey and the wind and the smell of evergreen were good medicine. Frank kissed Jane and his daughters and bid everyone a Happy New Year and then he quietly disappeared and went upstairs to his room and the narrow cowboy bunk. He finished his drink in the darkness listening to the wind, and then he closed his eyes and waited for his thoughts to turn into dreams.

CHAPTER TWENTY-FOUR

Frank would have given anything if he could have awakened from his New Year's sleep and found it was all a bad dream and that the forbidding events that would forever scar 1929 and tattoo painful memories in the mind of America had never happened. He would have given up the town house on Cherokee Avenue. The servants. The shiny limousines. The suite at the Ambassador. Maybe he would have even given up Woolaroc, his beloved sanctuary in the Osage.

Instead, he opened his eyes, reached for his spectacles, and discovered that 1930 and the Great Depression had dawned stark and sober and cold as ice. Frank felt like the empty whiskey glass on his nightstand. He reviewed the events of 1929 in his mind and he knew that for much of the nation the good times were over. The Jazz Age was lost. That was the reality that came shrouded in plain brown paper. No fancy wrap and no silk ribbon. The confetti and the noisemakers were gone. The champagne had lost its bubble.

Frank didn't even imagine that before 1930 was over he would have drafted a letter of resignation for the Phillips board of directors. But there was much that lay ahead that no one, not even Uncle Frank, could anticipate.

Over the course of this new decade the face of the nation would change. Smiles would turn to frowns; friends would become enemies. In the 1930s suicide rates increased, Hoovervilles sprang up, and breadlines formed. It was a time for apple peddlers and dust storms that rolled across the nation's heartland like black smoke. Odd jobs, hard times, and hungry years followed. Crops failed and so did people. Businesses went belly-up. Life became hard to swallow.

Frank felt the nation's economy begin to crack during the summer of 1929 when steel and automobile production declined. Soon the whole economy started to show signs of weakening.

As early as January 3, 1929, the year of "the St. Valentine's Day Massacre," Wyatt Earp's death, and Herbert Hoover's inauguration, Frank, wary of the rising market yet wanting to believe the political slogan "Four More Years of Prosperity," called for control of the oil industry through legislation designed to curb excessive production and compel cooperation among the major firms with regard to oil prices.

In early January, a *Wall Street News* article reported:

> Frank Phillips, president of Phillips Petroleum Co., in a public statement, says that voluntary cooperation in the petroleum industry is too low to keep pace with the problems of oversupply, and the trade needs the assistance of appropriate laws which it can invoke at the proper time to prevent crises. He further declared that voluntary cooperation permits a small minority to prevent the orderly development of pools which the majority desires, and the industry clearly needs the assistance of legislation which can be invoked by a majority of producers interested in a pool to require full cooperation by all . . .

The January 1929 issue of *National Petroleum News* published more of Frank's thoughts about the state of his beloved industry. Frank was quoted as saying:

> The oil industry has been giving increasing attention during the past year to conservation and the economic problems arising from over-production of raw material and over-expansion of refining and marketing facilities. Some progress has been made; however, much remains to be accomplished. . . . Although the character of legislation and regulation now applicable by federal and state governments to railroads, other public utilities, or banks, would not be in any respect appropriate to the oil industry, it is interesting to note—in view of the widespread and natural prejudice against government intervention in industry—that such regulation does not necessarily mean injury to the business in-

volved. On the contrary, the industries referred to have been
enjoying a more stable prosperity for some years, and their
securities are in high favor with the investing public.

Again, Frank was giving off mixed signals. His published remarks
advocated one stance while his actions were totally contrary. One day he
declared there was far too much oil production and too many refineries
and retail outlets, and the next day his company would buy more leases,
drill new wells, expand the firm's refining capacity, and build or acquire
more gas stations.

A month after Frank's call for restraint and conservation in the industry, Phillips Petroleum acquired the Wilhoit Oil Company, headquartered
in Springfield, Missouri, and its chain of forty-six service stations and
thirty bulk gasoline stations located in thirty-one cities. By the end of
1929, Phillips would have also acquired the Winters Oil Company, Kansas
City; State Oil Company, Lincoln; Morrison Oil Company, Denver; Han-
cock Oil Company, Minneapolis, as well as a number of smaller concerns.
The Borger refinery was enlarged, a new refinery at Kansas City was
authorized, and Phillips 66 fuel was available at 6,750 outlets across the
nation, as compared to 1,800 outlets at the end of 1928. Philfuels Company, the wholly owned subsidiary, was becoming a fount of considerable
revenue and a stable source of income.

Frank claimed the fierce competition Phillips faced was the cause of
the company's continued expansion and growth. His critics, of course,
disagreed and said Phillips Petroleum was only interested in cornering the
market and regulating prices to suit itself.

Besides growing in corporate size and strength, Phillips was also adding more employees. Some went to work in the Texas panhandle, some in
the Oklahoma and midcontinental oil patch, and still others in offices
spread from Bartlesville to New York. Phillips also continued to draw
many of their best young engineers and execs from the University of Kansas, the school that yielded Boots Adams and Paul Endacott, two of
Frank's most talented upper management prospects. Another University
of Kansas product attracted to Bartlesville and Phillips Petroleum was
Stanley Learned, a bright engineer who began earning wages in a paint and
wallpaper store in his hometown of Lawrence, Kansas, at the age of
twelve. In 1924, Learned was hired as a surveyor for the Phillips engineering department. By the close of the decade he had proven his worth and

paid his dues by trudging with the surveying crews over the prairies of Oklahoma, Texas, and Kansas in search of new oil lands.

Yet another University of Kansas recruit was William Wayne Keeler, brother-in-law of Boots Adams and the grandson of George Keeler and Nelson Carr, two prominent Bartlesville pioneers. Bill Keeler became a full-time Phillips employee in 1929 after working summers for the company in the Kansas City refinery. Young Keeler, like his sister Blanche Adams, had been raised with Cherokee as his first language because of his Indian grandmothers. Destined to follow in the footsteps of his grandfather Keeler, who helped bring in the Nellie Johnstone—Oklahoma's first commercial oil well—Bill Keeler, valedictorian of his high school graduating class, was awarded a Harry E. Sinclair science scholarship to study engineering at the University of Kansas. But when Sinclair Oil became embroiled in the Teapot Dome scandal, the scholarship money was cut off and Keeler, like Boots Adams had done several years before, dropped out of college and went to work for Uncle Frank.

Other bright young men and more and more women were also joining Phillips Petroleum. "We loved working for Uncle Frank," recalled Izola Moore, who started as a secretary in the engineering department in 1929 for $75 a month but almost immediately took a pay cut to $68 when the economy began to sour. "I didn't mind taking the cut so much. We were just glad to have jobs to go to every morning. I'd come down to the Phillips Building and there would be Uncle Frank arriving in his chauffeur-driven automobile. He'd talk to us in the elevator and say, 'Oh, you're one of my girls,' and he'd squeeze us a little bit. He was friendly and everybody loved him." As bright new faces continued to appear at Phillips Petroleum, some of the old hands left. Or at least they began to relinquish some of their authority.

L.E., Frank's brother, who helped him found the company and who stuck with him from the beginning in both the banking business and the oil patch, scaled down his responsibilities and gave up the number two spot in the Phillips chain of command. In 1929 L.E.'s health became a concern, and after much discussion within the upper echelon of the Phillips corporate ranks, it was decided he should spend more time with his banking duties in Bartlesville and as a director of the Federal Reserve Bank in Kansas City. When General Patrick J. Hurley, an old friend of both Frank and L.E. and the newly appointed Secretary of War in Hoover's Cabinet, offered him the post of governor-general of the Philippines, L.E. politely declined. He needed to take it easy and not take on any strenuous jobs.

Frequent excursions abroad with Node, and fishing trips, from the Sea of Galilee to the West Indies, were the best prescription.

On April 16, 1929, just a day after a special session of Congress met at Hoover's request to deal with the mounting problems of the nation's economy, the Phillips stockholders and directors gathered in Bartlesville for their annual meeting. Following a brief discussion, the board decided to "retire" L.E. from his post as vice president and general manager and elect him as the first chairman of the executive committee. They appointed Clyde Alexander to take his place.

The scuttlebutt during gin-and-poker sessions on lazy afternoons in the locker room at Hillcrest Country Club and around the office water coolers was that if Alexander continued to play his cards right, he'd wind up the second president of Phillips Petroleum Company when Frank decided to throw in the towel and retire to the ranch. It was as clear as bootleg gin to everyone who was privy to the gossip—from caddy to roughneck—that nobody else, especially any of the Phillips family offspring, stood in Alexander's path to the presidency.

For the time being, though, Frank Phillips was still very much in command. There was no question about that. As 1929 progressed and L.E. learned to adjust to his new role in the company, Frank became involved with other changes and departures and shifts both within the corporate structure and beyond the scope of Phillips Petroleum.

In June, just as Hoover signed the Agricultural Marketing Act in an effort to bring financial relief to American farmers, Frank—by investing through Frank Phillips and Associates, a private corporation—became part owner of a magnificent old New York institution—the Waldorf-Astoria. Considered to be one of the grand hotels of the world, the Waldorf-Astoria was located on fashionable Fifth Avenue and was destined to be razed in order to make way for the Empire State Building. The Waldorf was purchased by a corporation called Empire State, Inc., for $16 million, a price which included both property and the hotel building. Empire State, Inc., was formed by John Jacob Raskob, creator of General Motors; Coleman and Pierre S. Du Pont of the Du Pont family fame and kin to Eugene Du Pont, a director of Phillips Petroleum; Louis G. Kaufman, one of Frank's New York friends; and Ellis P. Earle, president of Nipissing Mining Company, New York, and another Phillips Petroleum director. Alfred E. Smith, former governor of New York and the Democratic presidential candidate who had been soundly beaten by Hoover in the 1928 election, was picked to head the corporation.

Frank inspected the hotel in July and paid particular attention to four handsome chandeliers which hung in the Tap Room. When the hotel was gutted in preparation for demolition, Frank was given the ornate lights. He promptly shipped them to Oklahoma and had them hung from the ceiling of the Great Hall at the Woolaroc lodge.

On October 1, 1929, the first trucks filled with demolition teams rolled up to the Waldorf. Construction of the Empire State Building commenced on St. Patrick's Day 1930 and proceeded at a breathtaking pace. The framework rose at a rate of four and one-half stories a week. In only one year and forty-five days, including Sundays and holidays, the world's tallest building was completed. The great limestone and steel structure rose 1,250 feet above Manhattan and contained 10 million bricks. Total construction cost was more than $40 million. On May 1, 1931, the Empire State Building was formally opened by Al Smith. President Hoover pressed a button in Washington, D.C., which turned on the building lights, and all the world could see the elegant Art Deco finger which jutted into the clouds and gave hope to the jobless souls selling apples in the huge building's shadow.

The new Waldorf-Astoria, another Art Deco treasure of limestone and light-colored brick with a granite base, was built in 1931 on Park Avenue with only St. Bartholomew's Church separating it from the Ambassador, Frank's New York residence.

Back in Oklahoma, the four fancy chandeliers Frank first spied dangling in the old Waldorf Tap Room that summer of 1929 burned brightly at Woolaroc, where they illuminated family gatherings, business meetings, cocktail parties, poker games, and the comings and goings of countless guests and associates of the Phillipses. Whenever anyone asked about the huge light fixtures, Frank liked to joke that they were all he ever got out of his investment. He called them "million-dollar chandeliers" and laughed while explaining that they were "the most expensive chandeliers in the country."

During that summer of 1929, as final preparations were still being made to erect the world's tallest building, Frank was doing more than running his oil company and picking out light fixtures for the ranch.

Art Goebel and the *Woolaroc* were back in the public eye once more. While promoting Phillips Petroleum for Frank by skywriting across the skies of the company's marketing territory, Goebel attempted to establish some transcontinental flight speed records with the *Woolaroc*. The airplane was modified by the Travel Air mechanics back in Wichita, and Goebel

buzzed off in search of more fame and glory. Instead, the aviator found
that the modifications severely impaired his field of vision to the front,
which not only made takeoffs difficult but generally made the *Woolaroc*
dangerous to fly. He encountered several problems and had a few close
calls, including a near-collision with a water tower in the foggy mist near
St. Louis. Goebel and Frank decided to have the plane restored to its
original appearance and put on public display at the F.P. Ranch. But first
Goebel and his trusty airplane would make one last circuit—a farewell
flight.

In a letter dated July 20, 1929, which was published in a booklet
called "Over Land and Sea: The Story of the *Woolaroc*," Frank explained
the rationale behind the last flight of the famous monoplane:

> My interest in aviation has always been prompted by the
> tremendous commercial advantages which flying permits,
> and it was for the purpose of aiding in a small way with the
> development of "air-consciousness" in the minds of the
> American public that I sponsored Colonel Arthur Goebel
> on his Hawaiian flight.
>
> It is not necessary to more than mention the part that
> flying has come to play in our lives. Trade and commerce
> have adapted air travel as an integral part of its daily opera-
> tions. Personal transportation by air is making great strides
> forward. This metamorphosis is the more interesting be-
> cause it is taking place in our own lives—before our very
> eyes.
>
> This intense interest is manifest by 403 requests which
> have come to me for a farewell tour of the *Woolaroc* and a
> closer contact with Arthur Goebel, its intrepid pilot, since
> the announcement that the plane would be permanently re-
> tired and preserved in a hangar on Woolaroc Ranch near
> Bartlesville, Oklahoma.
>
> In answer to these requests, Colonel Goebel has kindly
> consented to fly the *Woolaroc* on a last flight to various
> population centers of the United States. In this way, thou-
> sands of people will be enabled to see and study this now
> famous trans-oceanic airplane.
>
> At the close of the tour the *Woolaroc*, with all of its origi-
> nal equipment, will come to rest in a permanent, fireproof

hangar where it will be faithfully preserved and protected for the benefit of future generations.

Joining with those who will have the opportunity to see the *Woolaroc* on its last journey, permit me to express my appreciation of and gratitude to Colonel Arthur C. Goebel in making the tour possible—and to Art Goebel, the pilot, for his historical achievement with the ship.

News stories, including the August issue of *U.S. Air Services Magazine,* described the farewell tour and mentioned the "permanent hangar of stone" Frank was building for the airplane, the first public acknowledgment of what would eventually become the Woolaroc Museum.

As Goebel winged from city to city in the *Woolaroc,* and plans were being finalized for the Empire State Building and for the stone hangar for the airplane, Frank and his lieutenants watched the stock market prices rise. And rise. And rise. By September the stock price index peaked at 216, the climax of a three-year bull market. Then the market dropped off sharply, surged up again, dropped once more, and this time did not edge back up. By the end of September, the market sagged lower and lower. And lower.

Despite slight fluctuations, most investors didn't appear to be too concerned. There was a chorus of voices from all the eternal optimists. Convinced that American business and industry was too powerful and diverse to be adversely affected by the stock market, they assured everyone that there was nothing to worry about and the prices were only enduring a temporary setback. In New York, Frank continued to watch the market like a hawk, and every day he received the latest intelligence reports from Fern and the directors during luncheons at the Recess, Bankers, and Lotus clubs or at late-night conferences conducted in the Phillips Ambassador suite.

From all appearances Phillips Petroleum was enjoying its best year ever. Annual motor gasoline sales had jumped from 10 million gallons in 1928 to an amazing 100 million by the end of 1929. The company was also turning out 800,000 gallons of natural gas liquids each day. That was far more than any of their competitors and more than even Phillips could use in its own refinery and marketing operations, which meant that at least two-thirds of the output could be sold to other companies. Even more important, thanks to the successes of the refinery and retail marketing

operations, the company's earnings were soaring to a record $13.2 million
—more than twice the almost $6 million Phillips reported in 1928.

In October the clouds forming over Wall Street turned menacing and
the cocky talk around financial circles changed to whispers of concern. An
unexpected selling spree sent the stock market prices plummeting, and
although the prices temporarily recovered, Frank spent most of his time
visiting with key analysts, who continued to reassure him that the bottom
had been reached and full recovery was near.

"I know of nothing fundamentally wrong with the stock market or
with the underlying business and credit structure," said a confident
Charles E. Mitchell, president of New York's National City Bank, on
Tuesday, October 22, upon his return from Europe. But not everyone
agreed. Just days before, there were heavy withdrawals of capital from the
United States as Great Britain boosted its interest rate to 6.5 percent.

The same afternoon that Mitchell offered his statement of support,
Frank studied the situation and felt comfortable returning to Oklahoma in
order to preside at the third annual Cow Thieves and Outlaws Reunion
and attend his niece's wedding in Bartlesville. After several conferences,
including a session with Bill Skelly, Frank and Dan caught a train bound
for home by way of St. Louis, where Frank visited with some of his directors and picked up George Vierheller, the famed zoologist, who gave
Frank advice about his wild animals at the ranch and was an invited guest
to his outlaws picnic at Woolaroc.

While Frank was in St. Louis, the steady decline in stock market
prices that had started since the peak back in September continued and
there were signs of panic at the New York Stock Exchange.

On Thursday, Frank and his party arrived in Bartlesville in time to
learn that there was a collapse of stock prices at the exchange. Almost
13,000,000 shares were sold on October 24, and big bankers and wealthy
investors such as Mitchell from National City Bank, J. P. Morgan & Co.,
Guaranty Trust, Chase National, and John D. Rockefeller himself tried to
buoy up the market by buying. But even the gods of Wall Street could not
turn around the collapse and check the market's fall. Throughout the day
Frank broke away from his guests at the cow thieves party to go to the
lodge, where he kept the telephone lines open to the New York office and
Fern Butler, who was monitoring the situation minute by minute.

In New York, there were wild rumors about troops being called out
and speculators committing suicide. Over the next few days it became

apparent that this time the big bankers had not been able to stop the avalanche.

Nonetheless, life went on. Frank attended Martha Jane and "Twink" Starr's wedding on Saturday, October 26, and afterward went with Jane and the rest of the family to the reception at L.E.'s house on Cherokee. He spent Sunday at the ranch resting for the week to come. Frank had a feeling it would be one of the most difficult times of his life. He was correct.

Monday came and went with little or no change, but then the next day, October 29, 1929—a day known forever as "Black Tuesday"—the New York stock market crashed with a fury heard around the world.

After weeks of watching prices plunge there was a panic sale of more than sixteen million shares of stock, all sold at declining prices. Many stocks closed at less than half the value they showed that morning. It was without a doubt the most catastrophic day in the market's 112-year history and served as the forerunner for the Great Depression. From this day forward all of America would stop growing until December 7, 1941, when the bombing of Pearl Harbor would mobilize U.S. resources.

Frank spent the morning of October 29 in emergency conference with the Phillips executive committee. The agenda was brief. Frank explained that although the Phillips Petroleum stock continued to be solid and retained its value, the "little fellows outside do not know that." If the stockholders joined in the panic and sold their Phillips stock, Frank believed they would one day be sorry. He didn't want to see that happen and neither did the executive committee.

After he heard from L.E., Alexander, Wing, and the others, Frank laid out a plan of action. He proposed that each of the six thousand Phillips Petroleum stockholders of record immediately receive an official communication from the company which would offer them reassurance about Phillips and the stock. The executive committee agreed.

Sixteen telegraph operators were recruited from Tulsa and Wichita and given their orders. They were told to get out the word and get it out as quickly as possible. "Do not sell . . . hold on to your Phillips Petroleum stock . . . the value is still there." The telegraph operators, under the guidance of C. O. "Pop" Shirley, Phillips communications supervisor, worked nonstop from eight o'clock in the evening until after four o'clock the next morning. They accomplished their mission. Every stockholder received a wire and the anticipated massive dumping of Phillips stock was avoided.

While the telegrams were going out across the nation, Frank went to Tulsa for a quick huddle with Waite and then caught a train bound for New York. Frank wanted to get back in the thick of things.

One newspaper reporter learned of Phillips Petroleum's decision to dispatch wires to thousands of stockholders and summed up the maneuver this way: "Very human sort of rich men are the Phillips brothers. There are many millionaires in this world but only a few of them deserve to be. Frank and L. E. Phillips we number among them."

At first many thought the Crash of '29 and the Depression that followed would not linger but would disappear after a while. The nation had lived through stock market panics and depressions before. But this one was different and lasted longer than any this country had ever seen. The days turned to weeks and then months and finally years and the Depression hung on like a bad cough.

Before 1930 was very far along more than 4 million Americans were out of work, and by December 1931 that number had climbed to an astonishing 13.5 million, or one-third of the nation's labor force. But in the first few weeks of the decade, while the economy continued to sink, commodity prices dipped, and the national income collapsed, Frank went about his business as usual in New York and Chicago and Bartlesville. And in Washington, D.C., where he and Jane stayed at the Mayflower Hotel and on January 21, 1930, dined at Secretary of War and Mrs. Hurley's residence along with President and Mrs. Hoover.

Frank genuinely liked Hoover, a native Iowan who was born the son of a Quaker merchant and blacksmith in 1874, less than a month after the hordes of grasshoppers descended on the Phillips homestead in Nebraska and prompted the family's retreat back to Iowa. Like Frank, a self-made man, Hoover was orphaned when he was ten years old and spent several boyhood summers at Pawhuska, where his uncle, Major Laban J. Miles, was appointed Osage agent in 1878. Some of the old-timers around Pawhuska remembered the little orphan boy who liked collecting rocks and exploring the Osage Hills, a hobby that blossomed into his mining and engineering career.

Dinner conversation around the Hurleys' table centered on an important international naval conference held that afternoon in London, where the United States, Great Britain, and Japan agreed on sizes, ratios, and schedules for expanding their fleets, while Italy and France turned down the major provisions. Hoover also confided to his devout Republican friends that in less than two weeks he planned to name Charles Evans

Hughes to succeed William Howard Taft as Chief Justice of the Supreme Court, a move that the Senate would confirm before Valentine's Day.

Dinners with the President and cabinet officers, or with a growing circle of celebrity acquaintances in New York and across the country, meant a lot to Frank and Jane. The mix of social and business functions provided not only a change of pace but important opportunities for picking up vital information and lobbying on behalf of the oil industry.

There were also periodic trips back to Bartlesville, to check on operations in the headquarters office and tend to family matters, such as Jane and Frank's wedding anniversary on February 18, 1930. That afternoon, following a series of executive meetings, Frank glanced at his watch and quickly ordered his car to appear at the front door of the Phillips Building. He had to reach the town house at the stroke of five o'clock, the exact time he and Jane were married thirty-three years before and an hour Frank always made it a point to spend with Jane on each anniversary.

In Oklahoma it was still difficult to tell that the nation was bogged down in a debilitating economic quagmire. Four days before Frank and Jane marked their anniversary, L.E. and Node, in a show of optimism for the company and the country, embarked from San Francisco, bound for Manila and a three-month holiday in the Philippines.

While L.E. was away, Frank, facing mounting competition from companies that maintained refineries in the Phillips marketing territory, launched the most ambitious single project Phillips Petroleum had ever undertaken.

Frank knew that the competitive plants were fed crude through their own pipelines running from Oklahoma, Kansas, and Texas. But the Phillips primary refinery was in the Texas panhandle, far from the major retail centers of Kansas City, St. Louis, and Chicago. There was no way Frank could continue to ship refined gasoline and meet competitive prices.

The king of the independents found himself in another pinch, but as usual he wasted no time in finding a solution.

To supply the company's products to the midcontinent, it became evident that Phillips needed a pipeline. Frank brought his top executives together and decided to build a gasoline pipeline from Borger, Texas, to East St. Louis, Illinois. That meant laying 681 miles of pipe—no small feat. Gasoline had never been transported any great distance by pipeline, and several reputable engineering firms told Frank that what he proposed couldn't be done profitably because of a waste of gas through evaporation. Frank told them all to go to hell and went to work.

First, he formed the Phillips Pipe Line Company to construct and operate the line. Clyde Alexander was chosen as president. Stanley Learned, one of Frank's young and aggressive Kansans, still in his twenties, was selected to supervise construction. Phillips engineers designed a wasteless welded eight-inch line, novel in design and all part of a closed system which would not expose the gasoline to air from the refinery in Texas until it reached the storage tanks on the banks of the Mississippi River.

Despite the state of the nation's economy and the criticism he received from some engineers, Frank was determined his pipe project would be completed. Survey crews ventured out and Frank searched for a way to finance the revolutionary new system which would alter the shape of gasoline transportation and help Phillips Petroleum survive the lean years of the Depression.

The morning of April 22, while he was hunting for the financing, waiting for the survey teams to complete their work and for the pipe to be designed and delivered, Frank ruled the roost at the annual directors and stockholders meeting in Bartlesville. Immediately afterward, he flew by company airplane to Ponca City, where the entire town and hundreds of visitors and VIP guests turned out to celebrate the dedication of the Pioneer Woman statue, a massive seventeen-foot, six-ton bronze of a sunbonneted pioneer woman and her son.

The statue, created by Bryant Baker, was meant to be a memorial to the gutsy women who dared journey into what became Oklahoma and settled America's last frontier. The idea for the monument came from E. W. Marland, the prosperous oil tycoon, who for many years had been a close friend of Frank Phillips.

Marland was the fabulously wealthy wildcatter who built a celebrated $2.5 million mansion with fifteen bathrooms, three complete kitchens, and Waterford crystal chandeliers to light the ballroom. Plans for the mansion had been drawn up four years earlier, in 1926, about the same time Marland first came up with the idea for the monument to pioneer women. He had announced he was going to finance an international competition and would select an entry from one of the country's twelve best-known sculptors.

Marland's wife, Mary Virginia Collins Marland, died in 1926. Shortly after his wife's death, Marland raised some eyebrows around Ponca City when he began showing—for the first time in public—amorous interest in Lydie Roberts, his late wife's niece, whom Marland had legally adopted, along with the girl's brother, when Lydie was ten years old. Soon the adoption was annulled, and in 1928 Marland and Lydie married. He was fifty-four and the new Mrs. Marland was twenty-eight. The marriage shocked and scandalized even some of his upper-crust friends.

A slender, charming young woman with dark brown hair, touched with an auburn cast, and a dark, almost olive complexion, Lydie Marland was shy—painfully shy. But her husband was just the opposite. He was so successful at finding crude oil and so adept at spending money it was said Marland had a "nose for oil and the luck of the devil."

But shortly after his marriage to the retiring Miss Lydie, Marland's devil's luck really ran out. The House of Morgan brought him down.

John Pierpont Morgan, Jr., head of the Morgan banking empire and the son of the legendary financier, J. P. Morgan, who died in the Phillipses' suite at Rome's Grand Hotel in 1913 just short days before Frank and Jane arrived, had the distinction of bringing Marland's oil baron career to an abrupt end.

Marland's extravagant lifestyle and casual management approach made his oil company especially vulnerable. The House of Morgan lent great sums of money to Marland—around $30 million—to feed his expensive habits and fuel his refinery and expand the retail marketing operations. As they gave him more funding, the bankers also gained control of Marland's board of directors. They began to tell Marland what to do and what not to do. He didn't like what was happening, but he had little choice. When his company showed a loss, Marland shrugged and said it was nothing to fret about. He felt a good part of the blame fell on his partners, the eastern establishment bankers whose loans he used to expand his oil operations into California, Mexico, and Canada. But when there were additional losses in 1928, Marland's discomfort with the bankers turned to worry. Then the price of oil dropped and the overproduction situation that Frank Phillips had preached about in the newspapers was reaching the critical stage by the close of the year. Things couldn't get much worse. But they did. The great Oklahoma City field was discovered and there was even more oil. The glut of crude was too great for the refineries to handle.

All the while, J. P. Morgan was watching from the wings on Wall Street, biding his time and waiting for just the right moment to strike. That happened in May 1928, when Marland came to New York with his executive committee and learned that the bankers had called his hand. They were all against him. He was told in no uncertain terms that his oil firm needed a better business mind in the president's chair. He was expendable. Marland's flamboyant and carefree style of management was no longer appreciated or needed. Morgan, the "Wolf of Wall Street" as Marland called him, had won. He ousted Marland by taking control of the board of directors. Marland's operation was swallowed up by Morgan's smaller Colorado subsidiary, the Continental Oil Company. Morgan named the new company which was created from the merger—Conoco.

After he lost control of the firm, even Marland's board was forced out. His recommendations for a new president went unheeded, and the bankers picked Dan Moran, a tough executive from the Texas Company, as the president of the new company. Then the bankers made Marland an offer. He could remain as a board member and chairman of the company and draw a $75,000 annual salary under the condition that he leave Ponca City, since his presence could undermine Morgan's organization.

Marland told the bankers to stuff their offer where the sun never shines. He could never bring himself to leave the town where he had built his storybook empire. Marland's polo team was disbanded and he was forced to leave his baronial mansion. He and Lydie moved into a small gatehouse on the grounds, where he plotted his comeback, held on to his eternal optimism, and dreamed of being a multimillionaire once more. But the stock market crash of 1929 snuffed out those hopes and crushed plans for the creation of a new company.

In a few years he would resurface with a political career, but for now there was one last obligation to fulfill—the dedication of the huge bronze statue which was Marland's brainchild when he was still flush. Now the oil man was broke, but the completion of the project was secretly paid off by Lew Wentz, another Ponca City oil tycoon and, ironically, a bitter Marland rival.

So on that balmy April afternoon in 1930, surrounded by loyal friends, curiosity seekers, and guests from far and wide, Marland played out the cards dealt years before and honored his promise to Oklahoma by dedicating the Pioneer Woman monument.

Governor William J. Holloway declared a state holiday and showed up. T. E. Braniff presided at the dedication, and Frank served on the program committee. Will Rogers was on hand to add some levity to the solemn occasion. Public donations paid for the cost of the ceremonies, and bus and train lines offered special fares to out-of-towners interested in seeing the big statue unveiled. Many of the survivors from the land runs of 1889 and 1893 and their families came, dressed in pioneer garb. Some had traveled to Ponca City by wagon or horseback, just to ride in the parade and see the big statue unveiled. All in all, more than 40,000 people showed up.

President Hoover kicked off the ceremonies with a brief speech which was broadcast live from the White House. Secretary of War Hurley's remarks were also broadcast across the nation from Washington.

"Will Rogers was the principal speaker and the crowd got a big kick out of it," recalled Marion Cracraft, former oil editor for the Tulsa *Tribune*. "At the end of the actual dedication ceremony, there was a kind of tour of the mansion. I remember going out to the back of the house and there was no one there except Will Rogers on the steps leading down to the swimming pool and I remarked to Will, 'Well, this has been quite a day.' And he said, 'She sure has, she sure has.' "

Out by the huge bronze statue, a crowd of politicians, weather-beaten old Indian Territory pioneers, and Marland's last and best friends gathered. Over the din of conversation and laughter, Frank chatted with Marland and the bashful Lydie. Wearing a top hat and cutaway, Marland was smiling and seemed pleased by the large turnout for the dedication ceremony. Frank squeezed Marland's hand and wished him well. Then Frank gathered his entourage and left for the airport to return to Bartlesville. All the way home he thought about Marland. He remembered the years of parties and polo games and Marland's excitement about his plans for the mansion and the Pioneer Woman monument, and how it had all but disappeared now.

That evening Frank left Bartlesville in his private train car for New York. Sitting in a lounge chair sipping a nightcap, Frank could still see Marland in his fancy clothes, standing by the big bronze, surrounded by thousands of people. He pictured the smile on Marland's lips and the look in his eyes as though that single moment was all that was left and somehow it had to be saved.

He thought about nothing but Marland as the train moved eastward, where Frank had his own battles to fight and where there were more bankers and directors and others ready and willing to swallow up any careless wildcatter who got in their way.

A shiver ran up Frank's back when he considered the Depression and the price of oil and all the money he was borrowing to make his dream of a pipeline from Texas to Illinois a reality. The train roared on through the warm spring night, and Frank promised himself he'd never let anyone—not even a Morgan or a Rockefeller—grab the company he'd built.

CHAPTER TWENTY-FIVE

In the summer of 1930, the Depression had a stranglehold on the nation, but it had not quite grabbed Bartlesville, where Frank Phillips barreled ahead like there was no tomorrow.

Talk of the proposed Phillips pipeline was still hot news when the citizens of Bartlesville further learned that Frank was again expanding his corporate offices by adding two more floors to the existing Phillips Building, as well as constructing an eight-story wing and a ten-story office tower. For good measure, and while he was in an expansion mood, Frank also decided to go ahead and tack another story on the First National Bank Building.

To make sure some of their personal wealth was spread around, the Phillipses also turned their sights toward Iowa. After two attempts failed to obtain a Carnegie Library in Creston, Jane and Frank told their hometown friends not to worry. They'd just go ahead and build them one. The Phillipses bought a building site at the southwest corner of Maple and Howard streets for $10,000 cash and presented the deed, along with another $15,000 in seed money, and promises for more contributions, to the building fund. The library became a reality, and the Creston city council was only too pleased to name the new facility the Matilda J. Gibson Memorial Library, in honor of Jane's mother.

In June 1930, in yet another dramatic gesture partially designed to bolster the morale of his employees and show them that the king of the wildcatters was still solvent, Frank shelled out a half million dollars to have the Phillipses' Cherokee Avenue town house, built in 1908, remodeled and redecorated.

Both the donation to the library in Creston and the work on the town house were paid for with Frank's own funds and had nothing to do with Phillips Petroleum. Neither did the First National Bank expansion. But not everyone knew that. Many people—especially the general public—had a difficult time separating Frank Phillips from Phillips Petroleum. To them Mr. 66 and Phillips 66 were one and the same. Sometimes even Frank had trouble separating his personal affairs and those of the company. Still, the effect was that anything he did for himself—be it redo the town house or throw a big party at Woolaroc—would undoubtedly make his company look good. What was good for Frank had to be good for Phillips Petroleum and vice versa.

With that reasoning in mind, Frank went top dollar all the way on the town-house project and selected the hottest architectual talent available to do the job. For the master design work he hired Edward Delk, the same architect who designed Waite Phillips' striking homes—Philbrook in Tulsa and Philmont near Cimarron, New Mexico. The architect's reputation was growing and Delk was quite the rage in Bartlesville, where he also designed the original Hillcrest Country Club, John Kane's residence, and La Quinta, the Spanish-style hacienda mansion built on the edge of town for H. V. Foster.

Arthur Gorman, the local contractor who designed Woolaroc, was chosen as an associate or resident architect to oversee the work and make sure that all of the plans developed by Delk were faithfully executed. The town house was expanded to a total of twenty-nine rooms, including nine baths, four service kitchens, and seven bedrooms.

During the year of renovation and revamping at the town house— when small armies of plumbers, plasterers, and wood, tile, and marble craftsmen descended on the Cherokee Avenue abode—the Phillips family retreated to the Osage and used Woolaroc as their primary Oklahoma residence.

Part of the face-lift included expanding the library by moving the entire south wall and lowering the floor to give the room more height. A new fireplace was constructed, doors with heavy leaded glass which opened to the yard were installed, and the ceiling was decorated with a large circular floral design. Paneling and the original shelves for the more than two thousand books in the Phillips collection were replaced and two secret compartments were built into the new bookcases. The doors looked just like the wall panels, so no one would suspect that the compartments, each large enough to hold one person, were even there. Inside one compartment was a telephone and in the other was an annunciator microphone which connected with the intercom system running throughout the residence. Since kidnapping, including the snatching of wealthy oil barons or their families, was in vogue, Frank figured the secret compartments would be good for family members or servants to hide in and summon help if intruders ever invaded the Phillips home.

The kitchen and servants' area was improved and a dining room for the hired help was added. All of the woodwork in the main dining room was refinished and silk damask covering was applied to the walls.

Many of the rooms on the upper floors were also altered and redecorated, including a butler's bedroom, guest rooms, baths, and a third-floor ballroom, originally designed for entertainment, which would be used by the family as a recreation center and for storage.

On the second floor, where most of the family's bedrooms were located, no cost was spared. Especially when it came to redoing Jane's boudoir and bath. She chose a French decor—very light and feminine with an intricate festooned pattern of embossed wooden moldings and divider strips and a decorative molded-plaster cornice. In the huge walk-in closet were sliding wooden trays, drawers, shoe racks, and shelves. There was also a secret cabinet and a hidden jewel safe.

But Jane's pride and joy was her elegant bath. It was marble throughout, with the entire ceiling composed of square plate-glass panels. The architect's instructions for the artist assigned to decorate the mirrored ceiling were to paint a design on the back of the glass and then silver it over the top of the painting to highlight the artwork with a mirror background. He did just that and left a design with a lattice effect entwined with blue and pink flowers—a perfect color match for the pink marble used for the interior walls. Mirrors were also installed inside cabinets and above the marble lavatory. The floor, covered with a fur rug as snow white as Jane's hair, was made of marble blocks. For an extra touch of elegance —and decadence—all the fittings and fixtures for the tub and sink were gold.

By enlarging the library downstairs, Frank's bedroom was also expanded and the walls were canvassed and covered with a painted woodland scene. He kept the same size bed he had before the remodeling—a single bunk like the one at Woolaroc, only without the ranch brands burned into the wooden frame. As was the case with Jane's bath, more attention was paid to enhancing Frank's private bathroom. It was furnished with a steam cabinet to sweat out poisons at the end of a business day or after a party; a small refrigerator and bar; an ice machine; and next to the stool was a compartment containing Frank's electric cigar lighter and, for a practical touch, a place to hang an enema bag.

A handsome barber chair, with green upholstery to match the green marble bathtub, was installed. Every morning he was in town, Frank would rise early, start the day ("to clean my eyeballs out," he'd explain) with a single jigger of Johnnie Walker Black Label Scotch—no more, no less—and then sit in his barber chair while Fred Winchester, or another local barber, shaved him, gave him a facial, and trimmed what hair he had left. Once each week a young woman would show up to give Frank a manicure, and almost every morning Doc Hammond would slip away from his house and wife, George, and come up to Frank's bedroom to give him an osteopathic treatment while Dan laid out Frank's clothes.

Not a bad way to start each day for an ex-barber from a country town in Iowa.

Most days both Frank and Jane took their breakfast—usually soft-boiled eggs, bacon, toast, orange juice, half a grapefruit, coffee, and sometimes hot cereal—in bed. Occasionally they'd dine together in the sun-room, which, except for the library, was their favorite room in the town house. When the breakfast trays were cleared, Frank read the newspapers and, if he wasn't feeling well, conducted informal business meetings propped in bed, surrounded by executives and cigar smoke, while Dan scurried about emptying ashtrays and fetching pots of hot coffee.

After her breakfast, Jane finished letters and correspondence or planned her busy social calendar, which was always filled and often on the verge of being overbooked. Jane also enjoyed playing solitaire and bridge and frequently had enough guests to fill twenty bridge tables. Besides cards, she was fond of working elaborate wooden jigsaw puzzles and almost always had a puzzle in progress on a bridge table set up in the library or sunroom.

Throughout the house were framed family photographs and portraits, including a flattering oil on the second-floor landing near the staircase called "Lady Jane," a portrait by G. H. Barrett of an ermine-draped Aunt Jane in New York in 1926 when she was forty-nine.

The Oriental carpets placed in various rooms were purchased by the Phillipses at art auctions in Hot Springs, Arkansas. Jane's needlepoint pillows were scattered about the sunroom and elsewhere, much to Frank's chagrin. He felt that his wife doing any kind of handwork was beneath her station and the needlepoint reminded him of when his own mother and older sister had to take in sewing to help make ends meet back in Iowa.

After a dinner party, when Frank took the men off for cigars and brandy and Jane entertained the ladies with coffee and gossip, she would take up her needlework but remain alert to tuck it behind her if she saw Frank coming.

At the town house, Frank was as paternalistic with his servants as he was with his employees at Phillips Petroleum. By 1930 there were usually nine full-time servants—cooks, maids, valets, gardeners, laundresses, drivers—and as many as a half dozen part-time servants to help out with dinner parties. Although Frank paid them only the going wages of the day and an occasional bonus, the servants developed a lasting allegiance to him and the family.

Dan Mitani, always Frank's personal valet, continued to accompany his employer on all out-of-town trips as well as tend to Frank's needs at Bartlesville and at the ranch. In 1925 Dan married a local woman, Lee Savage Mitani, who was working next door at the L. E. Phillips residence. Soon after she and Dan married, Lee was dusting L.E.'s bedroom when she discovered a note from Node asking L.E. to dismiss Lee because she married Frank's valet and Node feared there would be an exchange of personal information about the two families going back and forth between the households. Lee was so incensed by the insinuating note that she marched to the family's breakfast table, confronted L.E. and Node, and told them there was no need for them to fire her, because she quit. Within three days, Jane Phillips summoned Lee to the town house and offered her a maid's position. She accepted and worked for the Phillipses for more than twenty years. When they first married, Dan and Lee lived in a third-floor bedroom which became a sewing room, but by the time the town house was remodeled the Mitanis had their own home, where they raised Lee's son from a previous marriage.

Shortly after the renovation of the town house, another Japanese servant joined the Phillips household staff. His name was Henry Einaga. Henry came to the United States in 1906. He arrived in San Francisco without a cent and had to take a dollar from the kindly ship's captain since it was illegal for a pauper to enter the country. Henry hid the dollar bill in his shoe for safekeeping and that night someone stole his shoes and his money. Not an auspicious start. But Henry's luck changed for the better.

While he was learning to speak English, he studied maps of the city with the street names printed in Japanese, memorized the street system, and got a job as a delivery boy. Henry stashed away some savings and looked for ways to better himself. After working along the California coast as a gardener, dental assistant, and railroad hand, he went to Kansas and hired on with both John Deere and Armour before buying an onion farm. He decided he was going to corner the market, but soon after he took over the operation, the bottom fell out of the onion market and he was forced to unload the farm and take a job as a butler with a Kansas City family.

From there, Henry moved to Bartlesville to work for Clyde Alexander, and before long Frank Phillips hired Henry away to work in the kitchen at Woolaroc. After about six months Frank moved him to the town house, where he was to remain for the rest of his life. Instead of staying in one of the servants' apartments built over the new six-stall garage at the rear of the property, Henry was provided with the third-floor butler's bedroom furnished with Jane Phillips' old bedroom furniture, including the big bed with a worn spot on the headboard, made when Jane leaned her head against it when she read and wrote letters.

A Buddhist with a black belt in jujitsu, Henry Einaga soon became the head servant at the town house and received his orders straight from Aunt Jane, who considered him her personal butler. Henry was responsible for keeping the first-floor rooms spotless, and the rest of the house was Lee Mitani's domain. Like Dan and the others, Henry quickly became part of the Phillips family. Among his other duties, he was an expert at flower arranging and also served as family photographer, taking snapshots and making motion-picture films of the children and grandchildren and guests at the town house and Woolaroc. On holidays and special occasions, the clan would gather in the music room and watch Henry's films projected on a screen set up in the dining room.

Through the years, as they served the Phillips family, Henry and Dan also became adept at mastering the family's quirks. But most of the permanent servants learned to put up with the Phillipses' personal idiosyncrasies. This included coping with John Phillips' penchant for booze and keeping him out of harm's way whenever he wandered over for a visit from his home—the big brick residence in young Johnny's name—located just across the street. They also remembered to keep fresh talcum powder sprinkled beneath the drawer liners in Jane's dresser and to care religiously for the pots of African violets she loved to see blooming in her bedroom and bath. Any servant who wished to remain in the Phillipses' employ for any length of time also quickly learned that Frank preferred onion sandwiches to munch on during late-night poker sessions on the third floor; that he was a stickler for seeing that the drapes were drawn at dusk; and that all the household bills were paid on time and in full. And pity the new

maid or helper who hadn't been told that Frank had absolute fits if he
spied a crooked painting or picture hanging on a town-house wall.

Several of the servants worked for Frank and Jane for many years,
and a few remained on friendly terms with them even after they stopped
working at the town house. One of those who always stayed in touch, and
in favor, was Matilda Barnhart, or Tillie as the Phillips family called her.
Tillie was the young woman from Missouri who had been hired as the first
governess for Jane and Mary when they were brought to Bartlesville from
New York. She left her job with Frank and Jane in 1920 after she met and
married Clayton Fisher, a driver for a local dry cleaners, but she remained
in close contact with the Phillipses.

That same year she quit her job as governess, Tillie went back to
Missouri to give birth to Lewis Byron Fisher, a healthy baby boy who the
two Phillips girls immediately decided was their very own living doll to
play with and cuddle every time Tillie came to the town house for a visit.
Immediately after her son was born, Tillie came back to work at the
Phillips residence, but soon she quit again to devote time to raising Lewis
and taking care of the family home on Jennings Avenue. Tillie's young
husband was given a job with Phillips Petroleum as a driver, first out in the
Burbank oil field, where he carried company mail, and later as one of
Uncle Frank's personal drivers. "For nine years, beginning in November of
'21, I carried company mail between Burbank and Bartlesville," recalled
Fisher. "The roads were tough and I wore out ten cars and trucks doing
the job."

In the early years of Phillips Petroleum, several different men from
the company garage, including Salty Sawtell, the driver who helped come
up with the Phillips 66 name, acted as Frank's chauffeur. Bruce Jones was
an early chauffeur for the Phillips family, and so was Charles Christian.
When Frank was ready to leave the office at the close of a working day, his
Bartlesville secretary, Marjorie Karch, formerly Marjorie Loos until she
married local dentist Ralph Karch, would call down to the Phillips garage
and summon a driver to take Frank to the town house or ranch.

After he became one of the Phillipses' drivers, Fisher's first assignment was to drive Uncle Frank on vacation trips to Colorado and Iowa.
For several years Fisher shared his driving duties with Homer Baker, a
quiet, mannerly man, and other times with Clarence Burton, a profane,
tobacco-chewing cuss. These three men—especially Baker and Fisher—
drove for Uncle Frank more than any of the others.

Baker, a Missouri native, came to Bartlesville when he was sixteen and tried a variety of jobs, including delivery boy, poultry dresser, and grocery clerk, until he found his calling in a garage where he worked in a mechanic's position. Baker was twenty-one years old when he was hired as the Phillipses' chauffeur, a job he kept from 1933 through 1936, and then returned from 1939 until 1941, when he went on to another job in the Phillips organization.

"When I began driving for Mr. Phillips in 1933, I started with a 1928 Lincoln limousine," said Baker. "I drove that car until 1935, when we got a new twelve-cylinder Cadillac. Mr. Phillips hated giving up that Lincoln. He never liked changing cars. He'd say, 'Changing cars is just like changing shoes—the new ones never fit quite as well.' "

Burton was an occasional driver for Frank, but a memorable one. A colorful character who started at Phillips Petroleum as a truck driver, hauling supplies and tools to Lot 185 and out to the Burbank field, Burton spit tobacco juice into a coffee can he kept under the driver's seat of the limo and had a fifth of either moonshine or bonded whiskey—depending on which one he could get—in the glove box. Every so often, Frank would take a swig from Burton's jug, but mostly he spent his time urging the salty country boy to drive faster. "Uncle Frank was a dear, fine man, but he was hardheaded. He believed automobiles should go faster than airplanes," said Burton. "There never was a car built that could go fast enough for Uncle Frank. One trip from town out to Woolaroc, he bawled me out thirteen times for not driving faster. Another trip out there, he had some gal along that he was trying to impress, and he kept getting on me to go faster. Finally I came out of a big curve and I never took my foot off the gas pedal. I tore through that ranch gate and speeded right up to the lodge and throwed on the brakes and Uncle Frank hit the floor. I knew right then and there I was fired." Burton was right, but after Frank's temper cooled, he was hired back. It wasn't the only time Frank fired Burton.

"Uncle Frank was up in New York City and was due to arrive on a Sunday afternoon at the airport, and since I was on vacation, my supervisor was supposed to have somebody else out there in the car to pick him up," remembered Burton. "Well, the other fella never showed up and there was nobody at the airport when Uncle Frank's plane landed." He learned from his supervisor that afternoon that Uncle Frank was about as angry as anybody had ever seen him and that Burton no longer had a job. Nevertheless, Burton stayed on and reported for work the next morning.

"I'd pick the old man up and he'd grumble and say, 'I thought you were fired, boy.' But then he'd go ahead and get in the car and I'd drive him. I wasn't paid for three months, but I kept on workin'. I had no place else to go. He'd grumble every time, but finally one day I was rehired and got my back pay. Uncle Frank never said any more about it."

Each morning, at a designated time, one of Frank's drivers would pull up in the circular drive at the north entry of the town house. Frank would emerge, pausing only to pick out one of his walking canes from the cylindrical holder by the door. If there was time and the weather was pleasant, Frank might walk part of the way to the office with the car slowly following behind.

When it rained, Frank would have the car pull up right on the sidewalk at the Phillips Building so he could get out under the porte cochere and stroll directly inside without getting a single drop of rain on him. This maneuver would have gone unnoticed since, after all, it was Uncle Frank, except for a cop called "Swede" who decided even the founder of Phillips Petroleum shouldn't be allowed to drive his limo on a city sidewalk.

Actually Lewis "Swede" Thompson was a likable chap who walked his downtown beat for years with his pet duck waddling behind him. Thompson would have sooner sat down to a roast-duck dinner than have provoked Uncle Frank, but the law was the law and as soon as he saw the limo pull up on the concrete sidewalk, Swede would walk over and hand Clayton Fisher, or whoever was driving, a parking ticket. Not a word was spoken. Clayton would simply nod, drive off, and turn in the ticket to be paid. It got to the point that whenever it looked like rain, Swede Thompson would hang around the Phillips Building.

Frank didn't mind a bit. That summer of 1930, and for most of the decade, he had more to worry about than beat cops who prayed for rain. There was business to be conducted; the Depression, like none the country ever experienced, was building a full head of steam and had to be reckoned with; and Phillips construction gangs—1,000 strong men—were laying a pipeline from the Texas panhandle to the Mississippi River, using machines which dug two miles of ditch, forty inches deep and eighteen inches wide, every day of the week.

Then there was the question about the future of Phillips Petroleum. Frank started thinking a lot about what was going to happen to his company after he was gone. Who would emerge to assume the mantle of power? Perhaps Alexander or Wing or even Dewar. Maybe one of the younger men struggling up the corporate ladder. It was clear that John Phillips was never going to mend his ways and become a responsible executive. Frank had sadly resigned himself to that fact.

Mostly because of Jane's influence, Frank had held out a glimmer of hope for John. He believed there was always a chance John might change his ways and get weaned off the bottle. But on most days, especially after hearing about another of John's wild nights, Frank realized it was only a matter of time before his son's troubled marriage would end and some serious steps would need to be taken to help John recover from his alcoholism.

While he was wrestling with the question of what was to become of John, the rest of his family—especially the children—brought Frank great happiness. He enjoyed his foster daughters, now blossoming into young ladies, and he cherished the beguiling Betty, mischievous Johnny, and baby Bobby—his three grandchildren, who called John "Dad" but always referred to Frank as "Father." Frank hoped they would survive the disruptions John's drinking and antics brought into their lives.

Beyond his own family Frank was always interested in some of the other young people in Bartlesville, especially the sons of employees, servants, and townspeople—boys of summer who were growing up around him. Frank kept his eye on the ones who he felt might turn into future leaders.

He looked them over to see if any had the spunk to go places with Phillips Petroleum. Through these lads, and the young execs at the company, Frank found a composite of the son that he always wanted and that John could never be. One of the young men who Frank believed showed real promise was Lewis Fisher, Tillie and Clayton's only child.

Starting during the Depression years, Frank made it a practice to give every schoolchild in Bartlesville a silver dollar and a sack of candy at Christmas, at first during visits he made to the schools and in later years at holiday assemblies held at the civic center. Lewis would get his silver dollar, just like the other kids, but then Frank would stop by the Fisher house on his way out to Woolaroc and give him something extra.

"Dad wasn't driving for Mr. Phillips then—that I remember," said Lewis. "He would call me over to the car, talk with me for a while, and ask how I was doing. Then, just before the car drove off, he'd hand me a gold piece. Not silver but gold." Lewis saw a lot of Uncle Frank after Tillie was given a special pass which allowed her son to swim in Clyde Lake whenever he pleased, and also when Lewis worked as a caddy and busboy out at the country club. "Mr. Phillips would walk into the main dining room out at Hillcrest and go right up to the edge of the floor and a hush would fall over the entire crowd. People were absolutely captivated by him. It was quite clear that Mr. Phillips had a commanding presence."

There were other Bartlesville boys who never forgot their encounters with that "commanding presence." Bill Jones was one of them. Bill was born smack in the middle of a string of nine children. His father was William Winfield Jones, known as W.W., a local realtor and a friend of Frank Phillips. Bill's aunt was Mabel McCreery, a Bartian who went to California to work as George Getty's secretary and in 1916—the year Bill was born—was elected a director of the Minnehoma Company.

Since W. W. Jones was well respected in town and was active at the Methodist church, where he served as Sunday-school superintendent, Frank trusted his judgment and valued his opinions. So whenever Frank heard about someone who needed financial help—a girl in trouble, a widow in distress, a family without a breadwinner—he would summon W.W. to his office and ask him to check out the story. If W.W. returned and reported that there was a legitimate need, Frank would write out a check for cash, with instructions that no one should ever know the source of the money. Frank and W.W. held their clandestine meetings and maintained their quiet system of charitable contributions for more than thirty years.

Bill Jones's first real meeting with Uncle Frank took place on Cherokee Avenue in front of Frank's town house. Bill was just a kid with a bicycle and a stack of magazines he wanted to sell.

"Frank Phillips was out in his yard and I came wheeling up on my bicycle and said, 'Mr. Phillips, do you want to buy a *Saturday Evening Post?*' He looked at me and said, 'Well, I might. What's in it?' I told him that I didn't know but I understood that it was a good magazine. Well, Mr. Phillips gave me a sharp look and said, 'You don't know what's in it? What's your name?' I said Bill Jones and he asked if I was one of W.W.'s boys and I told him yes. He looked me over some more and said, 'I'm surprised your father hasn't educated you any better than that. Come over here and sit down with me.'

"I sat down next to Mr. Phillips on the front steps and he took the magazine from me and proceeded to explain everything about it. He told me about the cover and he opened it to the table of contents and explained that. Then he showed me the stories and the drawings and told me about the advertising. He went through that entire magazine, page by page."

Frank and Bill sat together on the step for twenty minutes. And when Frank was through with his lecture on marketing, the boy knew more about that issue of *The Saturday Evening Post* than the editors. He also knew how to sell.

"Now, I'm going to buy this issue from you," said Frank. "And from now on, anytime you see me out in the yard, I want you to stop and try to sell me a copy of the magazine. I'll be fair game. But there are two things to remember—don't you ever ring the doorbell and don't you dare ever again try to sell me something without knowing what you're selling."

Even though the going rate for the magazine was a nickel, Frank tossed the boy a dollar—a new, shiny silver dollar—just like the ones he handed out to all the kids at Christmas.

"After that day I wore out two bicycles just riding around that big ol' house, waiting to see Mr. Phillips and sell him one of my magazines," said Bill. "Sometimes I'd catch him outside and I'd stop and sell him a magazine. I always knew everything about the issue I was selling and Mr. Phillips always gave me a silver dollar."

For many years after Frank's death there were grandfathers and grandmothers in Bartlesville who carried silver dollars in their pockets or purses, old coins worn smooth as slugs, memories of their youth when Uncle Frank, smelling of cigars and bay rum, reached out his hand and touched their lives during the Great Depression.

A great many lives—and not just schoolboys'—were touched by Frank Phillips during the summer of 1930. After weeks of dealing with nothing but worried bankers and concerned company directors in New York, St. Louis, Chicago, and Bartlesville, Frank had found little time for relaxation. His town house was disrupted and torn apart and so was his life, thanks to the Depression. Frank's big shows of optimism and bravado were fine, but the reality was that the nation was in the throes of economic turmoil and confusion.

All of the company developments and outside projects demanded Frank's time, and he was only able to manage a few getaway weekends with Jane and their daughters and a big July blowout at the ranch for 2,500 Phillips employees. But in August he found some days to slip away. He headed to Philmont, Waite's ranch nestled high in the mountains of northern New Mexico. While at Philmont, Frank found there was more to do than ride horses and eat trout suppers. There was business to discuss. Big business. But it wasn't the same as the meetings and the negotiation sessions he was used to in New York. It was an easier time. He was dealing with his little brother—the man he helped start in the world of business—

and they were under the cobalt-blue skies of the West, where everything seemed clean and fresh.

Before he left Philmont to join John and his family in Colorado Springs, Frank firmed up an agreement with Waite. After more than fifteen years it appeared the Phillips brothers were going to get back together in the oil business. At least on paper.

Although Waite had sold his oil company in 1925 for $25 million cash, it hadn't taken him long to return to the business. By the following year, he and R. Otis McClintock organized the Philmack Oil Company, and in 1927 Waite acquired large holdings in the Independent Oil & Gas Company, a thriving concern that had been organized by E. H. Moore in the boom days of the big Okmulgee field. Philmack merged into the Independent Oil & Gas Company, and the production, refinery, and marketing facilities were immediately expanded. Moore resigned from Independent Oil and R. C. Sharp, former president of Oklahoma Natural Gas Company, became president. Waite, a major stockholder, soon became chairman of the board of Independent Oil and a short time later was made chief executive officer.

Around the same time he was getting back in the oil business, Waite continued with his banking interests and also built several prominent office towers in downtown Tulsa, including the Philtower on Boston Avenue, a twenty-four-story ornately Gothic building that quickly became known as the "queen of the Tulsa skyline." Another design by Edward Delk, the Philtower's elaborate façade featured gargoyles and Waite Phillips' initials over the main entrance. Inside the fancy lobby were custom-made chandeliers, travertine marble floors and walls, and massive brass elevator doors, framed in marble, bearing the distinctive "WP" shield.

On a neighboring corner Waite built the Philcade, a nine-story Art Deco building of office suites and shops with an ornate lobby in the shape of a "T" for Tulsa.

The Philcade and the Philtower were connected by an eighty-foot brick-lined tunnel dug under the avenue by miners imported by Waite. The tunnel was built so supplies could be moved from building to building with ease, but also because Waite, like his brother in Bartlesville, was alarmed by a rash of kidnappings of wealthy men and wanted to be able to move about secretly and without fear of being whisked away by hoodlums.

Successful with all of his real estate, banking, and oil ventures, from the moment Waite became connected with Independent Oil, there were rumors of a merger with Phillips Petroleum. Now there was truth in the rumors. Talks in New Mexico between the brothers had gone very well. The merger of Phillips Petroleum and Independent Oil was a "done deal" and by late August the cat was out of the bag. The Phillips boys—Frank, L.E., Waite—were reunited.

In a joint statement released on August 29 by Frank and R. C. Sharp, the business world was told that the two firms "have agreed upon terms and conditions for consolidation" and would immediately combine their talent and resources. "The physical properties of the two companies are of such character as to effect an immediate reduction in capital expenditures, and are so located as to supplement each other without duplication, thus contributing to the natural economics of this consolidation, amounting to several million dollars annually." All that remained was for the directors to give their blessings.

After the directors of both firms had a chance to meet on September 2, news of the ratification of the merger was made public in an official statement issued by Phillips Petroleum: "Frank Phillips, president of the Phillips Petroleum Co., announces that directors of the Independent Oil & Gas Co. and directors of the Phillips Petroleum Co., at their meetings today, unanimously approved the action of their respective executive committees for a merger of the two companies on the basis of 76 shares of Phillips stock for 100 shares of Independent stock, representing the approximate book values of the two companies."

Newspaper headlines across the country told it all: "Battle of Brothers in the Oil Field Is Closed" and "Brothers Unite under One Business Flag Again."

The Oil & Gas Journal, a bible of the oil business, proclaimed: "The three Phillips brothers are regarded as among the smartest oil men in the industry. They have extensive banking facilities in New York, Kansas City, Tulsa and Bartlesville. The combination of Phillips and Independent forms a powerful organization in the hands of men of known ability and long experience."

No change in the management, headquarters, or employees of either of the two companies was contemplated. Waite did not plan or intend to ever become involved in the actual daily operations of the new company. He would retire once more and devote his time and energy to his real estate, banking, and charitable interests.

Years later, Waite's son, Elliott, explained that although Frank and Waite maintained the deepest respect and admiration for each other, the fact was they were too much alike to get along in business. "The reason," said Elliott, "is that if two men are riding the same trail, one has to be in front."

In essence, Phillips Petroleum Company acquired Independent Oil through an exchange of shares and consolidated two thriving oil firms into a concern with more than $316 million in assets, 3,600 producing oil wells, 54 gasoline plants, and 11,600 retail outlets. The merger also gave Phillips two additional refineries, one in Okmulgee and another in Kansas City. Phillips now leased 2.6 million acres in ten states, and the marketing territory was expanded from Texas to Canada and from the Rocky Mountains to Ohio.

On September 20, while oil and business writers were still telling the story of the big merger, there was further good news for Phillips Petroleum. On that date all existing records for opening-day gallonage were broken at the grand opening of a Phillips service station at the intersection of Clayton and Big Bend in suburban St. Louis, one of the heaviest traffic intersections in the world. Between the hours of 5:40 A.M. and 12:45 P.M. the station sold 70,154 gallons of Phillips 66 to more than 10,000 motorists, a new record which doubled the previous world's record for a single day's sales by one station.

The Phillips station was not only the largest in St. Louis but one of the largest in the world, with sixteen electrically operated pumps, and the most modern greasing, washing, and oil equipment money could buy. It took 104 men and women working the pumps, directing traffic, collecting money, greeting customers, and a fleet of nine Phillips trucks shuttling back and forth all day long from the service station to the bulk station three miles away just to keep the gasoline flowing and to serve the swarms of customers. By 9 A.M. police reserves were called out to handle traffic problems, and by 1 P.M. a double line of vehicles stretched out for blocks.

A large and spectacular "66" shield neon sign, fifteen feet in diameter, mounted on steel posts, was illuminated from sundown until closing time and was visible for miles. The St. Louis newspapers and radio stations spread the word far and wide about Phillips' latest addition to its string of retail outlets.

Word of the service station's success buoyed everyone's spirits back in Bartlesville. Especially Uncle Frank's. He and his oil company were flying as high as an eagle. Others recognized that as well, including his old friends and allies, the Osage Indians.

On September 27, 1930, a week after the big blowout in St. Louis and three weeks after the acquisition of Independent Oil became public knowledge, the Osage tribal council honored the man who had drilled hundreds of oil wells on their lands and helped rake millions of dollars into Osage hands. During a ceremony in the midst of the fourth annual Cow Thieves and Outlaws Reunion at the F.P. Ranch, Chief Fred Lookout—surrounded by 1,500 cowboys, outlaws, Pawnee Indian dancers, Osage tribesmen, and other guests—adopted Frank into the Osage tribe and gave him the name Hulah Kihekah, or Eagle Chief.

It marked the first time the Osage ever adopted a white person into their tribe.

Aunt Jane and the children and grandchildren were there. So were all four of Frank's brothers. Zack Miller, Pawnee Bill, Grif Graham, Henry Wells, and hundreds of painted Indians and booted cowboys quietly watched alongside Wall Street financiers and other guests.

The old-timers knew it was a solemn moment, and the crowd became silent when Chief Lookout and Frank faced each other and, in accordance with tradition, Lookout gave the newest Osage chief the gift of a pony, a saddle, and a single eagle feather. The saddle, handmade from the bone and skins of wild animals, had been in Lookout's family for more than a century. Frank was moved by the quiet but eloquent gesture and immediately following the adoption ceremony Eagle Chief presented Chief Lookout and Chief Madlock of the Pawnees with fine robes fashioned from the hides of buffalo which had run in the large Phillips herd.

After the exchange of gifts, the Indian drums began and Pawnees dressed in feathers and bells performed a stomp dance. Frank fed his guests one thousand pounds of Brahma beef, choice cuts of buffalo, great kettles of beans and peas, bushels of potatoes, and 3,600 bottles of soda water. It was a memorable occasion, and the visions of the Indian dancers, the old-time range riders standing next to the guests in neckties, and most of all the image of Chief Lookout handing him the reins of the Osage gift horse stuck in Frank's mind.

That autumn John Phillips, thanks again to his drinking, effectively managed to sever his relationship with Phillips Petroleum Company.

The episode had started as an innocent enough assignment from Frank. He asked John to go to Kansas City to meet with some of the eastern directors of the company and then accompany them to Bartlesville in Frank's private rail car. All the arrangements were made and John departed for Kansas City and celebrated his arrival in his usual manner—by trying to drink the town dry. Near closing time in the last bar he visited, John took a liking to a spider monkey, bought it from the bartender, and staggered back to his hotel.

The next day John was in a daze. He not only forgot about meeting the directors but wasn't really sure why he had come to Kansas City in the first place. He did manage to remember there was a train waiting, so he returned to Union Station in time to board the car, pour himself a stiff drink, and head south to Bartlesville, where Frank, Jane, John's children, plus a number of executives and employees were waiting at the depot to greet the directors.

As the train ground to a halt, the band Frank had assembled began to play and the crowd started clapping. The door of the train car opened, and a conductor placed a stepping stool on the ground for the passengers. Out waltzed John Phillips, as drunk as he'd ever been, with a monkey in his arms and not one director in tow. The band stopped playing. The crowd stopped clapping. Frank stopped smiling. It took only a few seconds for everyone present, even the children, to figure out what had happened.

Jane and the children were embarrassed. Frank was livid. As far as he was concerned, this was John's last screw-up. He gave his swaying son a public tongue lashing right on the spot. John was also quickly removed from the Phillips Petroleum board of directors and was never again associated with the firm.

The trouble with John and his drinking, as always, bothered Frank, but there were enough other problems rearing their ugly heads to keep his mind occupied. The Depression was starting to sink in and Frank was feeling its fingers tighten around the company. He was forced to admit that the economic troubles of the nation were finally affecting the firm's operations.

Phillips had been able to maintain a relative position of strength immediately following the stock market crash. The merger with Independent Oil contributed to the expansion of production and marketing facilities, and although several Phillips gas stations in East St. Louis sustained bomb damage that autumn of 1930 as the result of labor troubles, operations were running relatively smoothly and the company maintained its place as the largest oil producer in Oklahoma. But the merger and the huge debt which resulted from starting the pipeline project were beginning to pinch the Phillips pocketbook.

Phillips Petroleum's earnings were beginning to erode. From peaks of $21.4 million in 1926 and $13.2 million in 1929, net earnings dropped to a meager $3 million in 1930. Severe declines in prices of both crude and refined oil products occurred during the last months of 1930, bringing them to the lowest level since the company was organized in 1917.

As 1930 was ending, Frank became concerned about the tremendous financial obligations his company had to meet. He had worked vigorously to borrow money for expansion so Phillips Petroleum would not be stuck with great stores of raw material that couldn't be sold. When he recognized that transportation of Phillips' products from production sources to the market was too costly, he approved the pipeline project. He still believed in the decision to pursue both the pipeline and the merger with Independent. But the debt was staggering. It had all been a gamble, just like everything else worthwhile Frank ever did in the oil business. Now he was wondering if he had made the right decisions and if his company could survive and weather the storm.

Others, including the bankers who held the notes and the Phillips directors who were charged with maintaining fiscal order, were wondering the very same thing.

"It was a critical period in Frank Phillips' life and the life of the company," Paul Endacott said years later. "He was boxed in and needed an outlet. He went through with the merger in order to expand the company's capabilities and resources and he had good reason to build the pipeline, but all of it cost money, and money was in short supply."

Frank spent most of the last three months of 1930 at his battle stations in New York, meeting with bankers and board members, and with old associates from the oil patch, including Bill Skelly. Frank and Skelly had a few differences through the years, including a set-to in the lobby of Tulsa's Mayo Hotel back in 1929 when after a few too many drinks they squabbled over a mutual oil matter. The argument led to fisticuffs when Frank accused Skelly of being a liar. Skelly, a hulk of a man, jumped to his feet and charged. Punches were thrown, but no real damage was done and witnesses called the fistfight a draw. Frank and Skelly quickly settled their differences, and once the Depression struck, both men remained faithful allies, turning to each other for advice.

As he worked hard that autumn and winter to clear up his problems with the many banks he had borrowed money from, Frank would have gladly taken a Sunday punch from Skelly instead of slugging it out with a bunch of nervous bankers. For the most part he stayed close to his New York office except for brief jaunts to Chicago, where he conferred with company officers at the Blackstone Hotel. In New York, Fern's comfort, the Turkish baths, and Doc Hammond's daily treatments helped ease the

growing stress and strain, and receiving the wise counsel of the Baruch
brothers at weekend retreats helped. But mostly Frank felt alone and over-
extended. In Endacott's words, "boxed in."

Some of the bankers and even a few of the company directors were
being especially unreasonable, or so Frank believed. Much of their anxiety
was caused by the growing number of banks which were folding across the
nation, more than 1,300 since the Crash of '29. But they were also concerned about Frank's attitude. During meetings of the eastern directors
Frank grew more irritable and seemed to have lost some of his control and
confidence. Even at board meetings of the Chatham Phenix National
Bank, where Frank was a director and major stockholder, both Samuel
McRoberts, the chairman, and R. H. Higgins, senior vice president, who
also sat on the Phillips board, thought Frank appeared annoyed and overworked. A few directors even whispered that perhaps Frank might be
going to the bottle a bit more than was good for even a hard-drinking oil
tycoon.

It was about this time that Frank found that the Phillips line of credit
was drying up. The bankers were putting the squeeze on and it hurt. The
time had come for some action and it had to be quick. Frank and Obie
Wing began beating the bushes, spending most of their time either with
bankers or with those who had a banker's ear. There was no time to waste.

Frank became so involved in trying to untangle the company's banking problems in New York that he completely forgot about his own bank's
twenty-fifth anniversary in Bartlesville. Frank could hardly believe it was
on December 4, 1905, when he and L.E. first opened the Citizens Bank &
Trust Company. It seemed like only yesterday.

It was now called the First National Bank, and both L.E. and H. J.
Holm were present in Bartlesville for the directors' meeting where the
bank's quarter of a century of service was observed with a flurry of congratulatory resolutions and a luncheon. Fred Spies, the young man from
Creston who came aboard in 1907 as assistant cashier, was now vice president of First National, and W. C. Smoot, also on the Phillips board, was
president of the bank. Frank remained chairman of the board. In a wire
sent to Clay Smoot, Frank admitted he forgot the anniversary and he
apologized for missing the celebration.

Other banks around the country weren't faring as well as those in
Bartlesville. Bank closures were becoming routine. Then, on December 11,
the Bank of the U.S., a major private New York bank with sixty branches

and more than 400,000 depositors, locked its doors. Frank could feel the heat being turned up. The pressure was building.

On December 16, after another full day of meetings with bankers and stockholders, Frank joined L.E. for a late-night strategy session at the home of J. L. Johnston, president of the Lambert Company and a longtime Phillips director. But of more importance to Frank, Jack Johnston was also a good friend and someone who could be trusted.

Together the three men worked for hours developing a plan which they believed would shock the bankers, as well as those directors who were losing faith in Frank, back to their senses. The meeting didn't break up until after two o'clock the next morning, when Frank returned to the Ambassador for a few hours of sleep.

When Frank reached his office shortly before 11 A.M. on December 17, he was ready to put the plan hatched the night before to work. All that was needed was to get his thoughts down on paper. He wanted it all to be stated in the form of a letter. A letter to the Phillips board of directors. A letter which Frank knew would soon make its way into the hands of the bankers who were hurting his company by not lending Frank the money he so desperately needed to continue with the pipeline and other important Phillips projects.

Instead of using one of the secretaries or stenographers to take his dictation, Frank summoned Fern to his private office and together they drafted his thoughts. The letter ended up being two pages long, and although it wasn't the most polished epistle ever produced, it came right to the point.

It was, after all, a letter from Frank Phillips. A candid letter from Frank to the Phillips board of directors. In truth, it was Frank's ace in the hole—a wild card he pulled from nowhere just when he felt his back against the wall. When they received their copies, the directors got the point before they finished the first paragraph. Those who didn't know him any better were stunned. It was a letter of resignation from Frank Phillips. It read:

> After forty years in business I have realized for the past couple of years that I was fast approaching a period of retirement and have indicated to many of our Directors that I desire they give serious thought to the election of my successor within reasonable time. My age and condition of health will not permit me to continue much longer at the head of

an active, aggressive organization such as we are fortunate in having.

In placing my resignation in your hands I want you to fully appreciate that I have only in mind the best interest of the stockholders of the Company who have been most loyal since the inception of this Company. I can truthfully say that I have hardly had an unpleasant experience with a single stockholder and my Directors have been most cooperative and loyal and I further say that I am very proud of the achievements of our Company under most trying experiences almost since its inception which I attribute largely to the cooperation received from you, our stockholders and the public.

I have no apologies whatever to offer for the policies we have carried out to date and which we have agreed should be pursued in the future. In fact, I think they have been lifesaving policies. After a certain amount of money was invested in this Company it was absolutely necessary for us to expand, create volume, and develop certain divisions in the industry for the maintenence of our success. While some criticism may be made of our program, I cannot see how our Directors could have proceeded along any other line than has been pursued. We cannot control the economics of the industry, neither can we control the economics of finance.

While it has been necessary for us to continue policies already adopted, certain developments have taken place which leads me to believe that new leadership should be provided at the earliest possible date. When we were only a very small concern we borrowed from New York banks much more money than we now need and upon a statement which was not by any means comparable to the one we now make. We have always had certain bank credits over a long period of years, and it was natural for our management to rely upon or consider that these would be available again under a constructive program. Our balances over the last several years have justified anticipation that these credits would be renewed. We had assurance that such would be the case at least until we permanently financed certain projects now underway. We put our cards on the table with these banks and one of the leading banks made the statement to us that our line of credit was available. On the strength of this we solicited all our other depositories and their lines were freely granted, after which this leading bank withdrew their line so far as I can determine and in checking we find line of credit has been withdrawn by other banks, which of course impairs our credit with all other institutions as these lines of credit were all necessary preliminary to financing our new projects. This withdrawal of credit which may or may not have been prompted by competitors further confirms my opinion that your Company needs a new President.

Please do not misunderstand me in this action as I do not want to be classed as a quitter. My entire thought in the matter is for the best interests of the stockholders of this Company. It may interest you to know that I today have the largest number of shares of stock in this Company that I have owned since its beginning. I have not sold a share since it commenced to decline in the forties until during the last week have sold approximately thirty two hundred shares for members of my family to establish tax losses. My last speculative purchase was some thirteen thousand shares at $32½ in order to help clean up the syndicate in our last financing. This of course was bought in addition to my full rights which I took up. Am giving you this information as I do not want you to feel that I am letting down on the situation. My greatest interest is in our stockholders and my conscience permits that I make the statement that this has always been true.

I will remain with the Company in any advisory capacity until you can agree upon a desirable successor who can have the fullest confidence of the interests which we must depend upon to carry out the program absolutely necessary to our ultimate success. In choosing my successor I could not approve of any member of my family succeeding me, therefore I think the name of the Company should be changed to one more national in character. However I would not object to continuing sale of Phillips products which have met with such a big measure of success as is evidenced by the fact in a statement before me that sales in our St. Louis district last

month over the previous month increased sixty-eight percent, which I feel is due to the quality of Phillips products,
and our success in this line has heretofore been unequaled in
the industry.

I am leaving this week for Oklahoma to spend Christmas
with my family and will be back here before the first of the
year. I shall call a meeting upon my return to hear your
recommendations and I trust this letter will be kept confi-
dential until that time.

The letter was delivered to each of the Phillips directors, twenty of
them, counting Frank and the Phillips officers on the board. Now all the
cards were dealt and on the table. Within a few days Frank and Wing
departed for Oklahoma to spend the holidays with their families and see if
Frank's bold letter would get anyone's attention.

As far as Frank was concerned, he was taking the ultimate personal
risk. There were no guarantees that the directors wouldn't accept his resignation and start the search for a successor. He would have to wait and see
if he would lose everything he had spent years building or come out the big
winner. No matter what happened, Frank was convinced of one thing—it
was worth the risk.

CHAPTER TWENTY-SIX

Frank's gamble paid off.

He learned that, just a few days after Christmas when he returned to New York to run his traps and see what effect his letter of resignation had on the board of directors and, more importantly, the bankers. Just as Frank had hoped, the directors didn't call his bluff and accept his resignation. Instead they weighed the possibilities and decided the Phillips name and image was too closely tied to Frank to allow him to step down as president of the company. They knew full well that he was Mr. 66 and that Phillips Petroleum and Frank Phillips were one and the same in many people's minds. The tenuous days of the Depression were no time to muddy the waters by tampering with the corporate leadership.

Back in New York, during the last days of 1930, Frank resumed his meetings with directors and bankers. He was anxious to test the wind. At first it appeared sweet, just like a breath off the Osage prairie. On December 30, inked in parentheses following a notation in Frank's business journal about yet another meeting with McRoberts, chairman of Chatham Phenix, Fern happily noted: "Financing deal in better shape." It looked like the additional money Frank wanted to continue the pipeline project and keep the company afloat was going to come through after all.

On New Year's Eve, Frank took Wing and W. N. Davis, another Phillips vice president to see the future New York home of the new Phillips Petroleum offices in the Irving Trust Company building, an Art Deco skyscraper resembling a solid shaft of stone under development at the corner of Wall Street and Broadway. That afternoon Frank lunched at the Recess Club and bid happy New Year's greetings to the many familiar faces he saw there, including Charles Schwab and others who gathered to salute friends and say good riddance to the old year.

On New Year's Day 1931, the sweet wind suddenly turned sour. After dispatching Wing to check on financing deals with banks in Boston and Chicago, Frank prepared for several days of intensive meetings with his contacts at Chatham Phenix, New York Trust, and Chase Manhattan.

Those meetings were eye-openers. It didn't take Frank long to see that although his poker-playing strategy may have kept him at the helm of the company, he wasn't going to be able to grab the entire pot. There was a steep price to pay. Frank wasn't going to escape without giving his board of directors and the banks who lent his company money something in return. They wanted their pound of flesh too. For openers there would be a new Phillips finance committee to include Wing and board members Samuel McRoberts, Jack Johnston, and H. M. Addinsell, president of the Chase, Harris, Forbes Corporation. Frank could handle that. The other stipulations were the ones that were going to be difficult to swallow.

It was pointed out to Frank that the Depression was getting worse and that Phillips was already deeply in debt, a fact he knew only too well. The company had secured huge bank loans to finance the acquisition of Independent Oil and to start the products pipeline. Although no one wanted Frank to step down as president of Phillips, the bankers were concerned that a steadying influence was needed at the helm to help guide the company through the lean years that everyone knew were ahead. They wanted someone capable in oil industry matters but also strong enough to work side by side with Frank, who through the years had run the oil firm he founded like a czar. It was further explained to Frank that the bankers, and some directors, felt he was overworked and needed a change in perspective. They would feel more comfortable if someone they endorsed and trusted acted as Frank's personal adviser; they even had a candidate in mind—Amos L. Beaty.

A slightly pudgy but distinguished gentleman with balding head and gold-rimmed eyeglasses, Beaty was the former chairman of the board of the Texas Company. Often called "Judge" because he had also served as the oil firm's chief legal counsel, Beaty had become president of the American Petroleum Institute since his retirement.

Beaty also maintained strong ties with the banking community. The bankers Frank was dealing with not only liked Judge Beaty but said he was their choice to serve as their liaison with Phillips Petroleum. Beaty would be the bankers' watchdog. The Phillips directors concurred.

Frank had known Beaty for several years and begrudgingly admitted he respected the Judge's oil-business savvy. In the mid-1920s Beaty and Frank spent time together when there had been serious talk about a Texas Company–Phillips Petroleum merger. It was no secret then that the Texas Company wanted to control Phillips. Beaty and other Texas Company officials had even gone over some of Phillips' books and inspected certain properties. Frank vehemently denied that there was any truth to the merger talks and eventually negotiations fizzled and the deal was declared dead.

Beaty now was going to get his chance to at least influence the man at the top of Phillips Petroleum. Frank was humiliated. The very idea that he would have someone from outside the company sitting in on critical fiscal and internal strategy discussions was unthinkable.

On January 17, Frank met with several board members and after dinner conferred with Judge Beaty for several hours at the Plaza Hotel. On Frank's journal page for that date, there was another parenthetical note: "A terrible day due to unreasonableness of certain directors."

The directors proved they could be just as stubborn as Frank, and after he considered the alternatives, Frank decided he had little choice but to go along with their plan. He desperately needed more financing to keep his pipeline crews busy and also ensure that Phillips 66 remained fixed in the consumers' minds despite a highly competitive market. Frank reasoned that at least he hadn't ended up like his friend Marland, totally out in the cold with no company at all.

A good portion of the payoff for cooperating with the bankers came on January 21 when Frank and John Kane conferred with directors E. P. Earle and E. E. Loomis. They went directly from that meeting to McRobert's office at Chatham Phenix, where Wing and the Phillips lawyers were waiting. There, at midafternoon a little more than a month after he penned his letter of resignation, Frank joined the others in signing off on a $20 million financing deal. Now he had the funding to finish laying more pipe and to give his company some breathing room.

Shortly after securing the financial shot in the arm for his firm, Frank told a reporter from *The Wall Street Journal* that the effects of the Depression on the oil industry during the final quarter of 1930 "will be reflected in substantial losses to our company in common with many others." Frank, on the basis of the advice of Beaty and the board, also suggested that the "action of directors on future dividends must be governed by the company's earnings, economic conditions in the industry, and the general financial situation."

Frank was hoping to soften the blow he felt was coming. He knew there was a strong likelihood that the board would not be paying any dividends to shareholders unless a miracle took place. And during the 1930s miracles were about as rare as rain in the Dust Bowl.

As Frank feared, things didn't get any better. Around the country more and more people were losing their jobs. With the increase in unemployment, money became even more precious, and the end result was a drastic drop in prices. Because of overproduction and the fallout from the Depression, crude oil had dropped from a peak of $3.50 a barrel in 1920 to $0.95 a barrel in 1930. In 1931 the crude oil price shrank to only $0.22.

At the same time Phillips stock was taking a severe beating. At the beginning of 1931 the stock was going for a high of a little more than $16 a share. By the close of the year the stock had dropped to $3.00 and change. Such a drastic dip made life miserable for all the stockholders, including many employees who took part in the company's 1930 drive to sell stock to family and friends. When the drive started, the stock was selling for $32 a share. By the time the price reached the lowly $3.00 level, many workers were ready to hide out in the Osage in order to avoid angry relatives and neighbors who were left holding handfuls of stock.

During a Phillips directors' meeting at the Bankers Club in New York on March 3, a decision was made to pay no dividends to stockholders in 1931, the first time this had ever occurred in the company's history. This action more than any other cut Frank to the bone. The payment of stockholder dividends was very important to him. He was disappointed with the board's move but agreed that it was necessary. In a statement to the shareholders written March 31, 1931, for the 1930 annual report, Frank explained that the controversial move came "in view of the difficult times through which the industry in general is passing, and believing it essential to the best interests of the stockholders to conserve cash until conditions improve."

"The company is carrying out its program of gasoline and natural gas pipeline construction," Frank explained in another message to the stockholders. "Upon completion, this project will not only increase and stabilize earnings, but will extend the company's marketing area through additional outlets being acquired. The industry is going through a severe test and our policy is to complete all projects and conserve our cash position so that we may emerge from the depression in a strong position to produce, refine, transport, and market with maximum efficiency and economy."

Stirring words of inspiration to a bunch of unhappy stockholders who unfortunately would not receive dividends for a total of three years until the company resumed payments in 1934.

After three weeks of making excuses to shareholders, there was finally a bright moment for Frank. He returned to Oklahoma to take part in an official Osage adoption ceremony intended to complement and enhance Frank's informal passage into the tribal ranks held the previous autumn during the Cowthieves and Outlaws Reunion.

The formal ceremony was held in the Woolaroc dining room, where more than one hundred guests gathered. Wires or letters were received from those who couldn't attend, including, William Allen White, who was hospitalized at the Mayo Clinic, and Vice President of the United States Charles Curtis, a former senator from Kansas and a distant grandson of the Osage chief Pawhuska.

One cable read: "Say wish I could make it but can't. The Osages was always the smartest Indians in America. Now they show it again. There is one hundred and twenty million white men and they pick out the best one in the whole bunch to make him an Indian. Again they show their wisdom. Best regards to all Osages, including Frank, from the Renegade Cherokee —Will Rogers."

Frank's foster daughters, Jane and Mary, took off a few days from the Ogontz School and, accompanied by Fern, arrived by train in time for the doings. Most of the other Phillips family members were in attendance, as were several close friends, including Pawnee Bill and Zack Miller, Doc Hammond, Armais Arutunoff, and an array of Osage leaders and tribal members from Pawhuska.

Logs blazed in the huge fireplaces at either end of the dining room while guests took their seats at long tables covered with Frank's favorite red-and-white-checkered cloths. Every member of the tribal council came, garbed in full Osage dress. As Chief Fred Lookout and his wife, Julia, entered the room accompanied by Frank, the guests rose and cheered.

Following the luncheon, George Labadie, a member of the tribe and a Pawhuska businessman, served as toastmaster and introduced Chief Lookout, the head of the Osage Nation, and former Osage chief Baconrind. Both of their speeches were in Osage and were interpreted by Assistant Chief Harry Kophay, who in his nervousness stumbled over several of the words, causing the chiefs some embarrassment when they later learned of his mistakes. Although most of the guests present, and Frank himself, didn't know any errors in interpretation had been made, both chiefs wrote letters of apology to Frank.

In his letter, Chief Lookout explained that he had always called Frank "Little Oil Man." The name was fitting since the Osage called most whites whose names they had trouble pronouncing "Little" and then added a name to signify their occupation. Use of the word "Little" was in no way demeaning but was a name of distinction and honor.

The chief went on to explain that when he was asked to come up with a proper Indian name for Frank, he immediately thought of the eagle. The eagle was important to the Osage people, and Chief Lookout knew it must be important to whites, since they even put the bird's image on their money. "I belong to the Eagle Clan," wrote Chief Lookout, "so last fall when they wanted me to give you name, I gave you name Eagle Chief. In our Clan when the first son is born, they name him Eagle Chief. When they ask for an Indian name for you, I thought of my son, who has been dead when he just a year old, and that was his name, and so I gave my oldest son's name to you. In my speech I said to you that you have been a good help to our people, and a good friend, and my tribe have drawed a good deal of your money."

Although the first adoption ceremony at the outlaws reunion meant a lot to Frank, the formal event, which included Chief Lookout's presentation of a parchment scroll certifying Frank's membership with the rank of chief, was hard to top.

After the two Osage chiefs made their remarks and presented the scroll, Zack Miller spoke a few words and gave Frank a stretched robe, made from the tanned hide of a Philippine water buffalo. The robe was Miller's personal gift to his old friend. An artisan from the 101 Ranch had spent two full weeks in a prone position inscribing one side of the robe with pictures which told the history of the Osage Nation, likenesses of Frank and the two Osage chiefs, and, in both Osage and English, a condensed version of the resolution which read:

> Because of his great friendship for the Osage Indians and the part he has taken as a pioneer in developing their vast resources . . .
>
> This is to hereby certify that pursunt [sic] to Osage Council Resolution passed on the 13th day of March 1931 . . .
>
> We adopted Frank Phillips as an Honorary member of the

Osage Tribe of Indians and that he shall now be known
under the Indian name of Hulah kihekah (Eagle Chief).

The ceremony continued for more than an hour, but the crowd remained hushed, struck by the quiet dignity of the chiefs and the magic of
the moment. Near the close, Tulsa attorney T. J. Leahy, married to an
Osage woman, came forward to deliver a brief address and present the
chief's regalia, consisting of an elaborate headdress with a feathered train,
a beaded vest, an Indian shirt, a breechcloth, leggings, moccasins, arm-
bands, and bells. The items were a collective gift from George Labadie,
John Kane, Clyde Alexander, John Phillips, Dr. H. C. Weber, Doc Hammond, and Bert Gaddis.

As Leahy finished his talk, Chief Lookout placed the ornate chief's
bonnet on Frank's head, symbolizing the completion of Eagle Chief's initiation into the tribe. The time had come for Frank to speak. He got to his
feet, faced the audience, which included some of his dearest and closest
friends, and responded to all that had taken place that day.

"I am overwhelmed. I haven't the words to express my appreciation
of the honor which is being conferred. I accept this fellowship with greater
gratitude and pride than anything ever bestowed upon me. I fear I am not
worthy and wonder if this wise old tribe is not making its first mistake
today. This is a new inspiration, however, and I shall do my best to prove
worthy.

"Recent world history reveals the American Indian owned this continent with the Osages as leaders among American tribes in all progressive
activities, and owning the greater part from New Orleans to the north and
northwest. The Osages have always since been leaders in morals, education, and all uplifting activities of the American tribes. In speaking of
civilization and ancestry my good friend Will Rogers once said that his
forefathers were standing on the banks of the New England coast to meet
the people who think they founded American aristocracy. American aristocracy was already here. Before these white people came and by questionable methods obtained this country the Indians lived in happiness and
prosperity. Panics, hard times, and breadlines were never known. Millions
of people today are jobless and hungry. This country should be given back
to the real American aristocracy and its original owners which would give
the Indians a chance to bring happiness and good times out of chaos."

The crowd didn't stir. They sat in silence and ruminated over Frank's
words. He left the dining room, escorted by the tribal council to his up-

stairs bedroom, where they dressed him in the complete chief's costume. When Frank returned and walked in wearing the colorful feathers and beaded moccasins, he was greeted by loud applause.

Frank coaxed the Indian dignitaries to come outside and pose with him on the steps of the lodge so Frank Griggs, a Bartlesville photographer who recorded every major event at Woolaroc and in Bartlesville, could capture the moment on film. The stoic chiefs and several members of the tribal council agreed. Still in his chief's finery, Frank also posed for more photographs standing next to the stretched buffalo hide which told the story of his tribal adoption. These portraits were some of Frank's favorites, and they gave him pleasure during the dark days of the Depression which followed.

As he settled in for a long siege of ups and downs which would mark his life during the next several years, Frank gradually became accustomed to Judge Beaty, and a mutual respect developed between the two men.

Beaty was elected to the Phillips board of directors, and within four months he was appointed a member of the company's executive committee, the only outside director ever admitted to that group. When Beaty joined the board of directors, Frank, in an uncharacteristic move, sent out printed announcements about the newest director's election to all of the Phillips stockholders, no doubt a maneuver suggested by some of the other directors, including Beaty himself.

Many of the top executives at Phillips were unaware of the uncomfortable position in which Frank found himself. They didn't suspect that the kindly Judge Beaty was actually a representative for the bankers' interest. Most executives who saw him in Frank's New York office knew Beaty was a retired oil man and thought he was simply there working with Frank as a consultant. Even after he joined the executive committee and was given a small office adjoining Frank's in the Phillips Building in Bartlesville, most people thought nothing of it.

That was just fine with Frank. He didn't want anyone to know Beaty's true mission at Phillips or how much clout he had in company affairs. A few of the bright young men working for Frank, however, figured it out. Paul Endacott, still assigned as a vice president and manager of the Philgas division in the Philgas Company general offices in Detroit, was one of them.

Endacott was pleased that, although some branches of the company were struggling, Philgas, mainly because of distribution stations in the

Northeast and sales to consumers in suburban areas, was giving the company a much-needed boost in earnings.

"Natural Gas Without Pipe Lines" was how the company described the Philgas service. "Just like city gas" was what the Philgas customers called it. Throughout the eastern portion of the country and on the outskirts of many large cities, there was a growing number of Philgas customers who learned to live beyond the reach of city gas mains. Tank trucks carried Philgas, under pressure as a liquid, to the small storage tanks outside each customer's home. Large industrial users, served by direct tank car shipments, were also snapping up sizable quantities of Philgas.

When Endacott continued to spend money for installation of customers' tanks and the metal cabinets which enclosed the tanks, Frank took notice. He believed Endacott's expenditures were an apparent violation of the company order, dreamed up by the new finance committee as a cost-saving measure, not to spend one dime on capital investments. Endacott was summoned to Frank's office in New York to explain his actions.

When he walked in to see Frank and discuss the Philgas expenditures, Endacott was surprised to see Judge Beaty sitting at Frank's elbow. He was introduced to Beaty, and then Frank looked Endacott straight in the eye and got right to the point of the discussion.

"Young man, I know your operation is going along pretty well and that you've been able to send a bunch of cash back to Bartlesville," said Frank. "But I also know that you're directly violating my order about halting all capital investments. You're spending money on these tanks and cabinets and equipment and such. That's a capital investment and we just can't be having any more of that."

Endacott listened to Frank ramble on for more than twenty minutes about the importance of saving money and how everyone needed to be looking for ways to cut expenses. The young engineer spoke up. He explained that there were great advantages in owning the tanks which the customers tapped into in order to fuel their stoves. The system allowed Philgas to store gas in the customers' tanks outside their homes and not have to maintain huge storage tanks. Also, the installation fee for the tank and cabinet was actually more than the cost of the new equipment, so the company made money there. Besides, by owning the customer's gas tank, Philgas was able to ensure that their competitors couldn't come along and deposit their product in the tank.

"Mr. Phillips, I'd be glad to comply with your order and sell these items—the tanks and so forth—to the customers. Then if we sell the equipment to them it wouldn't be considered a capital investment. But I'll tell you this: if that happens and we change our current financial arrangements for this equipment and give up ownership, then we can't make as much money."

Beaty looked up from the Philgas financial report he was reviewing. He had heard and read enough. Everything Endacott said made sense and the bottom line of the Philgas report didn't lie. There was no need to fix something that wasn't broken. Beaty leaned over to Frank and said, "Frank, I think we're going to let this young fellow go ahead with what he's doing, aren't we?"

Frank had no recourse. He replied only: "Yes."

Nothing more was said; nothing more needed to be said. Endacott turned on his heel and left, aware that probably his job had been saved by someone who was much more than a kindly retiree doing some consulting.

Despite Beaty's presence and the Phillips directors' need to keep close watch over the company's fiscal matters, Frank somehow managed to maintain a front for the public and his employees, so there was never any question about who was leading Phillips Petroleum.

Frank also knew that his firm's close relationship with the Chatham Phenix crowd and Beaty wouldn't last forever. There would come a day when Phillips could break its ties with those directors, and he was right.

Over the next few years Phillips Petroleum was able to chip away at its debt. Starting in 1931, hundreds of Phillips employees were laid off and a practice of job sharing was started, with the company allowing only one employee per family. Salaries were trimmed and dividend payments halted. Frank made sure he was no exception. He cut his personal salary to a token one dollar a year.

A major event occurred in August 1931 that influenced the retirement of the substantial debt Phillips owed and helped Frank gain independence from the bankers. Frank's pet project—the Phillips pipeline—was completed. It ran 681 miles across America's midsection and connected the Phillips refinery at Borger, Texas, to the giant storage tanks at East St. Louis, Illinois, with a line branching from the main line at Paola, Kansas, to the company refinery at Kansas City.

The skeptics who had predicted that "the darn thing wouldn't work" were wrong. It worked just fine. A cheering crowd was present on August 5 at the formal dedication ceremony in East St. Louis, when Frank turned a spigot and the first gallon of gasoline from Borger flowed into the tank of Mayor Frank Doyle's automobile.

After the ceremony, a cub reporter, awed by one of the facts in Frank's history, cornered Frank in a hotel lobby across the river in St. Louis. "Mr. Phillips, is it really true you were a barber?"

"You're damned right!" roared Frank. "And if you get me a pair of shears and come up to my room, I'll give you a better haircut than the one you're wearing."

And Frank was feeling good. He was full of himself again. He found this rush of happiness and satisfaction when he thought back to his many sessions with the bankers. In oil circles, the pipeline had been a big joke. Frank nonetheless stuck to his idea. He borrowed the cash on short-term bank loans and arranged for long-time financing through a bond issue to be floated after completion of the project. He went along with the directors' desires to form a finance committee and he acquiesced when Judge Beaty was brought aboard.

When the work was finished, Wall Street, with the Depression bearing down, begged off when the bonds were offered. That's when Frank went back to the bankers who held his short-term paper. "Boys," he told them, "you're in the pipeline business."

They informed Frank they were not and that the pipeline was his baby. "You nurse it," one banker told Frank.

That is exactly what Frank did. Despite the conditions of the loan, including the addition of Judge Beaty to the board as the bankers' watchdog, Frank actually took the bankers' response as almost a vote of confidence. Without hesitation, he plunked down millions of dollars on the pipeline. It was a courageous investment and a smart one.

The Phillips pipeline was an engineering first—the longest ever built to carry gasoline. It expanded the Phillips midwestern marketing area enormously. It also changed the shape of gasoline transportation and helped Phillips Petroleum survive the worst years of the Depression.

The pipeline was designed to simultaneously transport several different Phillips products, a concept that many people in the industry questioned. *The Oil & Gas Journal* asked: "Could one pipeline carry such a volatile substance as liquid butane, as well as the heaviest grade of refinery gasoline and the materials . . . that lie between these two extremes, without danger of serious contamination?" Once again Frank's bright young geniuses had shown they could get the job done. They were confident their system would work, after the successful tests they performed on an experimental twenty-four-mile-long circuit laid out near Tulsa.

When the Borger–East St. Louis pipeline became operational, millions of gallons of gasoline of different grades were pumped through the line at the same time. One grade followed the other without mixing and various grades were drawn off at terminals built at Wichita; Kansas City; Jefferson City, Missouri; and East St. Louis. At those points, finished gasoline was then blended and shipped by truck, tank car, or river barge around the marketing territory, which even expanded into the congested Chicago area, where a huge Phillips 66 shield was erected near downtown. The glow of its lights could be seen for miles around.

The company was beginning to struggle out of the red and back into the black. Although in 1931, for the first time ever, Phillips Petroleum lost money, by 1932 earnings came to $776,000 and by 1933 earnings had creeped back up to $1.5 million.

Hefty bank loans needed to build the pipeline were paid back in full within three years, and by 1933, the bankers from Chatham Phenix— R. H. Higgins and Samuel McRoberts—were no longer factors in Frank's life. Both of them disappeared from the board of directors of Phillips Petroleum Company. Their bank, at one time one of the largest financial institutions in the country, was acquired by another firm and eventually closed. For the first time in several years, the lone wolf of the oil patch could breathe a little easier.

Judge Beaty, by now a close friend of Frank's, never tried to make a move to assume greater influence than he was originally given. Through the years he worked alongside Frank, the two men became close friends. Judge Beaty and his wife were frequent house guests of the Phillipses in Bartlesville and enjoyed dining with Frank and Jane on the Waldorf roof in New York. As Phillips Petroleum's business outlook brightened and the nation's economy began to improve, Beaty withdrew from the company. In 1935 the Judge stepped down from the Phillips board. His watchdog days were over.

But before his life would become more tranquil, Frank had to face those intervening years of the decade—from 1931 through 1935. During this period, marked by a continuous string of high and low points in Frank's life, he continued to pursue a lifestyle which he truly felt was the most appropriate for the general public's consumption. Image was always important to Frank. For that reason, Frank continued to hobnob with the best of them. He would go to circuses with the Ringlings; fly off for picnics in Iowa; dine with presidents, princes, and other big shots; spend weekends at Fern's place in Connecticut; entertain Will Rogers or Amelia Earhart or one of countless celebrities in his circle; or maybe even just disappear for a day or two, with Dan and Doc Hammond, for a good old-fashioned fishing trip.

"I suppose you might say that some of the things he did back in the 1930s were frivolous," said Endacott. "And I guess some of them were. But it was critical for Frank Phillips to keep his image up and always put his best foot forward. And during those unbelievably tough years, that's just what he did. He was kind of like a fellow whistling through the graveyard."

Frank found himself puckering up and doing quite a lot of whistling as the 1930s progressed. As Frank borrowed, bullied, cajoled, and finessed his way through those turbulent years, it was his habit to carry a small notebook and pencil wherever he went. During a meeting of one of the several boards he served on, such as the lunch sessions of the North American Aviation directors in the Chrysler Building's Cloud Club, a juicy morsel of business gossip was likely to find its way into Frank's little book. He even stuck his notebook and pencil in his pocket when he took his afternoon horseback rides at the ranch. When an idea occurred to him—a suggestion for the boys at the research laboratory, an improvement in service, the name of a promising employee who should be watched—Frank would jot it all down. Between the notebooks and the business journals Fern kept, Frank had a built-in system for jogging his memory.

But Frank didn't need his notebooks or his business journals to remember several of the events of the 1930s. Many of them were forever branded in his mind.

Family problems were no exception. There was a long laundry list of family conflicts and problems that continually cropped up and needed resolution. At the top of that list, naturally, was John Phillips. Frank certainly needed no notebook to remind him of John's constant woes.

After the humiliating episode with the monkey at the Bartlesville train depot in 1930 and his subsequent exit from the Phillips board of directors, John still found solace in an old companion—booze. As his drinking increased, Mildred's tolerance level finally topped out. The couple separated. That started the wheels turning to put their marriage out of its misery. On December 1, 1931, after fourteen years and three children, John Gibson Phillips and Mildred Beattie Phillips were no longer married —their divorce was granted.

Mildred kept the house on Cherokee, which was in her son's name, and stayed on in Bartlesville while Betty and Johnny went off to Garfield

School. John took to the road and spent most of his time living at the Ambassador Hotel in New York with his parents.

Frank was determined to get his son's mind off his shattered marriage and put some color in his cheeks. In January 1932 he put John on a train at Grand Central and shipped him upstate to Garrison, New York, a peaceful little town just across the Hudson from West Point. Garrison was the home of Bill Brown's Health Farm, named for its owner, Bill Brown, a former New York State boxing commissioner.

Brown's Health Farm had gained a reputation for being just the spot for the wealthy to dry out from too much Prohibition drink. John Phillips was tailor-made for the place. As part of the two-week alcoholic recovery program, attendants made certain the guests took brisk morning walks, ran fifty-yard sprints, and sweated their sins away in the steam baths and saunas. All the meals were low-calorie, and carrot juice was the toast at the cocktail hour.

After several days and nights at Brown's, John was feeling good and looking better. He was impressed with the program and himself. Instead of carousing or playing cards, he played catch with a medicine ball. When he had the urge to go to the bar to mix a stiff one, he went to the steam cabinet. Frank was so impressed with the reports he received that he joined John for the last five days of the program and signed them both up for another purge session for the next October.

In between visits to the health farm, Frank had yet another diversion planned for his son. After several meetings with Dr. Roy Chapman Andrews, the famous naturalist and explorer from the American Museum of Natural History, Frank arranged for John to accompany a museum team on a six-month-long expedition into the backwoods of South America to study the Andean region. Andrews, who would eventually direct the Museum of Natural History, was well known for leading expeditions throughout Asia and for his discovery of the remains of primitive mammals and humans and the recovery of fossil dinosaur eggs.

There's little doubt that Frank made it worth the museum's while in order to sweeten Chapman's interest in John and include him on the trip. No matter. Frank had found something else to keep John busy and on the wagon. It was worth any price.

John was to be part of an exploratory team led by Dr. Wendell C. Bennett, assistant curator of anthropology at the museum. On March 11, 1932, the eve of the expedition's departure, Frank and Jane hosted a buffet dinner for John and Bennett at the Ambassador. After the plates were cleared, Frank gave a brief speech and was surprised by Andrews, who said he wanted to make a presentation. Andrews carefully handed Frank a box containing a portion of one of his coveted dinosaur eggs discovered in 1923 at Flaming Cliffs in the center of Mongolia's Gobi Desert. There was also an exact replica of a complete egg and some desert sand from the site of the find.

In a formal letter of presentation, Andrews said his gift to Frank "is the only privately owned part of a dinosaur egg in the world . . . and is guaranteed to be ninety-five million years old." Frank immediately shipped his latest prized possession to Bartlesville with instructions that it be kept at his airplane museum, where a curious collection of odds and ends was starting to gather.

The South American trip was a definite high point in John's life. He thoroughly enjoyed getting out in the wilderness with Bennett and the others and roughing it. The expedition also met with success. That summer the New York *Times* carried stories each time the museum team made another discovery.

They uncovered significant Inca relics in ancient ruins at Lake Titicaca on the high Bolivian plateau. In one of the fortified sites discovered by the expedition there were remains of more than one thousand stone-domed houses. A few weeks later, working at a pre-Inca site in Bolivia, the team unearthed a massive stone idol measuring seventeen feet from head to foot with a four-foot base. The huge idol weighed eighteen tons and was several feet under the earth.

John made detailed notes of everything he experienced. "Hit the head of our monolith and got it pretty well uncovered before the day was over . . . Had to quit work for two days as the Indians all had to celebrate the Fiesta of San Pedro. It was really a marvelous sight though and I don't suppose many Americans have ever seen anything like it. There were fifteen hundred to two thousand Indians in the plaza from all over this district. Many of them had dressed in tiger skins and parrot feathers which they wore as shirts."

John returned to the States with bizarre tales and gifts, including a pair of shrunken heads—one male and one female—which came from the Jívaros, a warlike Indian tribe from Ecuador, notorious for shrinking heads taken in blood feuds.

Frank promptly sent the grotesque trophies to join the dinosaur egg at his museum at the F.P. Ranch while he sat and caught his son up on what he had missed while tramping through the countryside of South America.

During John's absence several family members and personal friends of the Phillipses had encountered problems which turned into real headaches. Not all of them, Frank explained to his son, could be remedied. Some were beyond fixing.

Such as the situation facing the Phillipses' old friends at the 101 Ranch. Less than two weeks after John left on his expedition, the 101 Ranch and everything on it was sold at public auction as part of a receiver's sale to satisfy debtors. The familiar ranch, where as a boy John camped with his father and listened to cowboys and coyotes howl at the moon, had gone broke. There would be no more turtle races, no more sharpshooting or bulldogging or yarn spinning at the fabled 101.

In the late 1920s things began to go bad for the Miller boys. In 1927 Joe Miller died of carbon monoxide poisoning, and two years later Frank was a pallbearer at the funeral of George Miller, killed in an automobile accident. Oil prices dropped, debts mounted, the Millers' circus failed and closed. Zack tried to keep the ranch going, but the tremendous indebtedness and nothing but the Depression ahead finally sank him.

Eventually the old ranch house was sold and the 101 was cut up and devoured just like a sheet cake at a country wedding. Another slice of western Americana had vanished. The demise of the 101 Ranch made Frank as sad as when he saw that faraway stare in Marland's eyes after he had lost his empire.

But more sadness was soon to follow. This time closer to home.

On May 23, 1932, John Gibson died. Frank's mentor—the spirited banker and entrepreneur, the man who dug iron mines in Mexico, harvested forests of mahogany in the Philippines, and built a coliseum for Chicago—was gone. The evening Gibson passed away his namesake, John Gibson Phillips, was acclimating to life in South America—exactly the kind of adventure the old man relished.

Jane was in New York with Frank when word came that her father was near death. The Phillipses and Dan raced for the train. During the night, as the train pulled into Ohio, the state where he was born eighty-two years before, John Gibson died in Bartlesville. The Phillipses took a plane from Columbus, and the next afternoon—Johnny Phillips' seventh birthday—they arrived in Bartlesville and went to the funeral home where the old man's body was being prepared for burial.

There was a private service at the Gibson residence at 717 Delaware, the house where Gibson entertained neighborhood children on the front porch with his poetry recitations. Then Frank and Jane, Gibson's widow,

Bertha, and his sisters took Gibson's body by train back to Salem, Illinois, for burial.

Frank was fifty-eight, about the same age as Gibson had been back in Creston when he sat down in Frank's barber chair and totally changed young Mr. Phillips' life.

Frank stood on the edge of John Gibson's grave. He comforted the cluster of mourning women and soothed Jane's deep sobs, but his mind was elsewhere. He was back in the barbershop trimming Gibson's whiskers and gobbling up the banker's stories of finance and business. He was back in a buggy with sweet Lady Jane peddling Gibson's bonds. He was back in Indian Territory eating dinner with Gibson at the old Right Way Hotel and scratching out plans to build a banking and oil empire. He was back on the deck of the steamship *Siberia* listening to Gibson recite his verse in a clear, loud voice that carried on the wind.

When the preacher said his piece and the gravediggers, off under the shade trees, put on their hats and reached for their shovels, Frank thanked John Gibson one last time and then he turned and walked away.

Frank went back to the world of the living and to a hornet's nest of business dilemmas that made him think he would have been better off if the directors had taken him up on his ploy and accepted his letter of resignation.

There was plenty of action all across the country as the summer of 1932 steamed up hotter than hell. Both the Republicans and the Democrats met in sultry Chicago to pick their presidential candidates. The GOP stuck with Herbert Hoover and Charles Curtis; the Democrats, at first deadlocked over a trio of hopefuls, finally went with Franklin D. Roosevelt and John Nance Garner. Out in Los Angeles at the tenth modern Olympic Games, Mildred "Babe" Didrikson starred in track and field and Clarence "Buster" Crabbe won a gold medal as a swimmer.

Frank had little or no time to think about politics or sports that summer. In East St. Louis, Illinois, the disputes between Phillips Petroleum and local union leaders, which had been festering for months like an angry sore, boiled to a head and exploded.

Phillips' problems in the area were tied to the pipeline which ran from the Texas panhandle to the terminal located near the Mississippi River. Back in 1931, as the pipeline reached St. Louis County, difficulties started because of a group of determined landowners who blocked construction for months by refusing to grant easements through their property. Not only were there long and costly court battles but some farmers posted obscene signs blasting Phillips and patrolled their land with rifles and shotguns to intimidate the company's digging crews. The individual right-of-way controversies were dragged out in the courts for months, costing Phillips thousands of dollars and causing the company to fall far behind in its construction schedule.

While the pipeline encountered trouble in St. Louis, just across the river in Illinois violence broke out near the new Phillips plant on March 13, 1931, when ironworkers from an East St. Louis labor union, objecting to the company's purchase of materials from a non-union firm, roughed up some Phillips workers and punctured their tires.

Two days later, about one hundred men, armed with shotguns and revolvers, ambushed six guards from the Chicago firm Phillips had hired to escort their workers to the Phillips storage tanks. A pitched gun battle left four of the guards wounded. The union leaders said they objected to the company's employment of non-union ironworkers and boilermakers on the tanks.

A week later the only two filling stations in East St. Louis that were owned outright by Phillips Petroleum were dynamited. The explosions not only caused havoc at the stations but shattered windows throughout the neighborhoods, and in one incident a thirteen-year-old girl was cut by flying glass. There was public outrage, rewards were offered for the culprits, but no one was apprehended. The stage had been set for more trouble.

More trouble started again in July 1932—almost a year after the pipeline and plant were opened. This time the problem was focused on the Phillips storage tank complex. When Phillips used the regular crew of the company which handled all painting and maintenance of Phillips properties to paint the thirty huge tanks, the union went beserk. Phillips officials in St. Louis pointed out that the non-union company was charging $5,700 to spray-paint the tanks as opposed to the $24,000 cost quoted by the union workers to paint the tanks by hand. The explanation was rejected, the work was suspended, and negotiations were launched.

But before any compromise could be reached, labor warfare erupted again. On July 25 a crowd of five hundred men, many of them armed, turned back a Phillips gasoline truck as it attempted to enter East St. Louis. More threats and attacks on Phillips drivers and employees followed. Some Phillips workers were pulled from their vehicles and beaten with blackjacks and pistol-whipped. The union's embargo on fuel deliveries also threatened to stop the retail sales of gas from Phillips stations.

East St. Louis Mayor Frank Doyle begged both sides to cooperate, but when the company requested police protection for the delivery trucks, Chief of Police Frank Leahy refused. He said he didn't have sufficient patrolmen to assign to the extra guard duty. The next day, however, at the instigation of the Chamber of Commerce and business leaders, Leahy changed his mind and swore in twenty-five special policemen to prevent further violence. Local business organizations offered to pay the additional salaries and the special force, backed up by regular patrolmen, started working twelve-hour shifts, standing guard at both the plant and Phillips gas stations. Despite the extra protection, there were more threats and violence.

Frank continually met with Judge Beaty and the rest of the executive committee during this period and also popped in and out of St. Louis to check with his local management team to see if the problems could be resolved. He was particularly concerned about the safety of his employees as well as the crippling effect the union unrest was having on the flow of Phillips products in and out of the East St. Louis terminal.

Frank wished he could bring in some Osage outlaws or a few of the toughs he hired years before to maintain law and order at his operations in the Oklahoma and Texas oil patch. They'd whip the union goons into shape in short order. But he knew that was unpredictable.

During his meetings in St. Louis, Frank made his thoughts quite clear. He wanted the problems solved and he wanted them solved immediately. The message came across loud and strong to the Phillips management team in East St. Louis: "Get the job done."

In late July and early August the newspapers published reports about a new guard force employed by Phillips to protect the plant and company workers. The newspaper accounts hinted that the new Phillips guards were not the standard-issue variety with neat uniforms and nightsticks. The new guards, the newspapers suggested, were none other than the infamous Shelton gang, a band of organized killers and thugs led by Carl Shelton and his brothers Earl and Bernard.

The Sheltons had been striking terror in citizens' hearts and making bold headlines since the 1920s, when their troops battled rival bootleggers with Army tanks and dropped bombs from airplanes during an incredible midwestern gang war which cost at least forty lives. Throughout the Depression years the Sheltons prospered from a variety of illegal activities while they carried on with their gangland feuds in southern Illinois.

A July 25, 1932, story in the St. Louis *Post-Dispatch* confirmed everyone's worst suspicions. "For two weeks Deputy Sheriffs of St. Clair County have been looking into a story that Shelton gangsters were employed as guards at the Phillips plant. With field glasses, at some distance from the establishment, officers reported they saw the following inside the fence: Carl Shelton, leader of the notorious Shelton gang; his brothers, Earl and Bernie; John B. ('Babs') Moran, Jack Britt, men named Wilders and Schmitke, and Eddie Walsh, described as a gang hanger-on." The newspaper went on to say that according to "an official source" the Sheltons had sent word to the east side unions to "lay off" the Phillips plant.

It was time to play hardball in East St. Louis.

Labor leaders urged owners of independent filling stations selling Phillips supplies to stop dealing with the company and affiliate with one of Phillips' competitors. At the same time there were new reports of the Sheltons importing some "armed men from Peoria" to aid in guarding the Phillips plant and workers. Tension continued to build.

On August 9, local authorities arrested Carl Shelton while he was riding with a Phillips Petroleum watchman. Two loaded revolvers, with the serial numbers partly obliterated, were found on the floor of the car. Shelton was taken in, briefly questioned, then released. The following afternoon, Monroe "Blackie" Arms, a feared gangster and Shelton associate, was arrested in East St. Louis in an automobile in which sheriff's deputies found a machine gun and a sawed-off shotgun. Arms told the officers he was "goin' squirrel huntin'." Two of Blackie Arms's companions—William "Bad Eye" Smith and Ray Dougherty—escaped.

That evening, Oliver Alden Moore, a popular labor leader and president of the Central Trades Council of East St. Louis, despite threats on his life consented to an interview with a St. Louis *Post-Dispatch* reporter. Wearing a revolver and guarded by four union boilermakers armed with rifles, Ollie Moore met with the reporter in a parked automobile in a quiet East St. Louis neighborhood. Each time another auto passed, Moore visibly flinched. His men craned their necks and watched the darkness for any signs of trouble.

Throughout the interview, Moore blasted the Sheltons and Phillips Petroleum, the firm he had been battling for almost two years. "Sure, I've been threatened," Moore said. "You see, the Shelton gang hasn't been making any money out of bootlegging recently and so they've been trying to make a racket out of labor in East St. Louis. But East St. Louis labor is not going to be intimidated by gangsters. Carl Shelton and his mob aren't big enough to make a racket out of labor here. Why, I was offered $30,000 by the Shelton gang if I'd move out and let labor alone. I turned that down. Word has reached me that the Shelton gang is out to get me, that they've imported six carloads of red-hots from Peoria to bump me off. Well, let them try it. They can't intimidate East St. Louis labor."

After taking the Sheltons to task and protesting Phillips Petroleum's decision to use non-union labor to build and paint the East St. Louis facility and tanks, Moore was satisfied that he had told the union's side of the controversy. Forty-five minutes later, back at union headquarters, Moore and his men decided to call it a night and headed for home. "Well, boys, I'll be seeing you tomorrow," Moore told the armed guards, and he started for his own automobile parked nearby.

At that moment, a large dark sedan drove out of the night with machine guns blazing from the windows. Witnesses said it sounded "like a motorboat running." Moore was struck twenty-seven times. He spun around like a top as the .45 caliber slugs hit him in the back, chest, and side. Within ten seconds Ollie Moore lay dead on the street. Two of his guards were wounded. No one could identify the assassins.

The next day, after a chat on the telephone with Police Chief Leahy, Carl Shelton strolled into police headquarters on his own accord. He was briefly questioned, seemed to have a plausible alibi, and was released. The authorities told reporters that the murder mystified them and said they didn't think the Sheltons had anything to do with it. "It might have been a personal quarrel," said Chief Leahy. "I don't think Carl had anything to do with that murder."

Frank, who had arrived in St. Louis the night before ("on account of labor troubles in East St. Louis," according to his daily journal), made sure that Phillips Petroleum quickly protected its public image. He bought a full page in the local newspapers and told the Phillips side of the story. The company's "Statement to the Public" spelled out the chronology of events that marked the two years of violence and discontent in East St. Louis. Phillips also disavowed any knowledge of or connection with Moore's murder. The statement read:

> In order to protect its property at its terminal, after consultation and advice from leading members of the business community, Phillips Pipe Line Company and Phillips Petroleum Company, through an agency engaged in such business, arranged for the stationing of guards within the enclosure around the Pipe Line Company property at its terminal

for no other purpose and with no other intention than the protection of the lives of the company's employees and the company's property. The management of neither of the companies was advised by the agency as to the personnel of the guards to be employed. That was a matter left entirely to the agency.

Only Wednesday evening, representatives of Phillips Petroleum Company, in conference with representatives of the labor unions, one of whom was Mr. Moore, were engaged in negotiations for a basis of working out the troubles . . . Under these circumstances, certainly it will be recognized that the death of Mr. Moore has been a serious blow to the carrying out of these plans. Surely no one regrets more than does the management of the Phillips Petroleum Company the unfortunate affair in East St. Louis Wednesday night.

Ollie Moore's funeral was a classic martyr's tribute. Six hundred union members from East St. Louis and neighboring towns gathered in front of the union headquarters, the site of their leader's murder. Led by a fourteen-piece band, the mourners slowly marched two abreast through a drizzle that stopped just as they reached Moore's brick bungalow, where the services were to be held.

Because the crowd swelled to more than 1,500, the funeral services were moved out of doors to a nearby vacant lot and the gunmetal-gray casket was placed on the grass beneath two trees. When the brief services were over, the crowd filed past the open coffin and a complement of six soldiers from Jefferson Barracks stood at attention in honor of the fallen World War I veteran, who left a widow and two young sons. At the cemetery the soldiers fired a volley over the grave and a regimental bugler sounded taps.

A union electrician who owned the automobile that was riddled with gunfire near the scene of Moore's murder outlined with white paint each of the eighteen bullet holes and painted in large letters on the doors and on the trunk: "Death Car." He then attached a sign to the auto's roof which said: "This car was fired on by the slayers of Ollie Moore, labor leader of East St. Louis and director of controversy with Phillips Petroleum Co." The man threatened to take the automobile from town to town and park it near Phillips service stations.

The opposition's plan to make Ollie Moore's murder into a cause célèbre died a quick and easy natural death. Frank saw to that. The company's published statement denying that Phillips intended anyone harm helped, as did the continuous effort to reach some sort of agreement with labor leaders. On August 17, less than a week after Moore was gunned down, Phillips Petroleum and the Boilermakers' Union hammered out an agreement that finally pleased both sides.

Meeting in a St. Louis hotel, representatives from Phillips and the union agreed that the work that had already taken place at the Phillips facility was done and couldn't be undone. But for all future construction in East St. Louis, Phillips said they would be sure to employ only union workers. Before the ink was dry, the union leaders withdrew all picketers stationed at the Phillips plant and company trucks immediately resumed delivery of gasoline to service stations.

Life and business went back to normal in East St. Louis. Ollie Moore's murder remained unsolved, and Carl Shelton continued as a king-pin in area gambling operations and other shady dealings until 1947, when the sixty-year-old gang leader was blasted out of a jeep by a hail of gunfire from a dark sedan on a country road near his farm. The gunmen, believed to be members of a rival gang, escaped.

Although, over the next several years, Phillips would continue to encounter some labor problems in the St. Louis area and also in Kansas City, where labor disputes led to the bombing of several filling stations, the resumption of production activity at the East St. Louis facility allowed Frank to breathe a little easier in the autumn of 1932, as the company announced its adoption of a five-day workweek for all employees throughout the Phillips system.

That October, just a few days before John and Frank were to leave for their second fitness round at Bill Brown's Health Farm, the Phillipses staged another dinner at the Ambassador to honor Dr. Andrews and Dr. Bennett and the Museum of Natural History. Among the guests was Jane's good friend Perle Mesta. The daughter of Bill Skirvin, owner of many oil wells and the Skirvin Hotel in Oklahoma City, Perle was the widow of manufacturing magnate George Mesta and was well connected in social, business, and diplomatic circles. She was considered to be "the Hostess with the Mostest" and served as the inspiration for the Broadway musical hit *Call Me Madam.* Jane liked socializing with the popular Perle, who, despite her party girl image, remained a faithful Christian Scientist and neither smoked nor drank.

But an aversion to tobacco and gin was not a problem for John Phillips. Much to Frank's disgust, John returned to his vices with renewed gusto even after the drying-out periods in South America and at the health farm.

In December, with Frank back in Oklahoma, Jane decided to celebrate the first anniversary of her son's divorce and throw a little New York bash. When she found herself with too many men on the guest list, Jane immediately put out the word that she needed more young ladies at her Ambassador suite that evening. One of those invited was Mary Kate Black, a young widow from South Carolina who was visiting friends in the city. Mary Kate came from an old Carolinian clan made up primarily of doctors. Her father was a doctor, and so were her three brothers. Mary Kate, naturally, had married a physician—Dr. Morton Hundley from Virginia. But her young husband was sickly and succumbed to double pneumonia on the couple's honeymoon trip to Switzerland. After living as a good and proper southern widow for almost three years, Mary Kate was ready to taste life again. John's divorce party was just what the doctor ordered.

"I'll never forget that evening," said Mary Kate. "And I'll never forget Jane Phillips. She was full of fun and had more wisdom than anyone I ever met. That night there was the party at their hotel and then we all went around to the speakeasies and clubs and Jane Phillips had her gold—and I do mean gold—checkbook. She'd write out a check wherever we went."

In the wee hours of the morning, after much champagne and laughter, it was obvious to everyone that John Phillips had fallen for the charming southern belle. He couldn't take his eyes off of Mary Kate.

"Finally, John looked at me and said, 'Mary, you know you can't dance very well and neither can I, but will you marry me?' I said that's the most I ever heard in one sentence in my entire life."

Although she was smitten by the curiously charming John Phillips, Mary Kate didn't give him an answer to his proposal. Instead she left the next day to see her ailing father and spend the holidays with family in South Carolina. While she was home, Mary Kate received a telephone call. It was Jane Phillips in New York.

"She told me that they were going to throw a huge New Year's Eve party on the roof of the St. Regis Hotel and that John wanted me to be there. She said that if I didn't come back for the party, John was planning to ask another woman. I told Mrs. Phillips to make me a reservation. I was coming."

The party was successful and so were John's renewed pleas for Mary Kate to throw caution to the wind and marry him. He told her that by now she should have made up her mind one way or the other. She agreed and accepted his proposal. They had a day to make the arrangements, and on January 3, 1933—after only two meetings at wild parties a month apart —John and Mary Kate were married in the heart of Manhattan at the Marble Collegiate Church, where a young man named Norman Vincent Peale had recently become the pastor.

"I wore a little ol' black-and-white dress I had with me," said Mary Kate. "That night we went to the movies and saw *Gay Divorcee.*"

John's marriage came as a complete surprise to Frank and Jane. They learned to accept Mary Kate but at the same time retained their relationship with Mildred, the mother of their three grandchildren.

"Right after we were married, John's father called me into his room at the Ambassador," said Mary Kate. "He said, 'You come here and sit on my lap,' and I did and then he said, 'I want to tell you one thing right here and now—my son doesn't work and, quite frankly, he doesn't have to.' Well, I just talked right back to him and said that John most certainly was going to work and then I saw to it that John went to Pittsburgh and got started in the steel business and then he became involved in Southwest Supply Company in Oklahoma. I wanted John to work and stay busy."

John and his new wife moved to a suite on the twelfth floor of Tulsa's Mayo Hotel, the popular oil man's hangout where Frank and Bill Skelly had their fistfight and J. Paul Getty once lived.

"I got along just fine with Frank Phillips," said Mary Kate. "He and I had this little game we'd sometimes play. Mr. Phillips would say, 'Dan, go get some money and bring it to me.' Dan would go get a pile of twenty-dollar bills. Then Mr. Phillips would take the stack and pick up one bill at a time and say, 'Now, Mary, does this bill end in an odd or an even number?' If I'd guess it correctly then he gave me the twenty."

But parlor games played with packs of crisp twenties weren't everyday events in 1933, not even for Uncle Frank, who for a time considered taking his daughters out of their private school because of the steep tuition. Banks were closing, jobs were scarce, and Okies were beginning to forsake the dust and hot winds that made western Oklahoma look like a landscape on the moon.

Frank still had more money than most, and what he had he shared with friends and family. "I always found him very generous with us," said Mary Kate. "And he was such an impressive man. He could talk to anybody—a country priest or some big corporate president. If a room had a thousand people in it, there would be a hush go over the crowd when he walked in. Frank Phillips had magnetism.

"He'd carry a cane and he'd say, 'Mary, it gives you style.' That was important to him—to have style. Frank Phillips was really something. Back in those years—those wild and crazy years—he was like a feudal lord. He was the lord of Oklahoma."

CHAPTER TWENTY-SEVEN

There were plenty of others who wanted to be "the lord of Oklahoma" and take over the reins of Phillips Petroleum Company. Clyde Alexander, vice president and general manager of Phillips, still considered himself to be the prime candidate. Obie Wing, the company's secretary-treasurer, remained Frank's golden boy and seemed a contender. A flock of young vice presidents and executives—Paul McIntyre, Howard Sherman, Stewart Dewar, and others—watched from the wings and jockeyed for position. They could only hope that someday they'd get their place in the sun.

But nobody—absolutely nobody—wanted to become president of Phillips Petroleum more than Boots Adams.

Working as assistant secretary-treasurer under Wing, the ambitious Boots, driven by his overwhelming passion to get to the top no matter what the price, did everything in his power to please Uncle Frank and charm Aunt Jane. At the peak of the Depression, Boots hit the road for Phillips, carrying a suitcase crammed with more than a million dollars' worth of company stock and bonds to be used for negotiating suitable marketing outlets. Boots crisscrossed the country, working twenty-hour days, pinpointing choice properties, analyzing potential sites, courting prospects.

In 1932 Boots and Frank were returning to Bartlesville by train after a lengthy stay in New York. That evening in the dining car, Frank told Boots that he intended to recommend to the board of directors that Adams be made his "assistant," a brand-new position.

"For a moment," Boots recalled years later, "I was aghast. For one thing, what I thought he meant by assistant was just a glorified secretary. And I had seen some of Uncle Frank's secretaries and assistants as they passed through our company. Why, some of them had lasted as long as six months in the job before they vanished from sight. I didn't want what I thought was a promising career to end that way."

Boots fidgeted in his seat as he worked up the courage to speak. "You know, Mr. Phillips, I can't take dictation or type very well."

Frank looked at Boots over the rim of his glasses and said, "I don't mean *secretary,* young man, I mean *assistant!*"

Boots replied that he wanted some time to consider Frank's offer.

"All right, you think about it. This is Saturday. You've got till Monday to make up your mind. That's when the board meets. But I want you to know something else. Nobody else on the board agrees with me that this is a good idea."

During what seemed like an endless weekend, Boots considered every angle and aspect of Frank's proposal. "Uncle Frank was a wonderful man, but the toughest taskmaster I have ever known," said Boots. "I figured that if I took the job, I didn't have a chance of staying with the company very long. But I also knew that if I didn't take it, I probably wouldn't be with the company even that long." On Monday, with his first cigar of the day smoldering in an ashtray, Frank had Boots standing tall before his gleaming desk. "Young man, you've had enough time to think. Do you want the job or not?" Boots had rehearsed his response for hours. "Mr. Phillips, you made up my mind last Saturday, but I didn't know it then. I'll take it."

Frank pushed the recommendation through his board of directors and announced that K. S. Adams was now the assistant to the president of Phillips Petroleum Company. Boots was given a new office, and for three weeks he didn't receive a single communication or directive from Uncle Frank. Then one Friday afternoon, Frank appeared in Adams' office.

"Well, what have you been doing for the past three weeks?" snapped Frank.

Adams reached into his desk drawer and took out a file. He explained it contained a list that he had drafted which included everything that he would do if he were the president of Phillips Petroleum.

"This took even Uncle Frank back a little bit," said Boots. "I handed him the yellow sheets and he took them and walked out."

That evening, Boots was home with Blanche and his children when Frank called and summoned him to the town house. Boots ran out the door and sprinted to Cherokee Avenue. Frank told his new assistant that he liked all of the proposals on the list, except for one—the suggestion for a budget committee for long-range financial planning to be headed by a company executive.

" 'I'll approve the whole thing as it stands with this one exception,' Mr. Phillips said, and he gave me back the list with the name of the executive I'd suggested crossed out and my name substituted as budget committee chairman."

By 1933, Boots was firmly ensconced in the Phillips Petroleum hierarchy. In a dozen years, the thirty-three-year-old brash college dropout rose from a lowly job as a warehouse clerk to become the number one assistant to Uncle Frank himself, with all the power and respect and glory that went with the position.

With more than its share of capable young executives like Boots Adams on the payroll, Phillips Petroleum was working hard to rise out of the muck of the Depression and return to the old levels of prosperity.

So was the rest of the nation. Sometimes the task seemed overwhelming. But Franklin Roosevelt was optimistic. F.D.R. and the Democrats had knocked the socks off the Republicans in the presidential election of 1932, and in March 1933 at his inauguration, Roosevelt—with a few words borrowed from Thoreau—told all Americans that "the only thing we have to fear is fear itself." It was time now for the "New Deal" and for bank holidays and all sorts of social and economic reform programs and projects.

Frank was not too pleased over the political turn of events. Even E. W. Marland, vengeful toward Morgan and the other eastern "economic royalists" that helped destroy his oil empire, had switched political parties and become an ardent New Dealer and won election to Congress in 1932 as a Democrat. By 1934, Marland would become governor of Oklahoma, succeeding the irrepressible William "Alfalfa Bill" Murray, the man who had shut down the state's oil wells in 1931 in an effort to stabilize prices.

Despite the national Democratic sweep in 1932, Frank's friend Alfred Mossman Landon was elected to his first term as governor of Kansas. Alf Landon was a millionaire oil man who had owned leases near the F.P. Ranch, and Frank believed Landon's political future was bright and offered some hope for the Republicans in the face of the growing number of public works programs Roosevelt had created. Still, even old Republican war horses like Frank Phillips went along with some of the Administration's plans designed to bring the economy around.

There was even talk of the oil industry unifying to help combat the ill effects of the Depression. Out in Beverly Hills, this news amused Will

Rogers. He observed in his nationally syndicated newspaper column that "Frank Phillips of oil fame was out the other day; said he was going to Washington. The oil men were going to draw up a code of ethics. Everybody present had to laugh. If he had said the gangsters of America were drawing up a code of ethics, it wouldn't have sounded near as impossible."

Frank knew Rogers' barbs were all in good fun. He looked forward to socializing with the popular comic, who by this time had become a close friend of a one-eyed aviator named Wiley Post. Both Rogers and Post spent time together with Uncle Frank at Woolaroc, playing poker, rolling dice, and swapping lies, and on one occasion Frank surprised the Bartlesville children during a school assembly by trooping out Will, who told them a batch of fresh jokes.

Post, the son of an itinerant Texas farmer, grew up in Oklahoma and at fourteen saw his first airplane at a county fair. Years later, after he learned to fly and he worked as a stunt parachute jumper, Post was forced to return to the oil patch to make ends meet. On his first day back on an oil rig, a flying bit of metal chipped off by a sledgehammer lodged in Post's left eye, which had to be removed when infection set in. Post was awarded $1,800 in compensation, and he immediately sank a good part of it into an old airplane which needed to be rebuilt. Soon he was giving flying exhibitions and carrying passengers throughout the backwoods country. The patch he wore over his glass eye became Post's trademark.

While flying as a Lockheed test pilot and for Oklahoma oil man F. C. Hall, Post won more than his share of prizes and recognition. In 1930 he came in first in the Bendix Trophy race and the next year he and Harold Gatty, an Australian navigator, set a world's record by circumnavigating the globe in eight days, fifteen hours, and fifty-one minutes. Their plane was named the *Winnie Mae*. In 1933 Post flew the *Winnie Mae* alone over the same the course, circling from New York to New York in seven days, eighteen hours, and forty-nine and a half minutes.

"What did I tell you about that little one-eyed Oklahoma boy?" asked Will Rogers. "He's a hawk, isn't he?" Frank had to agree that Post was indeed just that.

Post's flight with the *Winnie Mae* captured Frank's imagination and fed his enthusiasm for aviation. In 1934 when he learned that Post was prepared to go for yet another record—this one for height instead of for speed or distance—Frank agreed to act as his sponsor.

Post outfitted his plane to navigate the stratosphere and invented a suit of clothing calculated to maintain normal atmospheric pressure and oxygen content for the body in high altitudes, a prototype of some of the space suits worn later by modern aviators and astronauts.

On December 7, 1934, Uncle Frank's daily log recorded that he and a group of guests went to the Bartlesville airport "to see Wiley Post take off on stratosphere flight." Post taxied down the runway and started flying higher and higher. The crowd strained to see the little single-engine Lockheed airplane climb to almost 55,000 feet, higher than any human had ever flown before.

Although all four of Post's attempted stratosphere flights were marred by mechanical failures, the aviator did contribute to the conquest of space through the development of the pressurized suit and by becoming the first flier to ride the jet streams.

But even before he became an active sponsor of Wiley Post's stratosphere flights, Frank, coaxed by Billy Parker and Art Goebel, was an ardent aviation supporter. After a race or air show in Bartlesville, Frank hosted the aviators and guests at his ranch, including Amelia Earhart, the famed American aviatrix from Kansas. Besides his sponsorship of transcontinental air derbies, Frank offered trophies and prizes such as the Woolaroc Trophy presented at the National Air Races in Cleveland in 1932. Frank flew up for that event in his private plane, accompanied by a pet goat named Pete, which had been adopted as official mascot of the derby.

In September 1933, Frank brought Jane, their two daughters, John, and Elizabeth and Johnny to Chicago to attend the World's Fair and the International Air Races. Spread over four days, the air races were a major aviation event and attracted thousands of spectators. Major Eddie Rickenbacker, Congressional Medal of Honor winner and the top American ace in World War I; Clyde Cessna; Merrill "Babe" Meigs; and Florence Klingensmith, a twenty-five-year-old female flier from Minneapolis, were among those appointed to the advisory committee. Included on the list of well-known referees were Jimmy Doolittle and Billy Parker.

But of all the races scheduled, the premier speed event and feature race of the entire program was the one-hundred-mile unlimited free-for-all —open to any type of airship, with any type of motor—over a closed course. Frank Phillips was the sponsor. The veteran and rookie pilots alike considered the Frank Phillips Trophy Race to be a high-speed course classic.

To provide some incentive, Frank posted a total of $10,000 in prize money and a handsome Frank Phillips Trophy which stood four feet six inches tall. It was predicted that existing speed records for similar events would be shattered in the contest. Competing planes had to do 225 miles an hour to even qualify for the race and two of the outstanding speed demons of the air, Colonel Roscoe Turner and James R. Wedell, were among those contending for honors.

On the final day of the International Air Races, the Phillips party took their seats and waited for the big event. Frank wore a bow tie, a summer suit, a dapper white hat, and carried his cane. Jane and the girls wore white gloves and hats, and Jane had her fox stole draped over her shoulders. Before the races started Frank kidded with photographers who snapped shot after shot of the family as they watched the preliminary events.

Besides sponsoring the race, Frank also brought along the "Phillips 66 Hollywood Trio," billed as "a group of motion picture stunt fliers that has no equal." The trio—Frank Clarke, Paul Mantz, and Howard Batt—were veteran stunt pilots and thrilled the crowd with a demonstration of acrobatic skywriting. They flew at 10,000 feet above the city and scrawled the name Phillips 66 in mile-high letters for the millions of fair visitors to see. Late that afternoon came the Frank Phillips Trophy Race. All eyes turned skyward once more as the racing planes began flying the one-hundred-mile-long course.

Everything went smoothly until the race neared its conclusion and a tragic accident marred the entire show. Florence Klingensmith, the daring young female pilot and the only woman who entered the free-for-all race, ran into trouble. At the end of the eighth lap, just as she passed the pylon directly in front of the grandstands, a bit of fabric and scraps of wood were seen falling from her plane. She leveled off and headed straight out over the south fence of Curtis Wright Field away from the crowd. She flew southeast for three miles without losing altitude.

Suddenly the nose of her plane went down. The craft dove straight into a tree in the yard of a residence. Her body was found eight feet from the wreckage. She was wearing her parachute harness. Her parachute was found in the debris.

Frank was shocked and greatly upset by the crash and Klingensmith's death. His face was ashen and grave in the photographs taken at the awards ceremony after the race when Frank presented James Wedell, who had set a world speed record, with the trophy and his share of the winnings.

Newspaper accounts said Frank "appeared deeply affected by Miss Klingensmith's crash and death." That evening Frank released a brief statement: "I am trying to give money for the advancement of aviation, just as others give it for education and in other fields. I feel that such races as these advance the knowledge of aviation and its commercial application. Most of us merely give money or effort, which comparatively means nothing. Miss Klingensmith, also trying to advance the art of flight, made the supreme sacrifice."

That night, Frank didn't sleep at all. He sat up in his hotel room and cried. He didn't stop crying until close to dawn. He couldn't get the plane crash out of his mind. The image of the fabric stripping off the fuselage and the young woman guiding her craft away from the field stayed with him all night long.

The tragedy at the air races in Chicago made Frank aware how quickly a life can be snuffed out. He told family members that after the accident he felt very vulnerable. And for good reason. That autumn events took place which convinced Frank that there was more to fear than fear itself. Charles Arthur "Pretty Boy" Floyd was about to enter Frank's life.

A sneering, swaggering desperado born in Sallisaw, Oklahoma, Floyd —for a decade starting in the mid-1920s—left a bloody trail of murders, bank robberies, and kidnappings across America. Called "Oklahoma's Phantom Bandit," Floyd was better known by a nickname he hated— Pretty Boy—probably hung on him by some whores at Mother Ash's place, a popular brothel of the day in Kansas City.

Pretty Boy Floyd was not the type of outlaw Frank invited to his annual cow thieves reunions at the ranch. Despite the innocent nickname and the Robin Hood image he loved to cultivate, Floyd was a cold-hearted killer and the most notorious criminal Oklahoma had known since the days of Al Jennings and the Dalton Brothers. He took up a career of crime and violence at eighteen when he robbed a local post office of $350 in pennies. Over the next several years, Floyd robbed thirty midwestern banks and filed ten notches on his pocket watch to remind him of the number of men he had killed.

Floyd and one of his lieutenants, George Birdwell, staged so many bank robberies and kidnappings that Oklahoma's insurance rates became the highest in the nation. After the demise of John Dillinger, the title of Public Enemy No. 1 was hung on Mr. Floyd. On June 17, 1933, Pretty Boy became one of the prime suspects in the infamous Kansas City Massacre at the Union Station, where machine gunners blew away four lawmen and their manacled prisoner, Frank Nash, the former Osage outlaw who helped rob the train at Okesa many years before.

Only a month after the big shoot-out in Kansas City, while G-men were in hot pursuit of Pretty Boy, Oklahoma City oil man Charles F. Urschel was kidnapped and held for ransom after four men armed with machine guns burst into his home and grabbed their victim while he played bridge. The kidnappers' payoff came to $200,000 and was made in Kansas City. Urschel was released unharmed and law enforcement officers were now on the trail of the culprits—Harvey Bailey and Machine Gun Kelly.

That September, Pretty Boy, taking his cues from the Urschel kidnapping, figured it was time to try his luck at earning some fast bucks by snagging his own oil millionaire. He looked around and determined that as likely a prospect as any was Frank Phillips in Bartlesville. Floyd knew that, given the Depression economy, by kidnapping Frank or a member of his family, he could take in much more loot than he'd glean from several bank jobs.

Floyd began spending time in the Bartlesville area, checking out the Phillips family and learning as much as he could about their whereabouts and general living situation. Pretty Boy also decided to confer with Henry Wells, an outlaw who had operated out of the area for many years and had served hard time with some of Floyd's associates, including the late Frank Nash. Floyd confided in Wells and explained his plan, unaware that Henry considered Frank Phillips to be one of the finest men he knew. Wells listened to Floyd and, so as not to arouse his suspicion, answered some of his questions about the Phillips family.

After talking to Wells, Floyd and his operatives felt they were ready to test the waters. Late that same afternoon, Betty and Johnny Phillips were walking home from Garfield School to their residence on Cherokee Avenue—Betty with her girlfriends and Johnny with some boys from his class. When her group was only a couple of blocks away from the school a black limo pulled to the curb and one of the three men inside asked the girls which one of them was Betty Phillips.

Betty, who had always been trained to be wary of strangers, was suspicious but spoke up and identified herself. Suddenly two of the men leaped from the automobile and tried to grab Betty and pull her inside. She yanked away from them and ran home without looking back. After she caught her breath and explained what had happened, Mildred called Frank, who was out at the ranch for an afternoon horseback ride. He advised his former daughter-in-law to keep the children inside and lock all the doors. Then, before he could make his next move, Frank had a visitor. It was a breathless Henry Wells.

Shortly after his meeting with Pretty Boy, Wells left his hideout near Okesa and slipped across the valley to warn Frank about the kidnapping. Once Frank learned about the plot and found out who was behind it, he called Jane and Mildred and told them to prepare to leave at once for the ranch. Then he notified the local authorities and sent a car to pick up Jane, Mildred, and her three children. John and Mary Kate, who were visiting from Tulsa, were told they would have to stay put. Frank figured they could learn to endure a few days with John's ex-wife much easier than spending time with Pretty Boy and his chums.

"We were out at the ranch when ol' Henry Wells called on Mr. Phillips and warned him and said, 'You can't leave here till I tell you,' " recalled Mary Kate. "It was Pretty Boy Floyd and he was out to get a Phillips, but fortunately for all of us Henry Wells was Frank Phillips' friend. We all stayed in the lodge because it was so well protected, with the fence and the gate and all. No one could get through that gate. We stayed that way for several days—just stymied out there by those gangsters."

Grif Graham, Henry Wells, some of the ranch hands, and a few federal agents took up positions at the main gate and throughout the ranch property.

"It was before I went to work for Mr. Phillips, but I was around and I recall at the time that Grif was loaded for bear out at the main gate," said Joe Billam, a cowboy who eventually became manager of the F.P. Ranch.

To young Johnny Phillips, the entire episode was thrilling—just like a Saturday-afternoon matinee. "I remember it all so well," he said years afterward. "We had all retired one night about midnight and everything seemed so peaceful. Then about two o'clock in the morning a big sedan pulled up at the gate and blasted its horn. There were agents in Grif Graham's gatehouse and more hiding in the brush. I understand the car tried to ram the gate and everybody opened fire and the car got out of there in one big hurry. They never did catch them. Of course, we slept through the whole thing, and the next morning Father told us what had happened. We stayed on at the ranch for a few more days, and Henry Wells came around and said that he believed Pretty Boy and his bunch had given up on their plan and had moved on."

Frank's friendship with Wells had saved his life and possibly the lives of several members of the Phillips family. The Osage outlaw had double-crossed one of his own to protect the man whom he respected more than anyone.

"It was the only time," said Wells, "that I ever ratted on any of the boys. But I told Pretty Boy I'd rather somebody would have kidnapped my daddy than Frank Phillips."

The Pretty Boy Floyd caper again showed Frank how vulnerable he really was and it proved to him that the friendships he had with people from all walks of life—including rascals from the Osage—could be beneficial.

About this same time Frank learned that, in some instances, even long-standing relationships with valued employees weren't as solid or as reliable as he had once thought. When a moneymaking scheme—carried on for some time outside of the company—was uncovered involving four of Phillips' most respected employees, Frank was not only angry but saddened by the betrayal of trust.

Three of those in the questionable venture, which involved competitive oil field activities and diversion of funds in a direct conflict of interest, were top executives. Paul J. McIntyre was vice president in charge of production; Stewart Dewar was vice president in charge of land and geology; and Howard J. Sherman was the superintendent of the land department. The fourth employee, C. C. (Dean) Cummings—several notches lower on the corporate totem pole—served as a contracts and field claims representative. All four of them had worked for Frank since the early years of Phillips Petroleum, and the three senior executives were Frank's personal friends and socialized with John Phillips and the flamboyant young Bartians who made up the Halcyons.

"The rumors about those executives were getting around the country club," said Tom Sears, the former Hillcrest caddy. "They were building big homes, and what we understood is that they had actually formed some sort of dummy company and were scouting leases and buying oil properties for themselves."

When Frank finally decided something was awry, he quietly investigated the matter until he was satisfied that his suspicions were correct. His own employees were using inside knowledge to acquire valuable leases. They were taking in huge profits which, by all rights, should have gone to the company. Frank was very hurt. He was loyal to his employees and he demanded total loyalty in return.

"They were some of Uncle Frank's most trusted men," said Sears. "But when he had his chauffeur drive him by the houses they were building and he realized they couldn't afford those kinds of homes, even on a executive's salary, he knew the stories were true.''

All four of the men involved left the company. Dewar left first, followed by the others. For public consumption, the company called their departures "resignations." Dewar remained in Bartlesville, became an independent oil operator, and in 1943 joined the H. C. Price Corporation as a vice president. McIntyre, Sherman, and Cummings formed their own oil firm in Tulsa and discovered and developed several prolific oil fields in Oklahoma and Illinois.

The departure of four of Uncle Frank's valued employees, including three senior executives, also meant that more of the competition standing in Boots Adams's path up the corporate ladder was eliminated. As Boots figured it, now only Obie Wing and Clyde Alexander stood in his way. And he didn't have long to wait before even they would be gone.

Clyde Alexander would be the first.

For some time, Frank had been conscious that the Phillips vice president and general manager wasn't happy. Alexander was disagreeing with many of Frank's decisions and was becoming vocal about it. At the same time, Frank was none too pleased with some of the things Alexander was doing. When L.E. found out that Alexander was openly criticizing Frank's management style with others, including some of the company's directors, he became concerned and told Frank to beware—a palace uprising was in the making. By early 1934, the situation was growing tense.

"Clyde Alexander took the position that Frank Phillips shouldn't be running the company any longer and that the time had arrived for him to take over," said Paul Endacott, who returned to Bartlesville in March 1934 from his duties with Philgas in Detroit in order to head marketing research for Phillips. "The Depression was on and Frank was concerned about the company. I also think he may have been taking a little bit too much alcohol. L.E. was still sick and going back and forth to the sanatorium at Battle Creek. Judge Beaty was constantly at Frank's side. Then he found out that Clyde was actually contacting some of the directors and talking behind his back. It was a very tough time for Frank."

In the midst of his dilemma with Clyde Alexander, Frank's mother, Lucinda Phillips—Josie—passed away. She was eighty-four years old and suffered a hemorrhage of the brain while preparing lunch at her home in Gravity. Josie was standing by the kitchen sink peeling apples when she suddenly fell over dead in her sister's arms.

Frank dropped everything he was doing and immediately sped to

Iowa. Flowers filled the church in Gravity, and there were close to ninety telegrams and cables of condolence, including one from Waite and Genevieve, who were on a trip in Hawaii. All the other Phillips children were at the funeral.

The preacher spoke of Josie's ten children, and of their successes, and he recalled her husband, Lew, who had died twelve years before. He said she'd be remembered for her endurance. She had known the hardships of frontier life and survived crop failures, sand storms, blizzards, and grasshoppers. As the men at the graveyard began covering Josie's coffin with rich Iowa soil, Frank thought about how his mother had sacrificed for him and his brothers and sisters. A mound of flowers was banked over the fresh grave and Frank broke off a bloom and took it with him.

When Frank returned to Bartlesville, and to his troubles with Clyde Alexander, he kept his mother's spirit of survival in mind as he wrestled with critical company problems. In a little more than a week after Frank came home, the Alexander situation began building toward a fiery climax. Several people, including some Phillips executives, were present for the showdown. Billy Parker was one of them.

"I knew Clyde Alexander and Frank were not getting along too well," recalled Parker. "Clyde made decisions and Frank didn't like some of them. Then Mr. Phillips called in Clyde and told him he was planning on raising the stock dividends for the shareholders. Clyde disagreed. He said it was the first year we were really making money in a while, and that since everyone was working so hard, instead of passing it on to the stockholders he wanted to take care of the employees. That started it."

The grand finale came a few nights later out at Hillcrest. "My wife and I were at the country club that evening," said Parker. "I went down to the rest room. Frank and Clyde were there. Frank had drunk his share of martinis and he was blubbering, 'You son of a bitch, if you don't do what I tell you to do I'm going to fire you!' He said, 'I'm not going to give these people a raise, we're going to up our dividend.' About that time, Mr. Phillips saw me and he asked me to join them. I wanted no part of it, so I went back upstairs."

Witnesses claim that moments after Parker left, Alexander, outraged by Frank's stubbornness, bellowed, "You can't fire me!" Frank asked why, and Alexander responded, "Because I quit!" Then he reportedly drew back his fist and struck Frank on the jaw, knocking him to the floor. The battle royal was short-lived. The palace uprising was over. Alexander stormed out of the country club, into the cold night.

With the assistance of Billy Parker, Frank Phillips assembled one of the first and finest corporate aviation fleets in the nation. Pictured in one of the first planes in the mid-1930s are (left to right) Jimmy Thompson, secretary; Frank Phillips; Boots Adams, Frank's assistant and company treasurer; and John F. Kane, executive vice president and general counsel. *(Phillips Petroleum Company)*

With Frank Phillips, chairing a meeting of the Phillips Petroleum Company executive committee in the 1930s, are (left to right): George Oberfell, director of research; Boots Adams, R.F. Hamilton, Jr., committee secretary; and Don Emery, general counsel. *(Phillips Petroleum Company)*

Fern Butler with Woolaroc, the Osage stallion raised on the F.P. Ranch and presented to her by Frank Phillips in 1943. *(Nancy Marshall)*

Attending the International Air Races in Chicago in early September 1933 are members of the Phillips family (left to right): daughter Jane, Mrs. Frank Phillips, granddaughter Betty, grandson John, Frank Phillips, daughter Mary, and son John. *(John Gibson Phillips, Jr.)*

Frank sits amid a shower of gifts he received from around the world when he celebrated his sixty-sixth birthday, November 28, 1939. Bartlesville was filled with guests, some of whom had to sleep downtown in makeshift quarters. 25,000 people celebrated with Uncle Frank, and a grand parade topped the day's events. *(Phillips Petroleum Company)*

Mr. and Mrs. Frank Phillips posed for this formal portrait on the occasion of their fortieth wedding anniversary in 1937. *(John Gibson Phillips, Jr.)*

ABOVE:

Four generations of Phillips men gather for a reunion in 1946. Left to right: John Gibson Phillips, Jr.; John Gibson Phillips, Sr.; Frank Phillips; Frank Phillips III (oldest child of John Jr.). *(John Gibson Phillips, Jr.)*

LOWER RIGHT:

Frank and Jane's grandchildren, the only children of Mildred Beattie and John G. Phillips, Sr., are pictured in 1937. Left to right: Robert Beattie Phillips, Elizabeth Jane Phillips, and John Gibson Phillips, Jr. *(John Gibson Phillips, Jr.)*

Frank Phillips, successful banker and wildcat oil man, built a magnificent mansion at 1107 South Cherokee Avenue in 1909. The structure was renovated in 1930. It served as Frank and Jane's town house in Bartlesville for more than forty years. *(Phillips Petroleum Company)*

The living room at Woolaroc, country retreat of Frank and Jane Phillips, was the scene of hundreds of social gatherings attended by some of the most accomplished and well-known figures of the day. *(Frank Phillips Foundation, Inc.)*

Frank Phillips, outfitted in garb presented to him by the Outlaws and Cowthieves of the Osage Reservation at their annual picnic at the F.P. Ranch (Woolaroc) in 1928. *(Frank Phillips Foundation, Inc.)*

On February 23, Alexander was in Frank's office one last time—to hand in his resignation as an officer and director of Phillips Petroleum. He had ended a relationship—and a friendship—that spanned many years. Later that afternoon, Frank released the story of Clyde's departure to the press. According to the official newspaper version, Clyde resigned to "devote his time to personal interests."

Frank regretted that his long association with Alexander had ended in such a fashion. He was bitter over the conflict, but in a tribute to the good old days before their differences pulled them apart, Frank never changed the name of the body of water in front of Woolaroc. It remained Clyde Lake.

That April, during the annual meeting of the Phillips stockholders and directors, someone else who had been with Frank since the beginning left the company. L.E., approaching his fifty-eighth birthday, decided to call it quits and announced his retirement from Phillips Petroleum.

Continually plagued with several physical discomforts, L.E. would have more time for travel with Node. He would also collect wooden cigar store Indians and oversee his champion hog-breeding program at Philson Farms, his ranch retreat outside of Bartlesville. He'd continue to serve on the board of the Federal Reserve Bank in Kansas City and always maintained an active interest in the oil firm he helped found in 1917.

The entire upper-echelon structure at Phillips Petroleum was changing. Alexander and L.E. were gone, and several other top executives had left. Frank was feeling the strain and pressure of the times, and in the spring of 1934 when he became jaundiced and weak, he checked into the Fifth Avenue Hospital in New York for rest and recuperation. A complete battery of tests revealed that he suffered from diverticulosis, an intestinal disturbance caused by the inflammation of sacs, or diverticula, which doctors diagnosed had developed early in Frank's life.

After a few weeks of treatment, Frank's condition improved and he was able to return to work and meetings with Wing and others at the Phillips office at 80 Broadway on the thirty-seventh floor of the Irving Trust Building in lower Manhattan. From the windows Frank and Fern and the other officers of the company occasionally watched the shipping traffic pass the Statue of Liberty. They were within a few moments' walk of the heart of the Wall Street district.

Later that summer, after Frank settled in for long conferences over company matters with Wing and Adams and his remaining top executives, there was more bad news. Frank's favorite company officer, forty-three year-old Obie Wing, was very ill. He complained of an intestinal and stomach disorder. Frank thought maybe Wing had diverticulosis too, but in late August, when the illness grew critical, Frank insisted that Wing go to Johns Hopkins in Baltimore for treatment. Frank and Jane went with him. The diagnosis was not good. Wing had cancer, and the doctors said he didn't have long to live.

Frank was overwhelmed. Wing was the company spark plug—the one who convinced Frank of the value of a basketball team; the one who helped him judge bathing beauty contests at the ranch. He made Frank laugh and feel good about the future of Phillips Petroleum. Wing was the firm's reliable treasurer; he could be trusted. Frank had already kind of decided Obie would be the next Phillips president.

Wing was realistic about his state of affairs but managed to keep smiling. "Now I'm going to do what I've wanted to do all my life. I'm going to hire a colored boy, get a bucket full of ice and some bottles of beer, and we're going to play golf," said Wing when he returned to Bartlesville.

During the periods he was outside of the hospital, Wing resumed his duties at the office but spent as much time as possible at the country club, fulfilling his fantasy. The New Year's Eve party at Hillcrest was one of Obie's last flings, and he came dressed in white tie and tails, as full of life and hope as anyone.

For several months Frank was able to get his mind off Obie Wing's illness. Good tidings were in the air. Intelligence reports from Oklahoma City seemed positive, and suddenly, after an arduous battle to open up the sprawling Oklahoma City oil field, Frank sensed victory. It had been a long, and often frustrating process.

It all started back in 1928 when H. V. Foster's ITIO brought in the discovery well—Oklahoma City No. 1. The field—known for its wild wells, rivers of flowing crude, and almost uncontrollable flows of natural gas—made Oklahoma City, for many years a stepchild to the more sophisticated Tulsa, a roaring boom town. It was the first time a major oil discovery had occurred in a large metropolitan area, and it left the city and its citizens confused and reckless.

Then the Sudik No. 1—better known as the Wild Mary Sudik—broke loose on March 26, 1930, and gushed out of control for eleven days, threatening fires and showering oil as far away as Norman, eleven miles to the south. The well was finally tamed. That same year, a young Phillips geologist named Dean McGee was able to convince his superiors that

ITIO had barely scratched the surface of the Oklahoma City field. McGee recommended drilling deeper into the formation, which ran directly beneath the city.

McGee's ability to persuade the Phillips land department to buy potential drilling sites away from the discovery well was important. Not just because it was another gamble that paid off, but because it was the first time an oil company expanded a pool based solely on geological information. The last vestiges of the old school's hocus-pocus had vanished.

Phillips and several other oil companies leased acreage which extended into the city limits, and the field became one of the state's major producers of crude. As the field's growth reached city sidewalks, local citizens began complaining about the destruction of the neighborhoods and protested oil wells being drilled next to schools, churches, and hospitals. Ordinances were passed prohibiting future drilling, but the critics of the oil wells were quickly labeled "enemies of progress" by the oil companies, who ignored the zoning laws and kept right on drilling. Finally, in 1932, Governor "Alfalfa Bill" Murray, as he had done the year before to stop overproduction and raise the price of crude, ordered the Oklahoma National Guard into the oil field to enforce the city's rules governing the drilling of oil wells.

A new and comprehensive code to control overproduction was adopted, and it appeared that order had been restored. The troops were sent home. But problems continued. Despite the governor's actions, the situation in the Oklahoma City fields remained chaotic, and several oil companies continued to produce outside the regulations.

McGee, who had become Phillips' chief geologist, knew his recommendation about the field's potential was sound. McGee was present when Phillips moved deeper into the field and even drilled a pair of wells in a local garbage dump. The names they gave the two wells were appropriate —Ash Can No. 1 and Ash Can No. 2.

Within two years, McGee left Phillips and hooked up with Robert Kerr, another leader in the fight to open all parts of the community to oil development. The two partners organized a large independent oil company called Kerr-McGee.

In 1935, the Oklahoma City field was the second largest in the country and drilling had reached the boundaries established by the city fathers. Phillips was faced with several problems. It wasn't clear whether the company was smart to lease more land within the city. Also, there was a question whether the regulations restricting oil drilling would be changed. Once again, it was time to roll the dice.

An intelligence report on the situation sent to the Phillips executive committee stated that a proposed zoning-extension ordinance stood little or no chance of gaining voter approval. In addition, any more leasing and drilling would be tremendously expensive, since it meant buying the leases lot by lot and moving houses off the sites. "Well, that's that," Frank told the executive committee. "We're not interested in Oklahoma City."

But Frank's young assistant, Boots Adams, had different ideas. "I don't agree at all," said Boots. "I'm sure the extension can be won." Frank took a hard look at Boots. "All right, if that's the way you feel, you go down there and do something about it."

Boots left for Oklahoma City that night. With Obie Wing's health deteriorating every day, Adams had moved closer to the front of the pack. If he could be instrumental in turning around the Oklahoma City crisis, Boots felt he'd be assured of a secure senior executive post.

The Oklahoma City newspapers were opposed to drilling in the city and singled out Phillips Petroleum as the villain. The papers raked Frank over the coals for his disregard for shrubbery and the beauty of homes and neighborhoods. As he usually did when he was pushed, Uncle Frank bowed his neck and fought back. He had the backing of the other oil firms, but he was the marked man. The only way he knew to win was to try to influence the people who were going to vote for or against the drilling regulations.

"I own the oil under my leases in this town, and I intend to get it out!" said Frank.

Meanwhile, Boots and a small army of Phillips employees were busy beating the bushes to win support for passage of the zoning extension.

Some Phillips workers even went door to door through the thriving red-light district in order to drum up support and votes to pass the ordinance. They also spent millions of dollars buying up leases within the city limits in hopes that their side would win the election and they could get down to some serious drilling.

When Phillips found the local newspapers wouldn't even allow the company to buy advertising space, Boots set up shop in the Skirvin Hotel and began publishing a free newspaper called *The Truth About the Oil Extension.* Oklahoma Citians became avid readers, not only because the price was right but also because copies of the regular daily papers had a way of vanishing from porches and front yards whenever *The Truth*'s paper boys came around.

The Truth contended that the good citizens lined up against Phillips beneath the banner of civic beauty were well-intentioned but were, in fact, all wet. Frank's newspaper said even civic beauty is only turf deep and that the owners of mortaged homes had a right to take the oil from under the ground.

The combination of a well-oiled propaganda machine and lots of cold cash worked. When the election was held on March 5, 1935, Phillips and the other oil companies won. Within half an hour after the returns were counted, trucks loaded with drilling tools thundered down the city streets. By dawn derricks were going up. Many of the wells produced more than 20,000 barrels a day. The next year there was yet another election and another fight. Phillips won again.

Eventually, Phillips' rigs popped up across the city and even appeared on the lawn of the governor's mansion and in front of the state capitol building. Capitol Site No. 1, also known as Petunia because it was erected in the midst of a flower bed, was completed in 1942, one of twenty-six original derricks on the capitol grounds. Phillips drilled the well diagonally so that it produced from a spot directly below the center of the capitol building. Petunia produced faithfully for forty-four years, yielding 1.5 million barrels of oil and 1.6 billion cubic feet of natural gas during its lifetime.

Boots' victory proved to be a milestone in the company's development. The Oklahoma City field was a godsend for Phillips. Once again, Frank's gambler's luck, aided by some solid geology, turned the trick.

Frank and his family were vacationing on the beach at Boca Raton, Florida, that March of 1935, when the first election was held. Jane's personal diary for that date noted: "Frank won his election in Oklahoma City —to drill in city limits." When he received the telephone call with the good news, Frank lit a victory cigar, had Dan break out the drinks, and stayed up until after midnight celebrating. But all too soon the glow of the Oklahoma City victory would be diminished. Obie Wing's hopes of beating his cancer were fading. He was sinking fast. Obie was operated on at St. John Hospital in Tulsa in April 1935, and for the next week, notes in Jane's diaries reflected the sad results: "Obie's condition very grave." "Obie not so well." "Obie no better." "Obie worse."

On Good Friday, Fern flew in from New York to visit Wing. Frank met her at the airport and they drove to Tulsa. Two days later, Jane wrote:

"Obie bidding everyone goodbye." Frank and Jane and Fern and Boots and many others went back and forth between Bartlesville and Obie's bedside in Tulsa. Jane and Fern comforted his wife, Oral, and the Wing children. Finally they too said their goodbyes to the young man Frank loved like a son. On April 24, at 7:41 A.M., Obie's suffering ended.

Oberon K. Wing's death was mourned throughout the Southwest. Included in the list of honorary pallbearers were bank and railroad presidents as well as Phillips directors, oil company officials, and old friends. Jimmy Gullane, the stocky golf pro from the country club who helped Obie work on his southpaw drives, was an honorary pallbearer. So were H. V. Foster, Judge Beaty, L.E., Waite, and John Phillips. Former executives Stewart Dewar and Paul McIntyre were taken back into the fold for the day, as was Clyde Alexander.

Boots Adams was a pallbearer. He had learned much about the business during the years he worked under Obie and would genuinely miss the man whose meteoric rise was enviable. In only eleven years Wing had become a Phillips officer and director as well as a director of the First National Bank.

Boots grieved over Obie's death but he also had to be thinking that, coupled with his role in the successful Oklahoma City oil-field fight, the death of Wing left the door to Uncle Frank's office standing wide open. All Boots Adams had to do was walk in.

Shortly after Wing's death, Frank cleared his head by going on a fishing trip to Guaymas, the northwestern Mexican city on the Gulf of California. Billy Parker flew Frank, Dan, and Charles Lemp, the Phillips director from St. Louis, to the Arizona border, and then the party motored down the Sonoran coast to the Miramar Beach Hotel, where they were joined by Wiley Post.

Aunt Jane, who had also been quite fond of Obie and remained close to his widow, Oral, used her busy social schedule to take her mind off Wing's death. In May, during a trip to Creston with granddaughter Betty, Jane was involved in more tragedy. It happened while she was visiting her childhood friend Minnie Hall Thomas, married now to Joseph Thomas, and their daughter Jane, named for Jane Phillips and the same age as Betty.

One afternoon, shortly after Jane and Betty arrived in Creston, Joe Thomas was burned to death when his automobile and a gasoline truck exploded during a severe lightning storm. Jane handled the funeral arrangements and consoled her friend for several days until she had to rush back to Bartlesville to complete plans for the wedding of her foster daughter Mary.

The wedding took place June 8, 1935, in the Phillipses' suite at the Blackstone Hotel in Chicago. Mary was twenty-one years old and was marrying Marcus Low, son of Noretta Low, a widow and one of Jane's Bartlesville friends. Mark, who had a football scholarship to Northwestern University, was forced to quit school during the Depression and worked both as a soda jerk and as a cleanup man in a service station. Before he allowed Mary to wed, Frank made sure her young man got a sales position with Phillips Petroleum.

About fifty guests—mostly family and close friends, including Fern Butler—attended the wedding. Betty, soon to be a student at Monticello, a private school for young ladies in Godfrey, Illinois, near St. Louis, and Jane, enrolled at Webber College, served as bridesmaids. Mark's brother, Dean, was the best man, and Uncle Frank, wearing a fancy suit and a wide grin, gave his daughter away.

Jane and Frank spent the rest of the summer in their usual manner, shuttling between their homes in New York and Oklahoma, with occasional visits to the Lido Club, their favorite seaside resort just outside New York.

They were relaxing at the Lido Club that August when word came of yet another tragedy. Two of the Phillipes' dearest friends had been killed in an airplane accident. Will Rogers and Wiley Post were gone.

The cowboy-philosopher and "that little one-eyed Oklahoma boy" died when their plane crashed near Point Barrow, Alaska. Billy Parker had warned Post before the fateful trip not to trust the craft he and Rogers were taking on their aerial vacation. "It's too nose-heavy and will go right down in one of those lakes if an engine conks out even for a moment," said Parker. But Post didn't heed Billy's words of advice. He told Parker he was determined to go to Siberia and hunt tigers. He never got his chance.

The August 15 Associated Press bulletin from Alaska told the story: "Death, reaching through an Arctic fog, overtook Will Rogers, peerless comedian, and Wiley Post, master aviator, as their rebuilt airplane faltered and fell into an icy little river last night near this bleak outpost of civilization."

The entire world was shocked and went into mourning. The New York *Times* devoted four full pages to the event and ran an editorial about both Rogers and Post. Singer John McCormack, a friend of Rogers for many years, wrote: "A smile has disappeared from the lips of America, and her eyes are now suffused with tears. He was a man, take this for all in all, we shall not look upon his like again."

The Phillipes couldn't believe it when the sad news reached them out at the Lido Club. A reporter from the New York *Evening Journal* caught up with Frank and filed his reaction: "Frank Phillips, backer of Post's stratosphere flights, was so stunned he could hardly speak when found today at the Lido Club . . . and informed of the deaths of Post and Rogers."

Frank could barely mumble in a shocked whisper: "Terrible, terrible. It is so stunning. I just can't say anything. It's hard to realize such a thing could happen."

Will's widow, Betty Rogers, had been in Maine when the accident occurred. She returned to New York, and Frank came in from the beach to console her and the family. He met with them at the Waldorf-Astoria, and when it was time for the sorrowing family to catch the Pennsylvania Limited and head for the West Coast to meet Rogers' body, Frank accompanied them in a private car to the train station.

On his way back to the Lido Club that evening to rejoin his own family, Frank considered all the tragedies he and Jane had faced over the past few years. Air crashes, friends' businesses and banks bankrupt, labor violence, divorce, alcoholism—all these things punctuated the Phillipses' lives. Then there were the deaths that had to be reckoned with: George and Joe Miller were gone and with them the 101 Ranch; John Gibson, Lucinda Phillips, Obie Wing, and now Will Rogers and Wiley Post joined the others.

The 1930s had not been easy years for anyone, including the Frank Phillips family. Yet when he felt low or just wanted to get away from life's pressures, Frank knew he could always return to his home in the Osage— Woolaroc. For pleasure, he knew no better place than his ranch, where Claude Johnson or one of the other hands kept the saddle horses groomed and ready for long rides. Back in New York, the Phillipses lived high and handsome—dining at fine resturants, spending evenings at the theater or the Stork Club or Bill's Gay Nineties. But at the ranch, they got back to the earth and mixed with friends and associates from all walks of life— millionaires and working folks. To the Phillipses, the size of a person's bank account mattered very little.

As Frank put it to a reporter in Omaha in 1935 when he was asked about his personal wealth: "Why, I can't eat any more ham and eggs for breakfast than you, and it's a cinch I can't take any more money with me when I die."

The ranch was also ideal for conducting business—as long as Frank was able to throw in a good dose of fun for seasoning. "If I can get a man to the ranch," Frank liked to say, "I can make a deal." He seldom failed.

Prone to jokes and exaggerations, Frank was particularly fond of hosting tenderfoots who had never slipped their foot into a stirrup, let alone ridden a stallion through a herd of cattle or past some nasty buck llama that liked to spit a thick mucus, which resembled chewing tobacco, at the guests.

"I have a hell of a time down here with those eastern bankers and investors," said Frank. "You should see them. They're always thoroughly primed with stories of hangings and shootings and scalpings. But I put 'em at ease right off."

Many times, Frank would offer his new guests big glasses of milk. The startled visitors usually expected a cocktail and were surprised by the rather tame beverage. Once everyone had his glass in hand, Frank would raise his and say, "Bottoms up." After they all had a big slug, Frank asked how they liked it. Was it sweet, or fresh, or cold enough? After a few more minutes of milk sipping, Frank would finish his drink and inform his guests that the milk they had just enjoyed wasn't from one of Frank's cows at the dairy barn but had come straight from his buffalo herd. After that remark, the room usually became quiet except for the sound of glasses being put on the table. Of course, the milk was actually from a cow, but Frank never told them the truth. Sometimes he'd even invite the guests to go out and help milk the buffalo at the stroke of midnight, since that was the only time the big animals were willing to be milked. The buffalo milk escapade became Frank's own version of a snipe hunt.

Frank once had a little fun at the expense of Elliott Roosevelt, one of F.D.R.'s sons, when he came to visit the Phillipses at their ranch. It was no secret that Frank was a strong Republican, but he liked young Elliott, twenty-five years old at the time, primarily because of his work to establish a United States–South America air race. Also, Frank understood that even though Elliott was the son of the President and served as an officer of the Young Democrats of Texas, he was outspoken on political issues and at times expressed an anti-New Deal point of view.

When he arrived at the airport, Elliott was warmly greeted by Frank, who brought him out to the ranch for lunch. On the way, Elliott mentioned that he'd enjoy stalking and shooting a buffalo. Frank said that could be arranged, and once at the lodge, he had the young man pull on a pair of chaps and a cowboy hat. After some target practice by Clyde Lake, Roosevelt followed Frank and some other guests to a pasture where a small herd of buffalo were grazing.

Young Roosevelt was excited when Frank handed him a .30-30 rifle and told Grif Graham to have one of his boys drive the herd past so he could take a shot. The big shaggy beasts came rumbling by and Roosevelt shouldered the rifle and fired. A miss. Frank cleared his throat and Elliott wiped the sweat from his face. The cowhands rounded up the herd and drove them in front of Roosevelt for a second try. Another miss. Without a word, Frank took the rifle from the chagrined Roosevelt and handed it to Grif. "Here," said Frank, "see if you can do any better."

For the third time, the buffalo ran past. Grif took aim at a big bull—about a 1,600-pounder—running at a good clip at least seventy-five yards away. He squeezed the trigger and put a bullet behind the big beast's ear. The buffalo dropped dead in its tracks like a load of bricks.

Humiliated, Roosevelt trudged back to the lodge with the others for a lunch of barbecued elk and beer. He could hardly stand to look up from his plate, he was so embarrassed by his poor marksmanship. About half-way through the meal, Frank began giggling. Soon he was laughing so hard he was in tears. When the red-faced Elliott looked up, Frank finally admitted that the rifle he had used for the buffalo shoot was loaded with three bullets. "And the first two," roared Frank, "were blanks."

Although he never returned to hunt more buffalo, Roosevelt wrote the Phillipses a gushy thank-you letter and promised he'd never forget their hospitality and the charm of the ranch.

Neither would all kinds of other well-known guests—politicians, actors, artists, executives, clergy—who made their way to the Frank Phillips Ranch to enjoy an evening or a weekend at Woolaroc. The guest list over the years was a Who's Who of the greats from Hollywood, New York, and the capitals of the world.

Kansas governor Alf Landon came to Woolaroc for early strategy meetings after the Republicans chose him in 1936 to lead their fight to reclaim the White House from Roosevelt. Despite a vigorous campaign that included four national tours, Landon suffered a crushing defeat and even failed to carry his home state. The loss ended his national political career, but he remained active in Kansas politics as well as in the oil business and continued to come to Woolaroc to visit with the Phillipses.

Rudy Vallee, the singing idol of the Depression era with the nasal twang and varsity manners, came to Woolaroc. As did Aimee Semple McPherson, the colorful evangelist whose well-publicized kidnapping in 1926 turned out to be nothing more than a secret tryst. Herbert Hoover and Harry Truman visited the lodge—at different times. Other guests of distinction included movie stars Greer Garson, Wallace Beery, Eddie Cantor, and Mary Pickford; singer Marion Talley; Jimmy "Thirty Minutes over Tokyo" Doolittle; big-band leaders Paul Whiteman and Vincent Lopez; sculptor Bryant Baker; author Edna Ferber; as well as many regulars such as Will Rogers, Wiley Post, Roy Chapman Andrews, Perle Mesta, Tom Mix, and Pawnee Bill.

Richard Gordon Matzene, a London-born photographer and art collector, was a frequent visitor at the ranch. Dave Rubinoff, the violinist and band leader who rose to fame on Eddie Cantor's radio shows in the early 1930s before he became a featured soloist on his own national radio program several nights a week, came to the ranch many times to serenade the Phillipses with his violin.

Frank formed a strong alliance with several prominent Catholic leaders, for political and business reasons and also because he genuinely liked several of them, such as Archbishop Francis Spellman, the powerful head of the Catholic Church in New York who became a cardinal in 1945, and Bishop Francis Clement Kelley, the spiritual leader of Catholics in Oklahoma. Both men, and several other archbishops and bishops, visited Woolaroc through the years. They'd pull on Indian headdresses and brightly colored blankets or a ten-gallon hat and pose for photographs in front of the lodge.

"Powder River Jack" Lee and his wife, Kitty Lee, were friends of Frank and Jane and enjoyed their time spent at Woolaroc. Powder River Jack had toured with Buffalo Bill for fourteen years and worked with Will Rogers and Tom Mix in Wild West shows. The Lees liked to compose melodies for Frank and Jane, and one year for Christmas they wrote a ditty called "The Five Phillips Brothers" and sent copies to Frank, L.E., Waite, Ed, and Fred.

Parties at Woolaroc were always interesting affairs. In the dining room servants laid out china from Marshall Field's on checkered tablecoths. Over one of the fireplaces Frank hung a fish he called his mascot. It was a tarpon Frank caught off the Texas coast, and he liked to show it off to dinner guests. He told everyone it weighed sixty-six pounds and was sixty-six inches long—the perfect dimensions for a Phillips trophy.

Outside in the foyer, Dan or Henry stood on the stone floor and mixed drinks at a table used as a bar. Above the table hung a painting entitled "The Old Soak," a portrait of a well-seasoned Irish immigrant with one hand holding a small glass and the other an unmarked bottle. The character's eyes were entrancing and followed the viewer from any angle.

Frank acquired the painting from the Calumet Club in New York after the Depression closed its doors. An Indian drum, also stationed in the foyer, was used to summon guests and family to meals, and a pair of cigar store wooden Indians standing across the bar were gifts to Frank from L.E.

As guests gathered in the Great Hall, the aroma of barbecued buffalo drifted from the kitchen. Oftentimes, male guests would get a drink and try to find an excuse to go upstairs to Aunt Jane's room, where all the walls were covered with photographs of the famous and not so famous who came to visit. Almost everyone of them was a photo of a man—"my boyfriends," Jane was fond of saying—and if someone fell out of favor with her for one reason or another, his portrait was almost certain to be pulled from the wall and banished to a discard pile. In Jane's photo gallery were portraits of presidents and cardinals, movie stars and oil tycoons. Clayton Fisher, the Phillips driver, even made the gallery wall, alongside Truman, Spellman, Doolittle, and Hoover. The criterion to get on Aunt Jane's wall wasn't the size of a bankbook or personal fame but whether she liked you.

On the balcony, where memorable poker games were played, Frank mounted a Texas longhorn head, a gift from Amon Carter, the prominent Fort Worth publisher and businessman. The head was wired for light and sound, and when Frank pushed a button, the steer's eyes lit up like glowing red coals, smoke puffed from its nostrils, and the voice of Will Rogers came out of its mouth.

Directly below the big steer head was a set of longhorns which measured nine feet ten inches from tip to tip, and hanging from the balcony rails were alpaca rugs John Phillips brought back from his expedition to South America. Frank also kept a long tusk from a narwhal, a whalelike mammal of the Arctic seas, on the balcony and it became an object of curiosity. If someone asked about the tusk, especially a lady visitor, Frank was apt to tell them, with a perfectly serious face, that it was actually a petrified penis bone from a whale.

Below the balcony—and the chandeliers Frank took from the original

Waldorf-Astoria—was the Great Hall and all sorts of visual delights. On the walls hung oil paintings of Indian chiefs and western scenes, mostly from Santa Fe and Taos artists, including a painting by Oscar Berninghaus of some cow ponies tethered outside a saloon on a snowy night. It was Jane's favorite, and she asked that it always be left hanging in the northwest corner of the Great Hall. There were also many Indian blankets, several of them purchased by Frank during his visit to the Grand Canyon in 1926.

The scores of animal trophy heads which covered the walls represented some of the wild animals which were unable to acclimate to the ranch and others were trophies presented by Frank Buck, Martin and Osa Johnson, and John Ringling North. Lloyds of London, which insured the huge collection, rated the horns and heads in the Woolaroc lodge as the most complete collection of its kind in the world.

On either side of the main entrance to the lodge, hung paintings of Wiley Post and Will Rogers, and stuck high on the same wall to the left of the door was a playing card. It was put there by Blackstone, a famous magician who came to Woolaroc to entertain some of Frank and Jane's guests. During his act, Blackstone asked Jane to pick any card at random from a deck. She selected the queen of spades. Next Blackstone asked her to put the card back in the deck and to shuffle the cards and then throw them against the wall. Jane did as she was told and all the cards fell to the floor except one—the queen of spades. It magically flew up to the wall and stuck there. Frank gave strict orders that it should always remain in that spot as a reminder, he liked to laugh, that even Jane could be fooled.

Scattered around the big room were stuffed chairs and sofas, elaborate cattle horn furniture, and ashtrays made from elk, buffalo, and rhinoceros hooves. And, of course, several vases filled with peacock feathers.

Then there was the Phillipses' piano. When they first purchased the Steinway, its mahogany veneer gleamed just like the piano in the music room at the town house. The workmen uncrated the new piano and placed it in a prominent place in the Great Hall. It was a Steinway Duo-Art, which meant it could be modified to be played manually or by using the Pianola rolls.

That evening a group of guests arrived at the lodge. After they got their drinks they moseyed toward the new piano gleaming in the late afternoon light. When Frank appeared he reached into the large bark-covered cigar box, pulled out a plump stogie, and grabbed a handful of wooden kitchen matches. After biting off the tip of the cigar, Frank casually walked over to his guests, took a kitchen match, and struck it on the side of the piano. The match's abrasive sulphur head left an ugly scar in the polished mahogany. Everyone in the room gasped in horror, but Frank didn't say a word. He just lit his cigar and puffed away.

Throughout the evening, Frank repeatedly struck the kitchen matches on the piano, each time leaving another scar and his guests a bit more shocked. Finally, near the end of the evening after no one present mustered the courage to ask what he was doing, Frank spoke up.

"You know, I can't help but notice that you all have been giving me some hard looks every time I strike a match on this Steinway," said Frank. "Now let me tell you something. It doesn't matter one bit what this mahogany looks like, because come tomorrow, it's all going to be covered with bark."

The next day, true to his promise, Frank's workmen applied slabs of pine bark to the entire piano. It was the same Arkansas pine bark veneer used for all the frames for the paintings in the lodge and in Frank's airplane museum, an original process Frank liked so much he had it patented.

Over the years the bark-covered player piano got a good workout. During the holidays and. at parties, scroll after scroll was fed into the Steinway. For hours at a time songs like "Rhapsody in Blue," "For Me and My Gal," "Embraceable You," "Thou Swell," "Ain't We Got Fun," "When My Sugar Walks down the Street," "Bye-Bye, Blackbird," "Sentimental Journey," "Tea for Two," "September Song," and, of course, Frank and Jane's memorable favorite, "Let Me Call You Sweetheart," poured out of the lodge, carried across Clyde Lake to the distant meadows and beyond. For Frank, the music was sweet as angel's songs and his ranch was heaven on earth.

"This isn't all a dream about something," he told a reporter from the *Kansas Stockman* in 1937, "but it's a place where I can get back to nature. The great difficulty with the American people today is that they are getting too far away from the fundamental things in life." Frank tried his best to practice what he preached that year but he wasn't always able to get back to nature. Both business and family matters kept Frank busy and on the road.

The year 1937 began with the announcement of young Sara Jane Phillips' engagement to Frank Begrisch, member of a prominent New York family involved in the real estate business. Another Frank and Jane were soon to be wed. Frank and Jane Phillips hosted six hundred guests during an engagement party for the young couple at the town house, filled with poinsettias, roses, and jonquils. Out-of-town guests included the Frank Bucks and the George Vierhellers. All the women attending the ceremony were given orchid corsages and the men donned gardenia boutonnieres. Frank wasn't particularly pleased about Jane's plans to wed, although he reasoned that his daughter was of age and Begrisch was a fine young man. But when young Jane and her mother told Frank the wedding would take place on February 18—the same date the Phillipses were married—he was won over and gave his blessing.

The wedding ceremony was held in the Phillipses' apartment at the Ambassador at 5 P.M.—the exact day and hour that Frank and Jane Phillips were married forty years before in Creston. The apartment was decorated with Easter lilies, and Sara Jane wore a white satin gown with a long train and carried a bouquet of white orchids and lilies of the valley. Betty Phillips was her only attendant, since her sister, Mary Low, was pregnant with her first child and could not attend.

Jane and Frank Begrisch remained in Manhattan and later moved to Rye, New York. Frank Begrisch later became a director at Phillips Petroleum, a post he held for twenty-four years.

Frank was pleased that both of his foster daughters were happily married. His son was still a worry to him, but at least Mary Kate was there to help him when he took too much drink. John and Mary Kate would soon be leaving their home at the Mayo Hotel in Tulsa to take up residence in Dallas in the Governor's Suite at the elegant Adolphus Hotel.

John's former wife, Mildred, in a twist of irony, in 1937 wed Dean Cummings, the former Phillips employee who left the company a few years before along with execs Dewar, Sherman, and McIntyre. Cummings, Mildred, and Frank's three grandchildren, soon moved to Tulsa. Although her marriage to Cummings was considered an insult by some, Mildred remained in Frank and Jane's good graces and she and the children continued to spend time in Bartlesville, visiting the Phillipses. As it turned out, her relationship with Cummings was doomed from the start and their marriage soon ended in divorce.

On June, 13, 1937—in the midst of the continuous family comings and goings—Frank observed the twentieth anniversary of the founding of Phillips Petroleum Company.

In an anniversary issue of *Shield*, the company magazine, Frank wrote: "Twenty years ago, a state clerk's pen scrawled a signature on a crisp legal document. It was the charter of incorporation of the Phillips Petroleum Company. Looking back across the years . . . I am moved by a feeling of deep gratitude and a sense of solemnity. Gratitude to all who have so ably played their parts in our progress. Solemnity, because our great growth has brought with it a vastly heightened responsibility to employees and stockholders."

Paul Endacott was shifted over to yet another assignment, this one as manager of the newly created employee relations department, unveiled in 1937 as a result of the National Labor Relations Act. This measure, considered to be one of the most drastic legislative innovations of the decade, was enacted in 1935, and was better known as the Wagner Act. Designed by the Roosevelt administration as an integral part of the New Deal, the Wagner Act was intended to be an economic stabilizer for the nation. The Wagner Act was the bill of rights for unions. It set forth unfair labor practices by employers, established a code of fair practices, promoted self-determination and collective bargaining for workers, and stimulated the growth of strong unions.

It also frustrated the hell out of Frank Phillips and kept Endacott so busy negotiating labor contracts and establishing new company policies and plans that he had to check into the hospital and have his stomach pumped. "It was a difficult period," said Endacott. "It was a time of transition for the company. We were moving away from Frank Phillips' personal rule to the adoption of policies and rules. The firm was getting bigger and he had to change and give up some of the reins. That was totally against his grain.

"We also had all sorts of contracts to deal with and everyone was feeling their oats. It was a bothersome time. But the Wagner Act especially bothered Mr. Phillips. He didn't like all the new rules and he wanted more flexibility. He was having to give up doing things that he had done since before he drilled the Anna Anderson in the early years of the century. When the Wagner Act finally became effective, it tended to greatly isolate employees from management, and this was certainly contrary to the Phillips management style. The company had always been like one big, happy family. Mr. Phillips became very aware of this sudden shift and how this affected the relationship with his workers."

That summer, the company began issuing a monthly newsletter for employees called *PhilNews*. It was Frank's way of keeping in touch with the people who worked for him and letting them know about changes and developments, such as the implementation of a vacation policy and the company retirement plan. Endacott also helped Frank draft a president's letter which appeared in every issue.

"One of my greatest regrets is that our organization has grown so big that I can no longer sit down on the edge of a derrick floor, or some equally inviting place, and chat with each of you about the things that go on 'behind the scenes' in our company," Frank wrote in the first letter, which appeared in the August 1937 issue of *PhilNews.*

That fall, Frank felt the time had come to bring some order to his many acts of generosity, which included paying off all the church debts in Bartlesville and giving tens of thousands of dollars to the Boy Scouts and the YMCA. The Frank Phillips Foundation, Inc., was incorporated and the charter spelled out that the foundation's primary mission was to "aid public religious organizations, charitable organizations, preparatory, vocational and technical schools, institutions of higher learning and scientific research; to establish, maintain, conduct, assist, and endow public charitable, religious, literary, educational and scientific activities, agencies and institutions engaged in the discovery, treatment and care of disease; and its endowments and funds shall be administered exclusively for such purposes."

Frank wanted the Foundation to continue his policy of supporting the nation's youth, through philanthropic projects, and a special focus of the Foundation's generosity was the Scouting movement in America. "I'm a spoiled old man," said Frank shortly after the Foundation was formed and he became its first chairman of the board. "Money spoils any man. But it is a strange thing about money—regardless of how much you have—you can only wear one suit of clothes, sleep in one bed, and eat one meal at a time. I always wanted to be able to give money away, but I learned it was a hard thing to do and do right. You should never do for a person what he can do for himself. I made a lot of mistakes and enemies trying to give away money until I got some good, professional help."

While he was doling out personal funds to churches, Scout troops, and others, Frank was also trying very hard to convince a federal jury that he wasn't out to fleece the public. Frank and several other leaders of the nation's oil industry faced charges that they conspired to raise gasoline prices, in direct violation of the Sherman Antitrust Act. Their trial was held in federal district court in Madison, Wisconsin. The afternoon of the trial, the Madison newspaper carried a front-page photograph of Frank, holding a pipe and looking rather professorial as he conferred with a cigar-smoking John Kane, his senior vice president and chief counsel. The headline read: "Oil Magnates Face Federal Court."

The indictments, which were handed down in July 1936, charged that

Frank and the other defendants adopted, without authorization, policies and practices of the defunct National Recovery Act petroleum code and expanded and extended them to their own advantage. Originally forty individuals, twenty major oil companies, and four subsidiaries were indicted. Over the course of the trial, dismissals trimmed the dockets to sixteen firms and thirty individuals. The antitrust trial was the biggest involving the petroleum industry since the government dissolved the Standard Oil Trust.

During an address he made to the Tulsa Chamber of Commerce which was broadcast over radio on New Year's Eve 1937, Frank blasted the federal government as antagonistic and took several shots at the antitrust laws, which he said were crippling big business:

> I am not an advocate of government control of the industry. I earnestly believe that the various antitrust laws and other laws which have a stranglehold upon the industry should be so modified as to permit the industry to regulate itself. This regulation should have governmental sanction and should be supervised and policed by some governmental agency that will protect the public against monopoly and unfair trade practices; an agency with the power to enforce compliance with reasonable regulations adopted by the industry itself.
>
> I have faith in our form of government, and I am confident that able men at the head of government will find a way for government and business to cooperate on a rational, sensible basis—as friends, rather than as enemies . . . As we look over the past year, we see a disappointing lack of cooperation. Business recovery cannot be accomplished as long as it is the policy of our Administration to harass business, while at the same time demanding that we reduce unemployment and increase wages. Confidence cannot be restored as long as the Administration continues to break its promises to reduce expenses. It is impossible to legislate our way out of the Depression with unsound laws. Labor cannot establish the security of permanent employment by destroying or discouraging capital which gives them their jobs.

Within three weeks, a jury of twelve small businessmen and farmers in Madison deliberated for one hour and found all sixteen major oil companies and thirty officials, including Frank and Phillips Petroleum, guilty of the governmental charges that they conspired to fix gasoline prices.

Frank found another chance to take a shot at his tormentors that April when he told some reporters that "the present Depression is undoubtedly due largely to apprehension. Fear, inspired by an apparent atmosphere of hostility existing between leaders in government and leaders in business, is retarding industrial progress. Because of shifting policies and confiscatory taxes, there is little incentive to take the risk of capital expansions. Consequently, business is proceeding cautiously at reduced speed in the cloud of confusion and doubt."

In June 1938, Frank returned to Madison and entered a plea of "no contest" to the last three remaining counts on the antitrust indictments. He shelled out $30,000 in fines and flew home. When he returned to Bartlesville that evening, he emphatically stated that the plea was only because he wanted to avoid additional expense.

"I thought and still think that the conduct of my company in its efforts to help restore prosperity by correcting trade abuses was its patriotic duty and cannot be a violation of the law," said Frank. "After the government had invited cooperation and the industry had generously responded, it was shameful persecution to procure the indictments at Madison."

But despite the federal litigation and the wrestling with the Wagner Act which he had to put up with through the rest of the 1930s and into the 1940s, Frank was generally pleased with 1937, one of the most successful years in the history of Phillips Petroleum.

Later that year the United States was beginning to turn the corner on the Depression, and as the nation's economy recovered, Phillips' earnings perked up. Net income reached a record $24.1 million in 1937, and in the annual report Frank noted that "the year was the best in our history for both stockholders and employees. Profits were higher than ever before. Wages paid to employees, except executives, were increased and now are the highest ever paid by the company."

At year's end, Frank announced in the *PhilNews* that Phillips Petroleum and Metropolitan Life Insurance Company had developed a retirement plan for the employees to provide them with a life income after they reached the retirement age of sixty-five. The first Phillips employee to take advantage of the plan was Ben Barbee, the old patriarch of the Barbee clan, and one of the company's original workers who dated back to the days of the early Osage drilling when Frank's "Wonder Well" on Lot 185 gushed to life and caused the founding of Phillips Petroleum.

As he approached his own sixty-fifth birthday, that January Frank announced the latest list of new executive promotions and paid tribute to some of the recent retirements, including John Kane, who had served as a company officer and director since the beginning.

The big news was the promotion of Boots Adams. After Obie Wing's death, Boots had remained as the assistant to the president, but in 1935 Frank promoted him to treasurer of the company. Boots then also became a member of the board of directors and the executive committee. The man who started as a warehouse clerk was about to get a new title—executive vice president.

As a direct result of Boots' big step forward at least fourteen other employees received promotions, including Paul Endacott, who moved from serving as director of employee relations to assistant to the executive vice president.

Boots was only a step away from the presidency of Phillips Petroleum, and he wouldn't have long to wait until his dream came true.

Four months later, on April 26, 1938, during the annual meeting of the stockholders and the board of directors in Bartlesville, Frank recommended that K. S. Adams, the fast-talking young man from Kansas with the big ideas, be elected as the new president of Phillips Petroleum Company. Frank would become the company's chairman of the board and chief executive officer. The directors' vote was, of course, unanimous. Boots Adams, only thirty-eight years old, and with seventeen years' service with Phillips, was now one of the youngest presidents of a major corporation in America.

Phillips Petroleum Company would never be the same.

PART FIVE

UNCLE FRANK
1939–1950

"The memory is a living thing—it too is in transit.
But during its moment, all that is remembered joins,
and lives—the old and the young, the past and the
present, the living and the dead."

—From *One Writer's Beginnings,* by Eudora Welty

CHAPTER TWENTY-EIGHT

America was about to lose its innocence in the autumn of 1939. The "dishonest decade," as one poet described the 1930s, was soon to end. Terrible things were happening in the world. More than 17 percent of American workers were still out of work, while in Europe war clouds were thick and menacing. Adolf Hitler and the German Third Reich had already swallowed up Czechoslovakia and Poland. Words like "blitzkrieg," "panzer," and "Luftwaffe" were creeping into newspapers and striking terror in hearts and minds.

Most Americans, like Scarlett O'Hara, decided they could afford to worry about all that tomorrow. *Gone With the Wind* was about to premiere in Atlanta, and a former roughneck from the Osage oil patch named Clark Gable would soon be telling everyone that he didn't give a damn. Rhett Butler wasn't alone.

Americans by the thousands were burying their heads in the sand and escaping the cares and woes of the world. They fled to the movies and watched Shirley Temple, Tyrone Power, John Wayne, Errol Flynn, Bogey, and Karloff. They followed Judy Garland down the Yellow Brick Road in search of the Emerald City. They swung with Benny Goodman, Artie Shaw, and the Dorsey brothers. They were chilled by Kate Smith belting out "God Bless America." They forgot their past and present and flocked to the gaudy New York World's Fair to peek into the future and learn about the world of tomorrow.

In Bartlesville there was a party. A birthday bash. It would go down as the largest individual birthday celebration ever held in the United States. On November 28, 1939, Frank Phillips—"Mr. 66" himself—turned sixty-six years old.

It was a magical moment—not just for Uncle Frank but for all the Phillips employees and for all of those whose lives Phillips Petroleum had touched. Frank's sixty-sixth birthday was a significant milestone in his life and for the company. Through the years, he had continued to receive fan mail addressed to "Mr. 66, USA," and on occasion Frank still registered at hotels by signing only "66" in the guest books. It was only fitting that all the stops be pulled for the birthday marking his sixty-sixth year.

Planning sessions were held months in advance. Paul Endacott, now assistant to President Boots Adams, was placed in charge. He carried out his tasks as if it were a championship basketball game on the line, working until he was exhausted and then working some more.

"Boots and I and some of the others were concerned because of Frank Phillips' image and the fact that the company was still perceived as a one-man show," said Endacott. "There were some outsiders who really believed Phillips Petroleum would fall apart after Frank Phillips was gone. People didn't realize he was a symbol. We needed to change that image, and after talking it over, we decided the big sixty-sixth birthday salute was a fitting way to acknowledge his impact but also to let the world know that there was life after Frank Phillips."

Not a trick was missed. Gold coins, bearing Frank's likeness as "Eagle Chief" on one side and the Phillips 66 shield on the other, were minted; colorful name tags and badges, bunting and banners were produced; and a striking portrait of Uncle Frank by Raymond P. R. Neilson of the National Academy was commissioned and 50,000 color prints were distributed.

A formal pre-birthday dinner, hosted by Clay Smoot and John Cronin, the president and vice president respectively of the First National Bank, was held for the Phillips chieftain at Hillcrest Country Club. More than one hundred and fifty of his hometown friends came to honor Frank and listen to a stream of tributes and reminiscences from John Kane; L. E. Phillips; Burdette Blue, the head of ITIO; H. R. Straight, top executive from Cities Service; and Boots, the final speaker, who choked up as he paid his respects to "the boss" on behalf of himself and thousands of Phillips employees across the nation.

"I just want to say that there are a lot of swell fellows here," Frank told the guests. "As much success as I may have enjoyed, my riches are not in worldly goods so much as they are in friends—friends like you, my fellow townsmen, friends and neighbors, who have gathered here tonight.

Money is worth only what it will do for my friends, my country, my state, and the youth of the nation."

When he sat down to thunderous applause, a huge birthday cake—a replica of the First National Bank Building with sixty-six candles blazing —was wheeled into the room and the first of many rounds of "Happy Birthday" which he'd hear over the next several days was sung to Uncle Frank.

Frank's birthday fell on a Tuesday, and for several days before the big event, thousands of visitors began arriving in Bartlesville to take part in the celebration. They came by bus, by automobile, by train and airplane, and a few even arrived on horseback. Downtown Bartlesville was jammed as crowds milled around the Phillips Building and the First National Bank and stuffed into the lobby of the Maire Hotel, where a neon-lighted portrait of Frank beamed down on the guests, among them smartly dressed Easterners, well-heeled cowboys, and visitors from all over the world. The twenty-five members of the Kansas City Refinery Band brought their own cots and bedding and set up camp in the Maire's banquet room.

"There were people everywhere in town—sleeping wherever they could set up a cot," said young Johnny Phillips, by then a cadet, along with Boots' son Bud, at Culver Military Academy in Culver, Indiana, where Frank had been made "official father of the cadet corps" by the school administration. "The sixty-sixth birthday was pure love. People opened up their homes to help accommodate the crowds. It was the grandest birthday party anyone had ever seen and it was all for Father!"

John Phillips, Sr., and Mary Kate came up from Dallas to attend the preliminary parties and stay for the big event. Both of them were just starting to feel a little more comfortable about traveling and being around large crowds. It had been only a little more than six months since kidnappers tried to steal John away.

The incident occurred when John and Mary Kate stopped in Chicago after visiting Johnny at Culver. During the intermission at an evening performance of *Streets of Paris* at the Grand Opera House, John went to a rest room on the mezzanine, where he was approached by three men.

"Are you John Phillips?" they asked. John said he was.

"You're our man," said one of the men, and he pushed a gun into John's back and ordered him to walk ahead of them out of the theater. "We're going to take you for a nice little ride." As they descended the stairs, John bolted away and dashed back into the darkened auditorium and found his way to Mary Kate.

"We got out of there as quickly as we could," said Mary Kate. "We got some help in the lobby and then took a taxi to Mark and Mary Low's apartment and then back to the Blackstone. John called his father. We had guards protect us until we returned to Dallas."

The kidnappers vanished and there were no further incidents, but John remained cautious. Of course, a few drinks helped provide courage, and with all the police officers descending on Bartlesville for the sixty-sixth celebration, John figured no kidnapper stood a chance. He was right. There wasn't an ounce of malice in sight. Especially when it came to Uncle Frank and the Phillips family.

The night before Frank's birthday, as darkness fell hundreds of visitors continued to pass in homage before the official birthday portrait on display in the windows of the First National Bank. Every shop window in Bartlesville had one of the prints taped to the glass, and over at the newspaper they were getting ready to go to press with a special edition devoted to Frank. The sounds of saws and hammers still echoed through the streets as additional reviewing stands and seats were assembled and finishing touches were added to the floats for the birthday parade.

Final plans were in place for handling the heavy flow of traffic expected in the morning. At the Phillips Building, where after sundown the lighted windows spelled out "66" in numerals eight stories high, the crush of gifts, telegrams, and cards which arrived from all over the world overflowed Frank's private office on the seventh floor and filled several rooms. It took a score of clerks nearly two weeks to sort, classify, and arrange the more than 30,000 gifts and greetings.

There were thousands of wires and birthday cards, including a postcard the size of a conference table (reported to be the largest ever sent through the U.S. mails) from the Tulsa marketing division. Unusual gifts included a neatly bound book containing an appropriate verse from each of the sixty-six books of the Bible; a desk pen and pencil set built on a base of wood taken from the first Greeley County, Nebraska, courthouse, where Frank's father served as the first county judge; a "66" made entirely from crucifixes; and a registered Holstein yearling which nobody even tried to bring up the elevator to Frank's office.

The official program for the birthday party started precisely at 4:06 A.M., when a delegation welcomed more celebrants on both the regular train and a special train which arrived at the depot from Kansas City.

Members of the Jane Phillips Sorority, an organization of Phillips female employees that was founded in 1937 in honor of Aunt Jane, started serving breakfast at 4:30 A.M. for the out-of-town guests. The sorority members were devoted to Jane, and she liked to brag that "I feel the girls are all mine." Jane's girls kept the hot food coming throughout the morning as more special trains arrived, including one from Oklahoma City and the "Frank Phillips 66 Special" from Phillips, Texas, the panhandle town named for Uncle Frank. There were one thousand Phillips workers on the train plus the fifty members of the Phillips Band.

"I got up in the pitch dark and got all slicked up to meet that train from Phillips, Texas," said Tom Sears, the former Hillcrest caddy. "Gladys Markee was on that train and I was to meet her. She was my best buddy's girlfriend, but I had a date with her since he was away at college. Gladys was one of the leaders of the honor band and was she ever gorgeous."

Twenty marching bands from several states assembled in Bartlesville that day, including an employee band, a cowboy radio band, and a drum and bugle corps.

"I was a kid up in Nowata, Oklahoma, and our high school band was picked to march in Uncle Frank's big birthday parade," said Bob Cotton. "Were we ever proud! And thrilled! Why, we had never seen so many people in one place. People were everywhere! They were everywhere!"

Down at the train depot, Jack Leonard, the young son of Frank's favorite printer, was feeling his oats. A football player at the University of Missouri and a former lifeguard out at Clyde Lake, Leonard and several of his friends decided to get dressed like cowboys and ride their horses in honor of Uncle Frank's birthday. "It was just a spontaneous decision on our part to get all decked out in chaps and boots and hats and even gun belts," said Leonard. "I got my buckskin named Brownie and we all rode downtown to greet those birthday trains rolling in all morning long. We rode along the platform, and there were bunches of good-lookin' girls all over the place, just standing there waiting to be taken to the hotels. Why, we just rode up to them and had them swing right on sidesaddle behind us. These gals were all dolled up and there were plenty of limos waiting, but they were thrilled as hell to ride horseback." After depositing several beauties at the hotel—including Julia Stradler, chosen as the sales department "queen" from the Chicago office—Jack rounded up his compadres and they dashed off in time to take part in the birthday parade.

To start his big day, Frank wore a three-piece pinstripe suit befitting a banker and oil tycoon, went to his office, and wrote out a check for $66,000. Then he dictated a short note to be delivered to the news media:

"You may announce on my 66th birthday that I have contributed $66,000 to the Frank Phillips Foundation, Incorporated, toward the creation of an educational fund to be used for advanced education among children of Phillips Petroleum Company's employees only."

Philanthropy was nothing new for Frank. He had been giving away large sums of money to deserving causes, individuals, schools, Scout troops, and churches for years. In 1937, *Time* magazine reported how Frank, through a banker associate, quietly paid off all the indebtedness of the five major churches in Bartlesville—Methodist, Baptist, Episcopal, Presbyterian, and United Brethren—for a total of $63,000. Over the course of several years, Frank's Foundation pumped more than $1 million into Boy Scout councils in thirteen states. Frank also sent money to Father Flanagan's Boys Town and to the University of Oklahoma to establish a collection of books and source material on the history of the state and the Southwest. The year before his big birthday party, Frank—through the Foundation—had given $50,000 to complete the new Bartlesville Senior High–Junior College, which opened in 1940 as one of the most modern and complete school plants in the nation.

Beyond those worthwhile projects, Frank wanted to do more for the employees of Phillips Petroleum. For some time Frank, the man who had to quit school at the age of twelve, wanted to establish an ongoing scholarship program. When he learned of the birthday celebration he decided it would be the best time to show his appreciation for all the festivities planned in his honor.

Frank had long been a great admirer of Sir Cecil Rhodes, the South African statesman and financier, who left nearly his entire fortune for the endowment of the famous Rhodes scholarships at Oxford. Rhodes died in 1902, shortly before Frank's first sojourn in Indian Territory. Frank always remembered Rhodes' dying words: "So little done—so much to do." Those words took on more and more meaning for Frank as he grew older. He repeated them out loud that morning as he signed the check and looked forward to the special day before him.

That morning, as trains pulled into the station and the highways leading to Bartlesville became crowded with more guests, there were guided tours of the F.P. Ranch and the Woolaroc Museum and open houses at the Phillips facilities in town, including the new research building. Elevator girls in the Phillips Building wore their best Sunday dresses along with cowboy hats and boots, as did many of the young women from the Jane Phillips Sorority.

"I remember the really big deal was several of us girls actually sought permission from our superiors to wear pants to the parade that day," said Loudia Reed, a former employee in the Phillips legal department. "That was a major breakthrough for women back then, but we really wanted them to allow us to dress that way. We did, too. They gave us permission and we wore our pants to the parade."

By noon the entire parade route through Bartlesville was lined with people. State troopers and police officers from Oklahoma City and Tulsa were imported to help control the crowds and keep traffic moving smoothly. Not only was the entire day a holiday for all the schoolchildren in town and all Phillips employees but both ITIO and Cities Service closed their offices at noon. The bystanders waited for the parade to start and watched the helium-filled blimp supplied by the Amarillo sales division hover in the cool breeze directly over the Phillips Building, where it was moored. Starting at 12:49 P.M., a series of sixty-six aerial bomb salutes exploded over the city, signaling the official start of the air exhibition. As Billy Parker in his famous 1912 "pusher" airplane and an armada of aircraft, including company planes, flew overhead in formation, the dashing Art Goebel and Father Paul Schulte, "the Flying Priest of the Arctic," gunned their airplanes higher and higher into the cloudless sky. Father Schulte, who started his flying career during World War I before traveling thousands of miles in his plane as a missionary in the Arctic, wanted to give Frank a special blessing from the air. He flew to an altitude of 10,000 feet and made a 3,000-foot-long cross of smoke, while Goebel, in huge letters and numbers, wrote "FP" and "66" and then climbed to two miles above his audience and scrawled out a gigantic "Happy Birthday."

After the air show, Frank wanted to greet his thousands of guests before taking his place in the reviewing stand to watch the parade. Accompanied by Jane, Oklahoma Governor and Mrs. Leon C. Phillips, and Endacott, Frank ordered his chauffeur, Homer Baker, to drive the entire length of the parade route so he could say howdy.

The automobile Baker drove was not the usual limo but the $25,000 Chrysler, specially built for King George and Queen Elizabeth of England to ride in when they came to the New York World's Fair. The fancy car was shipped to Tulsa the day before and brought up to Bartlesville under police escort. After the birthday party the car was returned to Detroit, where it resided in a museum. "That car was a dandy," said Baker. "But the very first thing we did when it arrived was to take off the tires and put on some that were sold at Phillips service stations. We didn't want to miss that opportunity to push Phillips products."

And there was plenty of coverage. Newspaper and magazine reporters clustered around the reviewing stand, where the Phillips party, including John and Mary Kate, stood to watch the parade. A newsreel crew recorded the day's activities and more than fifty radio stations provided nationwide coverage of the event, including Elliott Roosevelt's twenty-four-station network based in Texas.

Next came the parade. There were sixty-six units, assembled in three sections. Everyone was to march by Uncle Frank as he watched from the reviewing stand.

The first section was a historical pageant with floats and displays which re-created Frank's early life, starting with his birth on the Nebraska frontier, through the years in Iowa, his entry into Indian Territory, and the founding of Phillips Petroleum.

Employees representing all twenty-one departments of the company made up the second section. They marched in the order in which the departments had originated, with Boots Adams leading the pack. "Old Fred," a workhorse who hauled pipe in the Burbank field for most of his twenty-two years, was in the parade, complete with his service badge. So was the Phillips 66ers basketball team, including some of its hot-shooting all-stars such as Bill Martin, the tough competitor from Blackwell, Oklahoma. The Men's Club, counterpart to the Jane Phillips Sorority, unveiled its new name—Frank Phillips Men's Club—during the parade, much to Uncle Frank's pleasure. Bringing up the rear of the employee section were some of the hands from the F.P. Ranch and a few of the animal residents, including a camel and some of the better riding stock.

The third section of the parade was comprised of all the special groups—Osage Indians in full tribal dress bearing a blanket, peace pipe, and tom-tom, all gifts for their Eagle Chief; pep squads; drill teams; civic organizations; Boy Scouts; Sea Scouts; and a bunch of colorful Cow Thieves and Outlaws astride their best mounts and wearing clean bandannas and the fanciest shirts they could beg, borrow, or steal.

But out of all the colorful parade entries, the one that was most dear to Uncle Frank's heart was a group of local youngsters disguised as circus figures—clowns, bareback riders, trapeze performers. It was the children of Bartlesville saying thanks to Uncle Frank for all the times he treated them to a free circus ticket. More than once, Frank was able to coax his old friends from the Ringling Brothers–Barnum and Bailey Circus to set up their big top on the edge of town. He bought every child in Bartlesville a ticket. Just a couple of months before his big birthday, when the Al G. Barnes and Sells-Floto Circus came to town, Frank again bought tickets for four thousand youngsters—from grade school through senior high, including Catholic students and the segregated blacks. He was still going through the stacks of thank-you letters from the children, about which he said, "I'm sure I experienced as big a thrill in receiving these letters as any the children may have experienced in attending the circus."

At the sixty-sixth-birthday parade, the children dressed in circus clothes passed the reviewing stand and waved at Uncle Frank. He smiled and waved at them and reminisced back to the time when his family had no money and he went to the circus for all his brothers and sisters and how he had come home and recollected the events for the entire family.

When the parade was over, Frank and his party and his thousands of visitors and guests jammed into the Municipal Stadium for the rest of the formal ceremonies. R. C. Jopling, Phillips' public relations director, acted as master of ceremonies, and a thousand musicians massed on the field serenaded the man of the hour. Once the last seat was taken, the bands saluted Governor Phillips with "Ruffles" and then broke into "The Star-Spangled Banner," with the singing led by Martha Atwood Baker, a former star with the Metropolitan Opera in New York.

Eldon Frye, editor of *PhilNews* and chairman of the employees committee which helped plan the party, brought Frank a brief greeting from the Phillips employees and spoke of the pride everyone felt about the company and their leader. "Uncle Frank, that parade should tell you better than anything I can say the many expressions of affections we have for you today and best wishes for many more birthdays," said Frye. "We are glad that you are 'just one of the hands'—our Uncle Frank."

Boots introduced the governor, who told the audience that "Frank is not one of those who have drifted to and fro, he chose to climb the high way instead of the low. I deem it a privilege, personally and as governor of this state, to be afforded the opportunity of paying respects to one of our most deserving and respected citizens . . . Nations are not great because of vast forests, tall buildings, fertile soil, and natural resources. They are great because of the men who till the soil, fell the trees, build the buildings, and develop the natural resources." He went on to say that it was his wish to talk about the "intangible spiritual force, that vision, faith, and push, that integrity which characterizes Frank Phillips." The governor talked about Frank's generous gifts to the Boy Scout movement, to education,

aviation, and other worthy causes, but he added that he didn't want to
stress them too much. "Frank Phillips has that other thing, that 'plus of
the soul,' " said the governor. "He has learned to live above the rim. It is
this thing that makes greatness." Governor Phillips also included a gracious tribute to Jane, "the lovely, faithful, charming wife who shares his
honors." This remark drew a long round of applause. He concluded by
offering birthday wishes from himself and the people of Oklahoma. "I
present you with the highest gift any man can receive—the respect and
gratitude of his fellow men—I congratulate you!"

There were other speeches. L.E. brought "felicitations from the family—the in-laws and outlaws." Police Commissioner Eddie Shields of
Tulsa gave Frank a police badge bearing the numeral 66. He also introduced Johnny, the diminutive celebrity bellhop from Phillip Morris who
offered his birthday greetings with a healthy "call for Uncle Frank." The
mayor spoke and so did a congressman and then General L. R. Gignilliat,
the head of Culver Military Academy, offered his salute to Frank.

When the last speaker had finished, a twelve-foot-tall birthday cake,
baked in the shape of the Phillips Building, appeared before the speaker's
platform. The giant cake was on a decorated wagon pulled by a team of
horses and was presented by Nona Farquharson, president of the Jane
Phillips Sorority, on behalf of all the Phillips employees. Tears crowded
Frank's eyes when he saw the mammoth cake and looked around the
stadium at the thousands of cheering Phillips fans. Deeply moved by the
birthday ceremonies—the air show, the parade, the speeches, the cake
presentation—he took off his hat, walked to the cluster of microphones,
and delivered a short and unprepared birthday talk:

> It's not my party—it's my birthday. It's *your* party, and in
> all these long years it's the swellest party I ever had. I al-
> ways have to brag about everything that I have anything to
> do with. I even brag about you, Governor, when I am away
> from home. I tell them that we have the most wonderful
> governor in the United States, and I mean it; I think he is
> doing a swell job. But certainly he hasn't done any better job
> and I don't believe he has done as good a job as you employ-
> ees have done in giving me this wonderful party.
>
> I am so chock-full, shaken down, and running over with
> enthusiasm—no, it isn't my enthusiasm, it's something else,

it comes up in my neck and gets down in my nose, and when I try to talk I stutter. . . .

But you don't know how proud I am of every one of you, and I have just reason. You have made the Phillips Petroleum Company. You have just used me for a symbol, that's all. I can't help it because I am sixty-six and have caught up with this number of yours. Yes, I have felt a little old at times, but when I saw this crowd and the parade today, the inspiration brought youth into these old bones, and I said, boys and girls, I am going to start out now and do something worthwhile—I've just been 'piking' along. As far as I'm concerned, it is you who have been doing the swell job. But it's a long day, and we have a lot ahead of us. I will be sixty-six till midnight, and I'm just going to go like hell from now until midnight. . . .

Again and again I want to express my appreciation to all members of this splendid organization who have been so wonderful, so gracious, and made the day so pleasant and so happy for me. As I said, I am just chock-full and am about to have a good old-fashioned bawl. Goodbye.

Frank's big party was far from over, though. That evening there were more open houses and cocktail receptions, street dances, and public balls featuring everything from complete orchestras and dance bands to marimba soloists and barbershop quartets. A variety show at the Civic Center presented the best fiddle players from the Osage, and "Powder River Jack" Lee and Miss Kitty played their guitars and sang lusty cowboy ballads. Lee explained to the crowd that "real cowboys don't yodel or whine through their noses, but sing right out from their lungs." That's just what Uncle Frank did when he joined in the singing with one of the quartets and helped them with several old-time tunes, including "Let me Call You Sweetheart."

Frank and Jane made every party and dance. They arrived at the new high school gymnasium, made possible by Frank's donation, just in time to see the start of the inaugural basketball game between the Phillips 66ers and the company's team from the Kansas City refinery. Outside, the night sky was filled with rockets and explosions, including sixty-six varieties of fireworks and a brilliant "Happy Birthday" and a 66 shield. The dancing and revelry went on until the next morning, when the walking wounded loaded happy drunks back on the train cars like so many cords of cut wood.

Bartlesville wasn't the only place they were celebrating Uncle Frank's sixty-sixth. Up in Creston, Iowa, there was yet another Frank Phillips party. Most of the southwestern part of the state poured into Creston to eat barbecue and sing "Happy Birthday" to the hometown boy who made good. It was a double-barreled celebration. Not only did the citizens of Creston honor Frank but they toasted the official start of the area's first oil well, spudded in on Frank's birthday by a Phillips Petroleum crew.

The Phillips exploration in Iowa, said one local newspaper, will result in "a well which all Iowa confidently expects will be the beginning of black gold riches for it—and all thanks to Uncle Frank."

Frank had always kept his ties with his friends and family in Iowa, and when his geologists determined that there very well might be crude oil deposits beneath the Iowa sod, Frank dispatched Phillips drilling crews to Arthur Brown's farm in Union County, a little more than ten miles south of Creston, to start oil and gas well tests.

Frank returned to Iowa to check on his drilling crew's progress about four months after they had begun. During a speech before the Des Moines Chamber of Commerce, Frank offered some bullish and optimistic talk about the state of the oil industry and its future in Iowa. "It is not unlikely that this great agricultural state may also become an oil-producing state," Frank said. "Thousands of acres in Iowa are now under lease and test wells may soon be drilled in what has probably been the world's biggest 'oil-less oil boom.' "

After his speech, Frank returned to Creston and thrilled everyone by circling over the city in his company airplane at an altitude of less than three hundred feet. Back in the old hometown he was treated like a conquering warrior. He visited each of the three Phillips 66 stations in Creston, shaking hands and chatting with customers and employees. The newspaper printed the headline in huge war type: "FRANK PHILLIPS COMING HOME," and Mr. 66 was the featured speaker at a banquet to honor the local basketball team, which had just won the state championship.

Frank played the oil game hard and long in Iowa but to no avail. By late 1939 the novelty of the drilling rigs set up on the farms was wearing thin. The stream of visitors every Sunday to see the oil derricks dwindled. Finally, the oil test well sunk by Phillips Petroleum was abandoned. No oil was found, nor was there any indication that any oil would ever be discovered. The community was disappointed and so was Frank.

The next year, on April Fools' Day, the Creston *News Observer* ran an editorial which lamented the failure to find oil but praised Frank for his efforts:

> There was a general hope that oil in paying quantities would be discovered. That hope is now lost. It was practically all we had to lose. Frank Phillips has also lost hope, in addition, to many thousands of dollars which he necessarily invested to make the test. He no doubt will be a good loser, and we should be equally good losers.
>
> Had oil been discovered, Frank Phillips would have been the recipient of enthusiastic praise and endless expressions of appreciation. He is entitled to exactly the same praise as though the project had proved successful. The citizens of this community do appreciate his effort, and we should make it a point to let him know that appreciation.

But back in November 1939, the disappointment of dry holes in Iowa was yet to come. In Oklahoma and Iowa it seemed like the entire world had come to help blow out Frank's birthday candles. The sixty-sixth-birthday celebration wasn't limited to Creston and Bartlesville. All around the marketing territory and at every Phillips office and installation, workers paused to remember their chairman. For the benefit of the field employees who couldn't attend the shindig in Bartlesville, radio stations out of Tulsa, Oklahoma City, Wichita Falls, and Amarillo broadcast sixty-six minutes of the party. Old roughnecks who had been pulling drilling tools for Phillips when Boots Adams was still a fraternity boy winked their toasts to Uncle Frank as they shared a pint of sour mash out in the lonely oil patch.

Phillips workers who held down the fort while everyone else came to Bartlesville remembered Uncle Frank. There were parties in St. Louis, Kansas City, and Chicago. Up in the New York office—high above Wall Street and Broadway in the Irving Trust Building—there was a celebration for Frank.

Someone was missing from the New York party, though. When they popped the corks and cut the cake, Fern Butler wasn't there. After almost twenty-two years as Frank's lover and a valued employee of Phillips Petroleum, Fern got her walking papers.

It was called retirement, but those close to Fern and Frank knew there was no other choice. Office politics and scurrilous rumors had taken their toll. The endless gossip about Frank and Fern and their long-standing relationship finally forced Jane Phillips to react. When Frank admitted he had been having an affair with Fern for decades, Jane's response was cool and not totally unpredictable. "When Father told Granny about Fern Butler, she said, 'Frank, those kinds of things are bound to happen. I can understand that. But Fern will have to go,' " recalled John Phillips, Jr.

Fern was to be replaced by Mildred Rowson, daughter of Ed Phillips, Frank's brother, who operated an automobile dealership in Okmulgee. Mildred was an ambitious young woman who had been hired by her uncle and was assigned to the New York office just a couple of years before. From the beginning she wanted Fern's job, and by mid-July 1939—just a few months before Uncle Frank's birthday party—Mildred had what she wanted.

Fern, only forty-six years old, resigned her post as assistant secretary with the company and manager of the New York office—the office she started for Phillips in 1920, just three years after she hired on as a stenographer back in Bartlesville. She quietly packed up her personal belongings and more than three decades of memories and was gone. "Fern was out and Mildred was in. It was that simple," said John Phillips, Jr.

"There was tremendous jealousy in that New York office—it was very competitive," said Nancy Locke Marshall, Fern's niece, who was a young girl visiting in New York when her aunt's "retirement" took effect. "It was clear to Fern that Mildred wanted her job. Also, Fern's relationship with Marjorie Karch, the secretary in Bartlesville, was going sour about that time. But the main problem stemmed from Mildred Rowson. After Fern left the company, at the mere mention of Mildred's name she would become absolutely rigid.

"There was some talk that Jane Phillips had her fired, but I'm not so sure that was the case. There was some real conniving going on and there may have been some pressure put on Mr. Phillips by Jane and some others. All I know is that in 1939, when Fern stopped working for Phillips Petroleum, she was at an absolute loss about what to do with herself. She had plenty of money, but she needed to keep busy. One of the stockbrokers she dated suggested she move into brokerage work, but she declined. She was tired of all that, and besides, it just wouldn't have been the same."

Fern had plenty of living options. She had maintained a comfortable apartment at 540 Park Avenue for five years and she continued to use her waterfront home near Westport as a summer residence. Also, Frank made sure she was provided with a handsome and regular income and he established a trust in her name. Although she was no longer in his life on a daily basis as she had been for many years, Fern always stayed in touch with Frank, mostly through telephone calls to Dan Mitani, and Frank kept tabs on her.

"From the day she left that office, I never heard Fern say one bad word against the company or anyone in the Phillips family," said her niece. "I'm sure that she never stopped loving Frank Phillips and Phillips Petroleum. They were, after all, her life."

Weary after spending her best years managing the Phillips operations in New York, Fern found refuge in her Park Avenue apartment, furnished with white carpets and draperies in the summer and navy-blue draperies and carpets in the winter. She also looked forward to going to her place in Connecticut for Lempie Krank's good cooking, long saltwater swims, and horseback rides on the bridle paths at the Fairfield Hunt Club. She had her friends and she had her dogs—the tenacious terriers she learned to trust more than any person. But still Fern missed certain things. She missed rising early and having John Graham drive her to the commuter train in the dim light so she could arrive in the city in time to confront Wall Street and the stock market. She missed the bright young executives, filled with news and excitement, bursting through the office door. She missed Frank's laughter and his rage. She missed the sweet smell of Havana cigars.

CHAPTER TWENTY-NINE

In one smooth motion, Frank brushed a match against the porch railing, brought the tiny blue flame to just below the end of his cigar, and sucked in rich smoke. When the cigar glowed like a campfire ember, he let the smoke roll out of his mouth and watched it catch hold of the breeze and disappear in the night sky. Behind him, the voices inside Woolaroc lodge were muffled. Off in the distant hills an owl called and nearby something—probably a raccoon or an opossum—ambled through the brush. Frank eased into a chair and worked on his cigar and waited for his eyes to adjust to the darkness. As the moon came up big and round like a boy's face and the stars began to arrive, Frank considered his life and his achievements.

He was sixty-six years old. From forty-four to sixty-six, he built a far-flung empire of oil. His $3 million company was now a $226 million firm engaged in all phases of the oil business, from producing oil at the well to delivering it to the consumer's car.

The Phillips 66 shield hung over the company's sales outlets in twenty-two states, and sales of refined products were more than a billion gallons a year. Shipments had begun through the new extension of the products pipeline from East St. Louis to East Chicago, Indiana. Philgas volume increased, and Phillips Petroleum had become a major producer of oil and gas, the world's largest producer of natural gasoline, the world's largest distributor of liquefied gas.

The original organization of twenty-seven people had grown to more than 30,000 employees, jobbers, and dealers. The handful of original stockholders had become 41,000. The research staff—Oberfell's "Whiz Kids"—now numbered more than 150 and operated out of one of the most modern laboratories in the industry.

The company's radio program, "Tune Up Time," was aired across the nation every Tuesday evening with "splendid entertainment . . . furnished by Walter O'Keefe, André Kostelanetz and his orchestra, Kay Thompson and guest stars." Phillips also sponsored news programs on 134 local radio stations, mostly in the Midwest, where its marketing efforts were centered. More Americans than ever were responding to the slogans "Phill-up with Phillips for Instant Starting" and "Phill-up with Phillips for Greater Mileage."

There was no question that Frank Phillips continued to rule the roost. He remained in charge of his own destiny as well as that of Phillips Petroleum Company. Boots Adams was the president and wielded tremendous power, but Uncle Frank, the senior officer as chairman of the board, was still *the* boss. And anyone who doubted that had a headful of cigar ashes.

Frank was in command. He wasn't ready to retire to the ranch. Not just yet. Not by a long shot. He stayed on the job. He remained active in management and policy decisions. He ran the executive committee like a drilling foreman. And he bossed the board of directors as though they were old cronies or Phillips executives handpicked for the job, which, in fact, they were.

Frank traveled the nation, speaking out in behalf of legislation to conserve oil and natural gas through prevention of waste, as well as the control of trade practices and regulations for the interstate shipments of oil. He preached against the "antiquated Sherman Antitrust Law" and called for the federal government to "cease their attacks upon business and cooperate with industry for the benefit of all." When there was some confusion about his stand concerning oil industry legislation, he published a booklet entitled "Sanctioned Self-Regulation" to clarify his position.

"We do not need government operation or management, but we do need a background of regulation that will enable oil producers to do their own operating in an effective and economical manner," wrote Frank. "Although the present need is only for adequate government regulation, it is conceivable that, if this regulation is too late in forthcoming, not mere regulation, but entire government control and management will result."

Frank was still feeling his oats. He was as feisty as ever—ready to take on the government or anybody else he felt posed a threat to the oil industry.

Although he kept a battery of physicians busy in New York, St. Louis, and Kansas City tending to a series of complaints ranging from severe headaches and deafness in his left ear to gastric distress and eye problems,

Frank managed to stay fit by watching his diet, taking long walks, and going horseback riding whenever he got a chance to slip away to the ranch. He also enjoyed daily osteopathic treatments from Doc Hammond. Compared to the rest of his family, Frank was in good form.

Jane's health was far from good. Frank worried about his Betsie and so did the rest of the family. She remained overweight and a compulsive cigarette smoker. But she tried to keep her physical ailments to herself and kept up a social schedule which would have staggered most healthy twenty-year-olds. John Phillips, Sr., wasn't in much better shape than his mother. Years of heavy drinking had sapped him. In photographs showing Frank and John dressed in formal coats, tails, and top hats, ready to make their traditional New Year's calls, Frank appears to be years younger than his own son. As the old cowboys put it, John Phillips had been ridden hard and hung up wet to dry.

And among Frank's siblings, L.E. was still battling the poor health that had bothered him since he was a young man. Down in Tulsa, Waite Phillips was another story. Waite had never felt better. Prosperity seemed to agree with him. Big money was a comforting tonic. "As I've always maintained, I've just been lucky," Waite said again and again when asked about the wealth he had accumulated over the years through oil and real estate transactions. "Just lucky, that's all."

Probably the most generous public benefactor ever to live in Tulsa, a city built by philanthropy, Waite felt duty-bound to distribute large sums of his wealth to the needy and for the public's good. Like most rich men, he was glad to be wealthy, but his wealth also concerned him. He spent a great deal of time brooding about what he ought to do with it. And he thought well and to good purpose.

Educational and humanitarian causes also interested Waite. During the depth of the Depression, he built Tulsa University's Petroleum Engineering Building, a youth center, new buildings for the American Legion and for the Community Chest, and a new hospital wing. He gave substantial monetary gifts to a variety of civic, cultural, religious, and educational organizations. Waite also donated three hundred acres of wooded land in south Tulsa to be used to build Southern Hills Country Club, which evolved into one of the finest golf courses in the nation.

After only eleven years of living at Villa Philbrook—and with their two children, Elliott and Helen Jane, grown—Waite and Genevieve decided to dispose of their palatial Tulsa mansion. They moved into the penthouse on top of the Philcade Building and donated their seventy-two-room residence with the surrounding twenty-three acres to the city to be used as an art museum. Being an astute businessman, Waite also supplied revenue sources—the Beacon Building was donated to provide for the upkeep of the mansion; the Elliott Building to maintain the grounds and gardens.

"The only things we keep are those we give away," said Waite. "All things should be put to their best possible use."

Such was the case at Philbrook. The villa was transformed into a stunning public showcase known for many years as the Philbrook Art Center and later as the Philbrook Museum of Art. Many of the original rooms on the first floor were left intact, while the upstairs and lower-level rooms were turned into galleries. The permanent art collections at Philbrook spanned several centuries and included a premier Native American collection which greatly pleased Waite. "Oil fortunes were made out of the Indian lands," he said. "I have a deep feeling of gratitude to the American Indian and I want to see his culture preserved."

Waite had the American Indian and his own love of the outdoors on his mind a short time later when, in one of his largest philanthropies, he donated 35,000 acres of his Philmont Ranch in New Mexico to the Boy Scouts of America and threw in $50,000 for developing and improving it. He later donated more of the sprawling Philmont acreage to the Boy Scouts, for a total of 127,000 acres, and also turned over the net income of his twenty-three-story Philtower Building in Tulsa to cover operational costs.

Waite liked what he saw when he visited the camp. During the summers he frequently returned to Philmont and quietly rode his favorite trail horse, Old Gus, through the campsites watching the Scouts at work and play. Sometimes he'd stop and ask the boys if they were enjoying their time in the mountains, but he made a point never to identify himself. Other times, Waite pulled up to one of the sites in an automobile and observed the activities for twenty minutes or a half hour before driving off.

A plaque, bearing Waite's own words, at the Philmont Scout Ranch perhaps best expressed his reasons for turning the ranch over to the Boy Scouts. It said: "These properties are donated and dedicated to the Boy Scouts of America for the purpose of perpetuating faith, self-reliance, integrity and freedom, principles used to build this great country by the American pioneer . . . So these future citizens may, through thoughtful adult guidance and by the inspiration of nature, visualize and form a code of living to diligently maintain these high ideals and our proper destiny."

In Bartlesville, Frank also continued to give away large sums of money through his Foundation, but he wasn't yet prepared to donate any of his properties to anyone, no matter how worthy the cause. His ranch in the Osage was still especially important as an entertainment and relaxation center. When the time came he'd be willing to part with his Osage land, but not just yet. The ranch was a showcase and offered a forum for Frank to share his views of the world, industry and business, and politics. As he grew older, Frank didn't become shy about letting his thoughts be known.

Frank not only made sure that he kept up with changing business trends but he stayed in the public eye in New York and back in Oklahoma. Even back home in Bartlesville, Frank made sure he was in the midst of the action when the old Odeon Theatre got a face-lift and a new name—the Osage. It was called "one of the slickest-looking movie houses in the Southwest," according to the newspaper, and on August 27, 1940, the owners hosted the world premiere of the musical *Irene* and arranged for the film's star, Anna Neagle, to come to the dedication ceremonies. Frank would act as her official host. Congratulatory wires came in from the top stars of the day—Carole Lombard, Tyrone Power, Henry Fonda, Lucille Ball, Ginger Rogers, Marlene Dietrich, Charles Laughton, Mickey Rooney.

Clark Gable's telegram read: "As an old-timer myself in Bartlesville my special good wishes are yours as the new Osage Theatre opens its doors." Spencer Tracy wired: "While we were working on 'Boom Town' Clark Gable told me so much about Bartlesville I almost feel I have been to your city."

Downtown Bartlesville was overflowing with people, but just in case someone wasn't sure of where to find the Osage that evening when the movie started, floodlights were set up to show the way.

In the afternoon, Miss Neagle and Frank joined Osage chief Fred Lookout at the dedication ceremony, at which the chief adopted the movie star as the first white woman in the Osage tribe and gave her an Indian name, which translated to "Princess of the Dance." As soon as the chief placed a headdress on Miss Neagle and gave her a large Osage blanket, Frank sidled up. "Now that we're both brother and sister in the Osage tribe, how about a little kiss?" Miss Neagle politely refused and was able to outmaneuver Frank without causing a fuss.

About that time, Art Goebel swooped down from the heavens in his skywriting airplane and, in an attempt to add a bit of color to the celebration, released a stream of pinkish-red smoke. Everyone cheered for the dashing Goebel, but when the cloud of smoke drifted to the earth and settled on the crowd, the cheers turned to murmurs of concern. It seemed Goebel's colored smoke stained everything it touched, including coats and hats. Fortunately none of the stuff got on Miss Neagle's pretty war bonnet, but Phillips Petroleum picked up the dry-cleaning tabs for hundreds of celebrants. Frank didn't mind. It was another good party. Later, when he entertained his Indian friends and Miss Neagle, still wrapped in her colorful blanket, at Woolaroc, he was even finally able to convince the reluctant actress that it was perfectly all right to allow a fellow adopted Osage to have a peck on the cheek.

A few months later, Frank threw another big party at Woolaroc, this time to entertain James Farley, formerly chairman of the Democratic National Committee and F.D.R.'s Postmaster General until he resigned both posts when he opposed Roosevelt's decision to run for a third term in 1940. Farley, who had reentered the business world, came to Bartlesville to convince Frank to buy the New York Yankees. Colonel Jacob Ruppert, the beer baron who purchased the Yankees in 1915, died in 1939, and Farley and some others felt Frank Phillips had the resources necessary to take over ownership of the ball club. Frank listened to their pitch but decided against putting any money into the Yankees. Already a dynasty with such greats as Ruth, Lazzeri, and Gehrig, the Yankees went on to more glory days with DiMaggio, Rizzuto, and Henrich. They remained under the ownership of the Ruppert heirs with Ed Barrow as president until 1945, when the club was sold for $2.8 million. But Frank never regretted his decision. After all, as he had pointed out to Farley, "I'm a National League man."

He was also a basketball fan, and the Phillips 66ers, led by such greats as Bill Martin, were providing Frank and thousands of loyal fans with enough thrills to last a lifetime. "We played a lot, sixty games a season, and most of the games were benefits, in small towns, raising money for good causes," said Martin. "Basketball became quite important to Phillips. We thought that we were gaining advertising we couldn't buy any other way and we were getting a lot of goodwill as a result of it. Mr. Phillips really didn't know too much about basketball, but he did know the difference between winning and losing, and he was a hard winner . . . or, I should say, a hard loser, a very gracious winner. He didn't like to lose, whether it was basketball or business or anything else. He used to come to our dressing room and it wouldn't take long to get the idea that in the second half, you ought to put out your very best effort. That was the kind of man he was."

Losing was no fun, but thankfully there were mostly winning teams coming out of Bartlesville. They were treated like royalty. In November 1940, Aunt Jane hosted a dinner for the company team—National AAU Basketball Champions—at the Ambassador just before one of their big battles at Madison Square Garden. The menu included buffalo, elk, deer, wild turkey, and pumpkin pie, and all the wild game was flown in from the F.P. Ranch.

That same month, the Frank Phillips Men's Club sponsored a huge dinner-dance at the Hillcrest Country Club in honor of Uncle Frank's sixty-seventh birthday. Frank wasn't able to break away from the New York office in time to attend, but he arranged to put in an appearance via the telephone. At 10:30 P.M., the telephone connection was made and R. C. Jopling announced: "Ladies and gentlemen, Mr. Frank Phillips speaking to you from Radio City, New York!"

Everyone in the room, dressed to the teeth for their chairman, leaned forward and listened for Frank's mellow voice to boom over the loudspeaker.

"Well, well! Hello, boys and girls. This is your Uncle Frank, back here in New York, just about crying my eyes out because I can't be with you. Makes me a little sad to be reminded that another year has rolled around and that I am no longer sixty-six. I still insist, though, that I am only sixty-seven years young—not old. Anyway, it makes me very happy to be remembered on my birthday. A year ago you gave me the swellest birthday party that any man ever had. The greatest thrill of my life was watching you marching in that great parade. It was the greatest tribute that was ever paid to me, and I shall remember it with deep and humble appreciation as long as I live."

To let his employees know just how much he did appreciate their attention, that fall Frank took the wraps off his ranch and museum and opened it for conducted tours several times each week. He also posted a statement in the entrance hall of the Woolaroc Museum which explained his reasons for opening the facility:

> This museum originated as a place of safekeeping for the
> many mementos given me by my friends throughout the
> world. In the hope that paintings by great American artists
> might educate and inspire, I since have added many of

these. They reflect the evolution of America, especially its Southwest, from prehistoric times to the present day.

As the adopted White Chief of the Osage Tribe, and as an observer of the final chapter of the long conflict between Red Man and White Man, I have known the traditions and viewpoints of both these peoples.

If the exhibits here can help to convince youth and future generations that America's hard-earned traditions and ideals must be preserved, my meager efforts will be repaid.

There was no question about the importance of the ranch in Frank Phillips' life. Ernie Pyle, the famed journalist, who went on to win the Pulitzer Prize for his front-line stories during World War II, witnessed Frank's deep attachment to the Osage and the lodge during a visit Pyle made back in 1939.

"When I visited the Frank Phillips Ranch, one of the oil company employees picked me up at the hotel to take me out," wrote Pyle. "It was my impression that Phillips was in New York, and I had no expectations of seeing him. You would not figure one of the richest men in Oklahoma would be easy to see anyhow. But when we got to the ranch, there were Mr. and Mrs. Phillips sitting on the front porch—waiting lunch for us! I guess that's Oklahoma hospitality or something."

After they finished eating, Jane gave Pyle a guided tour of her bedroom portrait gallery. She talked about each of the men in the photographs and explained the discard pile for those who died or fell out of favor. Pyle liked the outgoing and gracious Aunt Jane, but he was particularly taken with Frank.

"Phillips is tall and slender and looks younger than his sixty-odd years. In his pictures, he seems grim and tight-lipped. But in person, his face is friendly. He is easy to talk to and, unlike many rich men, pays attention to what you say."

Dressed in an immaculate riding outfit, Frank personally led the journalist on a tour of the property. "It's thirty-six miles around the ranch, which has fourteen miles of roads, and a half dozen lakes," wrote Pyle. "You can easily get lost just driving around the place. Right within the ranch you can fight a forest fire, go motorboating, shoot a buffalo, or lounge in Park Avenue style . . . Mr. Phillips spends about forty percent of his time here, about forty percent in New York, and the other twenty percent somewhere else, which may mean anywhere from Europe to California. He travels mostly by air in his own Boeing transport. Last year alone he flew about 20,000 miles. Mrs. Phillips flies, too.''

Pyle was shown Clyde Lake and the picnic shelter. He inspected the stables and looked at the herds of livestock and the surviving wild animals which had learned to adapt to the Oklahoma weather.

"Phillips is generous with his ranch," wrote Pyle. "His executives and many lesser employees come out regularly and ride horses. He gives gigantic parties several times a year. Phillips loves these big parties . . . Frequently there are one thousand guests. He has even brought out four special trainloads of guests from New York. For these big blowouts, they barbecue four buffalo, killed right on the ranch."

But of everything he saw during his visit to the F.P. Ranch, Pyle was most fascinated with Frank's Woolaroc Museum, especially the curious collection of art and artifacts; dinosaur eggs, shrunken heads, and gleaming saddles. "All around the walls are great oil paintings—mostly western scenes or portraits of Indians. Every picture is hung against a rug of animal skin, with the edges protruding a few inches. It gives a softening effect that is marvelous. It sounds like a hodgepodge, but it isn't. The thousands of items are expensively mounted and classically presented. The curator is Mrs. Donald Zulke, the ranch foreman's wife. She never had any museum experience. She and Mr. Phillips just figured out together how they wanted things done."

Not long after Ernie Pyle published his account of the F.P. Ranch, Frank determined that some changes needed to be made at Woolaroc. His "airplane museum" had been enlarged and a trophy room added, but still more space was needed. A large part of the memorabilia collection housed at Woolaroc came from Frank's sixty-sixth-birthday celebration, and the inventory of art, artifacts, animal trophies, and other collectibles was growing.

Jane kept meticulous books listing the donor, the gift, and the date it was received. It was a most eclectic collection. Besides the shrunken heads, dinosaur eggs, old stagecoach, and *Woolaroc* airplane, the museum contained Miles Standish's ashtray; Frank's complete Osage Indian outfit and the stretched buffalo hide which told the adoption story; an array of firearms including outlaw Al Jennings' old single-action Colt revolver; a stuffed alligator Boots brought back from Miami; a 1,200-pound block of anthracite coal; a fourteen-foot-long petrified tree root dug up near Hogshooter, Oklahoma; the historic medal of the Legion of Honor presented to an associate's ancestor by Napoleon after the Battle of Waterloo; an Egyptian fly brush; a machine gun which belonged to "Diamond Jack" Altarrie, one of Al Capone's lieutenants; a hair ball taken from a steer's stomach; and a mounted two-headed calf.

The museum truly was a hodgepodge of odds and ends and curios mixed with some fine examples of western art. Frank realized that professional guidance from someone trained in museum work could bring some semblance of order to the collections. It was no longer just a job for the ranch foreman's wife.

More material for the Woolaroc Museum was uncovered at the Spiro Indian Mound, a significant archaeological excavation located on the south bank of the Arkansas River in southeastern Oklahoma. Spiro Mound was discovered in 1933, and between 1936 and 1941, controlled excavation was carried out by the University of Oklahoma, with help from the University of Tulsa, the Oklahoma Historical Society, and WPA work crews. Many of the highly developed artifacts, including masks, pottery, utensils, and weapons, were distributed to various museums around the state, including Philbrook and Woolaroc. Frank Phillips furnished much of the funding for the excavation work and went down to the site to get a close look at busy diggers and on at least one occasion even tried his hand at excavating.

"I recall when Mr. Phillips first became aware of Spiro," said Paul Endacott. "A fellow showed up in the Phillips offices in Bartlesville. He just appeared there one day out of the blue. He was an old cowboy type, with grizzly sort of looks and he had an old beat-up and ragged hat. I went ahead and saw him and he came walking in with that hat filled with pearls. There must have been two or three hundred pearls in that hat. He said he'd been down digging at Spiro and came across some things that might be good for Mr. Phillips' museum. The man wanted to sell the pearls. I immediately brought that old-timer right in to see Mr. Phillips. It was soon after that that he became involved with the digs out at Spiro."

Once he became associated with the Spiro Mound work and saw that the collections at his museum were going to continue to grow and expand, Frank began to search in earnest for a professional museum director. The person Frank was looking for came limping into his life in 1940. His name was George Welles Patrick Patterson. He had enough Indian blood in him to also use the Apache name Kemoha. Most people called him Pat.

Pat Patterson was a profane and blustery Irish-Catholic Democrat who had nothing in common with Frank Phillips except for their mutual love of the West, especially the art and folklore. Born and raised in a circus family, Patterson was badly crippled with polio when he was only nine months old.

"The polio kind of slowed me down for any useful life in show business," said Patterson. "My father worked the flying trapeze and walked the wire. He sang and danced, the whole goddamn bit. I can remember my mother selling tickets and Dad performing. My sister, Marcella, was a performer too. They'd leave me with the elephants, with Old Maude to look after me while they worked. She'd protect me with her trunk. Never will forget those damn ol' elephants, especially when they had to move their bowels."

Pat Patterson was a colorful character and he was right up Frank's alley. Anyone who was raised under the big top was top-drawer in Frank's book. He could forgive Patterson's political affiliation just to hear the circus stories. Besides, Patterson knew what he was talking about when it came to putting together a first-rate museum.

"When we first met in 1940, we got off to a rather shaky start," recalled Patterson. "I was down at Norman at the University of Oklahoma. I had my degree in anthropology, and went on to graduate school and was doing some teaching. One afternoon I was throwing a little party at my apartment. I had nineteen girls up there and two campus cops and I was cooking a giant ham. The phone rang and this big ol' Texas gal answered it and I grabbed it out of her hands and yelled, 'Operator, is this goddamn call paid for?' Before she could answer I heard this voice roar back, 'You're damn right it's paid for!' You see, I was used to some of my former students calling me up collect. Anyway, the voice said that it was Frank Phillips on the line from up in Bartlesville. He said he wanted to talk to me and he was sending his plane down in the morning to pick me up. Now, when I heard that I knew it was some sort of a joke and I had this ham cookin' and all those gals waitin', so I said I'm sorry but I'm gonna be busy in the morning and I hung up on him and went back to my party."

The next morning, Patterson was teaching in his classroom when a proctor broke in and told him he was wanted immediately at the administration office. "I walked in there and they asked me why I hung up on Frank Phillips. An hour later I was on my way to Bartlesville. They brought me out to the ranch and there was Mr. Phillips sitting out on the porch, gazing out across the lake to his pastures. I sat down and he said he wanted me to go to work for him. He said he wanted me to build up his collections and make his museum a good one."

Patterson had been active at the Spiro excavations and had heard that if a person wanted to work for a tough customer all he had to do was get a job with Frank Phillips.

"I knew all too well the stories about the old son of a bitch, and I knew he was the one man I never wanted to work for under any conditions," said Patterson. "So I told him I was still in school doing graduate work and teaching, but he wouldn't have any of it. Finally, I told him I'd work for him for thirty days if I could get all the money I needed with no questions asked. I thought that would end the discussion right there. But it didn't. That tough old man looked me right in the eye and said, 'Well, don't just sit there—get started!' "

Patterson put in his thirty days, but he didn't return to school, because Frank had taken care of that. "You can't go back there," Frank told Pat. "I got you fired."

Patterson was shocked. He went to the telephone and called the university to see about his teaching post. They were waiting for Pat's call. "It was true, all right. I hung up the phone and went back to work. I stayed with Frank Phillips."

Frank's decision to commit substantial funds to improve the facilities and his hiring of Patterson as a museum director was the turning point in Woolaroc's haphazard development.

The museum building, constructed of native sandstone quarried from nearby cliffs, was again expanded. The collections were also enhanced. Glass display cases and the pine bark frames which originated at Woolaroc were added, and the museum's permanent collection was vastly refined and improved. Before long, the Woolaroc Museum would house more than 55,000 individual exhibits. As the old-time oil men would have said it, Woolaroc was becoming a real "Darb."

Magnificent sculptures and paintings from the western masters—Russell, Remington, Leigh, Sharp, Couse, and many more—were purchased and displayed. The finest American Indian handicraft—baskets, pottery, rugs, and blankets—from Pueblo villages, the Apaches, and the deep Southwest were also shown.

In one exhibit were Woolaroc's collection of western saddles and accessories, including the 101 Ranch "Wild West Show" saddle—billed as "The World's Finest" and valued at $10,000. It was studded with fifteen pounds of sterling silver and gold, and all sorts of precious jewels, including diamonds, rubies, and sapphires. The collection also included Buffalo Bill's saddle and guns, and the saddles of May Lillie, Tom Mix, Zack

Miller, and Teddy Roosevelt. A saddle that Henry Wells used while fleeing a sheriff's posse was in one of the cases, as well as the skull of the horse, complete with bullet hole, which was shot from under Wells by Grif Graham only a few hundred yards from where the museum was eventually built.

Woolaroc acquired other bits of Americana—the original bronze model by Daniel Chester French for the Lincoln Memorial in Washington, D.C.; portraits of Will Rogers, General George Armstrong Custer, Sitting Bull, and other great Americans; and an outstanding series of miniature portraits of the thirty-two Presidents of the United States, all hand-painted on ivory, which had been originally started by the artist A. J. Rowell under contract for "Diamond Jim" Brady. Woolaroc would continue the series.

About a year before his death, E. W. Marland, still financially strapped, asked Frank to help him by purchasing some of his paintings, tapestries, rugs, and bronzes which had been acquired by Marland in his more prosperous years. Frank agreed. He was pleased to be able to help out his old friend. Frank sent Patterson to Ponca City to inspect the collection and make Marland a fair offer. Woolaroc acquired much of the Marland collection as well as the twelve original castings from the "Pioneer Woman" sculpture competition sponsored by Marland in 1926, including the original bronze model which won by sculptor Bryant Baker. Besides the dozen bronzes of the "Pioneer Woman," Woolaroc eventually purchased from Marland five other major bronzes—"Pioneer Man," also by Bryant Baker, and four created by artist Jo Mora: "The Indian," "The Squaw," "The Cowboy," and "Belle Starr." All of them were given prominent places in the museum.

"There was a lot of controversy about Governor Marland—his politics and his second marriage and all," said Patterson. "And when he died in 1941, there weren't a hell of a lot of people at his funeral. But Frank Phillips was there. Rest assured of that. Frank Phillips stuck with him to the end."

Patterson always respected Frank, and over the years, as he continued to work as the museum director, the outspoken man with the exaggerated gait caused by polio learned to love Uncle Frank. Still, working for him wasn't always easy.

"That museum was a one-man show," said Patterson. "It was his own personal plaything and he hated anyone coming along and sticking their nose in it. We fought. Fought all the time. I swear he fired me at least once a week. I know for a fact he fired me three times in a single day! We'd have some knock-down-and-drag-outs over policy or what should be placed where and what we should be acquiring. I stood my ground and I learned that you had to do that with that tough old son of a bitch. He respected those who stood up to him if they believed they were right. But Frank Phillips was never one to insist that you jump to attention and click your heels for him. He knew he was in charge, so he didn't have to be reminded. If your boss is confident and sure of himself, then everyone will feel good about themselves. That's what made the ranch and the museum succeed. That's why it was such a beautiful place to work and such a beautiful place to be."

Another one who learned to put up with Frank's explosive temperament was his oldest grandson—John Phillips, Jr. During his summer vacations from military school, Johnny was given jobs as an errand boy in Frank's bank in Bartlesville and also at a Phillips 66 station in Tulsa. But in the summer of 1941, just prior to his senior year as a cadet at Culver, Frank summoned his grandson to the Phillips offices and told him that he wanted him to go to work on the ranch as a hired hand for the summer. It was the job Johnny had dreamed of since he was a small child.

Frank agreed to pay John, Jr., and his good friend, Will Parker—son of Billy and Cindy Parker—twelve and one-half cents an hour, plus room and board, for a forty-hour week. Before they marched out of his office, Frank promised the boys that if they worked hard all summer, there would be a bonus waiting for both of them. The two teenagers went straight to Woolaroc and were placed under the tutelage of John Seward, the tough Texan who was then the ranch manager.

The work was hard and the hours long and the boys loved their summer at the ranch. They especially liked mixing with the ranch hands— Joe Billam, who was soon to become the next ranch manager; "Huck" Swift, one of the best riders on the spread; and Claude Johnstone, manager of the horse barn. Johnny and Will earned every penny of their pay. They had a few dollars and lots of aching muscles and blisters and sunburned necks to prove it. There was little time to get into trouble. At summer's end, Frank called Johnny and Will back to his office. Their work was over and it was time to review their performance.

"Boys, I've gotten some mighty fine reports about your work and conduct this summer," Frank told them. "Now, I have a surprise for you. That bonus I promised. In the morning I'm putting you on a company airplane for New York. You can live it up for the next two weeks in my suites at the Ambassador. The rest of the summer is on me!"

The boys were thrilled. They left bright and early, arrived in New York, checked into the Ambassador, had a huge dinner at the Waldorf, and made the rounds of all the places a pair of sixteen-year-olds were able to penetrate. But after a few days of living high off the hog, Johnny and Will got bored, and lonesome. They also got a wild hair and decided to send the Phillips pilot back to Oklahoma to fetch their girlfriends. The pilot was at their beck and call and left as ordered.

"I'll never forget what happened next," said John, Jr. "I telephoned Granny back in Bartlesville to see if she'd go along with our scheme and she hit the ceiling. She said that it couldn't be done. Well, I had sort of expected that reaction, but then she kept talking and added, 'If you think for one minute that I'm going to stand by and send those girls up there to have all the fun, you're mistaken. I'll be on the plane with them! I'll be the chaperon!'"

Aunt Jane was as good as her word. She arrived with the young ladies, and for the next two weeks, Jane and her young charges took Manhattan by storm—eating at the expensive restaurants and shopping at the most exquisite stores.

"The first Saturday night Father got to the city, he was hosting a small dinner party at the hotel," said John, Jr. "He asked me what all of us youngsters were going to do that evening, and I told him not much, since we had spent all our wages earned as ranch hands and were flat broke. He said we should all come to the dinner and at least we'd eat well. Later we all entered the dining room and looked for our place cards and saw that our plates were all turned upside down on the table. Just ours and nobody else's. We sat down and turned over our plates and under each one was a crisp one-hundred-dollar bill. I looked at Father and he was laughing."

After dinner, Frank loaded the youngsters, flush once more, in his car and sent them off for a final fling. They danced the conga to the strains of Xavier Cugat and his band and later, at another hotel ballroom, were thrilled when Frank Sinatra crooned "I'll Never Smile Again" at Johnny's request and then joined the kids from Oklahoma for a short visit at their table. "That's what I remember best about Granny and Father," said John, Jr. "They were giving people and they enjoyed having fun and seeing others have fun too."

Pat Patterson also learned about the fun-loving side of both Frank and Jane. He became very close to the Phillipses. Occasionally he escorted

Jane to a few of the Bartlesville night spots, including the Clover Club, where she enjoyed playing blackjack. Sometimes Pat would go with Jane on a New York adventure.

"One afternoon, a couple of years after I went to work out at Woolaroc, Mrs. Phillips and myself had been to a big New York show and we went over to this fancy damn old place to have a drink," said Pat. "Never will forget it. We walked in there—just Jane Phillips and me—and the room is filled with all these society types. Well, as soon as the orchestra spies her, they break into 'Oklahoma!' Every eye in the joint is on us. Jane played it to the hilt. She walked real slow and looked around at everybody with her fancy eyeglasses. Everyone made a big fuss over us. I mean a big fuss. We got seated and everybody's still watching and Jane turns to the waiter and orders our drinks and then tells him to also bring her a big tub of hot water.

"He does just as he's been told and comes back with the drinks and the tub of steaming water. Well, Jane proceeds to kick off her shoes and peel off her stockings and plunge her feet straight into the water. Then she breathes a great big sigh of relief and looks around the room at everybody staring with their mouths wide open and she says, 'You should all try this —it is absolutely wonderful!' It was really something. They never made another one like Jane Phillips."

After Pat had been on the job at Woolaroc long enough to get the museum under control and learn the ropes, he would often join Jane and Frank for dinner whenever they came out to the ranch. "One night we were all sitting there eating and suddenly Mr. Phillips turns to me and says, 'You look tired.' I said that I was tired and that I planned to turn in early. He said, 'That's not what I mean—you need a vacation!' I said, 'Okay, I'll take a vacation.' We all kept eating. He cut his beef and took a bite or two and said, 'Well, where are you going on your vacation?' I said that I really hadn't thought much about it yet but maybe I'd go over to Arkansas and do some fishing. He said, 'By God, don't you have any ambition? You can fish right here at the ranch! Don't go off fishing! Why don't you go see some other museums and see just what the hell they're doing that we're not doing.' I said all right, I'd do just that.

"Mr. Phillips ate some more beef and poked at his potatos and said, 'When are you leaving?' I said, 'If you want me to go so bad, I'll leave in the morning!' He said, 'Good! here's your tickets.' He reached in his pocket and pulled out a string of tickets pinned together. I measured them. It was seventeen-feet long. A string of rail tickets and reservations at hotels and notes of entry and introduction to every museum on the east side of the Mississippi River."

Not long after the expedition back East, Frank financed a similar trip for Patterson, but this time he sent him to see all the museums west of the Mississippi. "He really cared a great deal about the ranch and the museum and he wanted to be sure that we didn't miss a trick when it came to making Woolaroc a first-class attraction," said Patterson.

As Frank and Patterson worked at shaping up the Woolaroc Museum, the angry clouds of war which already engulfed most of Europe and Asia moved like a threatening prairie thunderhead toward the United States. Phillips and the rest of the nation earnestly prepared for the inevitable— war with the Axis powers.

Actually, the research end of the Phillips business had been working day and night to prepare for the war effort and two key Phillips contributions to the Allied cause—high-octane aviation gasoline and synthetic rubber—were made possible because of research breakthroughs in the 1930s.

The company was a pioneer in the development of butadiene, an essential ingredient in the manufacture of synthetic rubber which came about when Phillips and B. F. Goodrich pooled their resources. Frank took great pride and pleasure in sitting down with David M. Goodrich to sign a joint agreement to form Hydrocarbon Chemical and Rubber Company, which later was named Hycar Chemical Company.

As a result of the company's continued success with the development of synthetic rubber and smooth-burning aviation gasoline, Frank was able to report in 1940: "No industry is in a better position to supply the nation's military needs. Requirements of petroleum products for contemplated airplanes, ships, and motorized war machines will be available long before such equipment can be built. The Company is equally prepared. New manufacturing facilities have been specially designed to produce the highest-grade aviation gasoline and ingredients. Raw materials are being converted for the manufacture of synthetic rubber. The Company has no investments of any kind in any foreign country; therefore, all facilities are available for any national situation that might arise."

As 1941 dawned and Roosevelt was inaugurated for his third term of office, the war clouds grew darker. By the springtime those clouds had become as dark as motor oil. In May, Roosevelt denounced the French who collaborated with the German occupation forces. He issued orders for the U.S. government to take into protective custody all French ships in U.S. ports, including the *Normandie,* the great luxury liner which Frank and Jane had sailed on, and also which Fern was on during her solo trip to England a few years before the war erupted. That summer, in rapid succession F.D.R. ordered all the German and Italian consulates in the country closed; vowed aid to the U.S.S.R. following the German invasion; froze Japanese assets in the United States in retaliation for their move into French Indochina; and launched a get-tough policy for Navy planes and ships to shoot on sight any Axis ships which moved into "the sea frontier of the United States."

The stage was set for the country's entry into the war. Phillips Petroleum, like most leaders in American business and industry, was center stage and only waiting for its cue. While waiting, Frank and brothers L.E. and Ed went up to Iowa to visit kinfolk and friends, pay their respects at the family burial plot near Gravity, and go back to the old Phillips farm. The Phillips brothers flew to Iowa in an Army Lockheed Lodestar loaned to Frank because of his new position as petroleum chairman of fifteen midwestern states for the Petroleum Industry Council for National Defense. Frank had been as proud as Jane's peacocks when Harold Ickes, the pugnacious Secretary of the Interior in Roosevelt's Cabinet, selected Frank to serve as one of the Petroleum Defense Coordinators in order to ensure that the nation's petroleum facilities were used efficiently with respect to the strong possibility of the United States entering the war. Ickes called for the first meeting to be held in December.

"I consider it my patriotic duty to accept these additional responsibilities and the hard work which is inevitable under the circumstances," wrote Frank in his August *PhilNews* letter to the employees. "My appointment places me in the same category as the many other employees of our Company who have been selected for service in other national defense assignments."

A few days after his visit to Iowa, Frank got another opportunity to speak about the war when he appeared before the Nebraska Historical Society. "Beyond all question," said Frank, "the present war will be won by the nation that has the largest supply of oil and makes the best-quality motor fuel. In war, as well as peace, oil is one of our leading industries. Perhaps it is the leading industry in war, if you consider that oil is our first column in national defense. Imagine, if you can, the calamity that would result should the supply of petroleum and its products be suddenly stopped. The national defense would crumble, for tanks, airplanes, and motorized equipment are helpless without gasoline and lubricating oil. Our navy would rust at anchor, and every wheel in the nation would stop.

"During the eighty-two years of its life the oil industry has grown from a medicine business to one of our largest and most important industries. It has continued its growth and progress even during depression years and has reached its present stature without shooting, plowing under, or otherwise destroying any of our natural resources and without one penny of help in the way of subsidy from our national government."

Frank remained in the New York office most of the autumn, except for quick trips to Washington to confer with Ickes. On November 28, Frank celebrated his sixty-eighth birthday by announcing that he and Jane were going to contribute $550,000 to the Frank Phillips Foundation. That morning when he arrived at the office, Mildred Rowson and Hy Byrd, the company's assistant secretary in New York, had Frank's entire office decorated with flowers. In the evening, Jane gave a party for Frank at the Ambassador and presented him with a bronze bust of Will Rogers sculpted by Bryant Baker. Will's widow was one of the guests. Frank and his entourage left for Oklahoma the next day.

Osa Johnson, the noted African explorer, came to Bartlesville in early December to appear at the Civic Center and lecture about her African adventures. She also was going to show her latest motion picture, *African Paradise.* Osa's appearances were part of an educational program sponsored by the Frank Phillips Foundation. Frank left the office early to introduce the attractive widow at her matinee performance for the schoolchildren.

But Frank wasn't able to attend Osa's show the next day on December 6. That evening he had to appear at a charity basketball game in Tulsa between the Phillips 66ers and the squad from 20th Century–Fox. The proceeds from the contest were donated to the YMCA fund for underprivileged children, and Frank was asked to make the presentation, an honor he could not refuse.

When it was time to introduce Osa, Frank ducked out of the game in Tulsa for a few moments and went to Waite's downtown penthouse atop the Philcade Building, where there was a telephone hookup to the Civic Center in Bartlesville. At 8 P.M. Frank's voice came over the loudspeakers. Frank told the overflow crowd about the charity basketball game and offered his apologies for not being back home for Osa's program. "The new Osa Johnson picture which you are going to see is one of the finest she ever made," Frank said. "When we saw it for the first time in New York, Aunt Jane said, 'I wish everybody in Bartlesville could see this picture.' Her word is law, as you know, so we made arrangements to bring Mrs. Johnson here and we hope you will agree with us that it is really a worthwhile entertainment.

"Now a word about Osa Johnson. As an explorer, hunter, and fisherman, she is without any equal. You have all seen the tremendous elephant head inside the entrance to my museum at Woolaroc. It is said to be the largest elephant head ever mounted. The trophy represents an extremely narrow escape Osa Johnson had when this elephant charged her in the jungles of Africa. She was able to save her life with only seconds to spare by bringing it down with her elephant gun. She's pretty too, and I am again sorry I am not personally in Bartlesville because at the last three performances I have been able to wind up my introduction of the speaker by stealing a kiss from her and I can't do that by long-distance telephone."

It was a hectic Saturday night, but Frank and Jane returned in time to have a nightcap with Osa out at the ranch. The three old friends sat beneath the trophy heads in the lodge and watched Dan feed the huge fire with armloads of cut oak. They talked about Osa's travel plans and the next stops on her hectic lecture schedule. She was due to leave the following day and travel all the way to the Hawaiian Islands. The Phillipses knew Hawaii well and they wished they could break away and go there with Osa to soak up the sun and feel the warm sea breeze.

By the next day—Sunday, December 7, 1941—Osa's plans were changed and her Hawaiian trip was abruptly canceled. That morning a massive Japanese striking force of 360 planes and scores of ships which had left their home waters on November 26, two days before Frank's birthday, attacked the U.S. naval base on Oahu, Hawaiian Islands. Their primary target was Pearl Harbor, where ninety-four U.S. Navy ships, including eight battleships, were moored. Within a half hour after the start of the attack, the *Arizona* was a burning wreck, the *Oklahoma* had capsized, the *West Virginia* was sunk, and every other battleship was damaged. Almost all the U.S. planes were destroyed on the ground. American casualties were 2,403 killed or missing and 1,178 wounded. This grim December Sabbath would be known forever as "Pearl Harbor Day." As Roosevelt said, it was "a day that shall live in infamy." The United States had been plunged into World War II. It would last for the next 1,365 days.

As soon as he could muster his troops, Frank left that Sunday for Washington. He was accompanied by Dan and R. F. Hamilton, secretary of the Phillips executive committee. Frank went to the nation's capital for the first official meeting of the Petroleum Industry Council for National Defense. Now the time had come for Frank and the other oil company executives reporting to Secretary Ickes to implement the strategy for the petroleum industry prepared weeks in advance.

While he was in Washington, all hell broke loose. Resentment against the Japanese was building. "Remember Pearl Harbor!" was the rallying cry. Dan Mitani, Frank's valet for more than twenty years, became the target of ridicule and scorn whenever he appeared on the street or in public areas. The hatred in the eyes of the men and women who jeered at Dan frightened him and concerned Frank. During the rest of their time in Washington and for Dan's own protection, Frank hid his valet in a room at the Shoreham Hotel. Dan eventually returned to his family in Bartlesville. But until the anti-Japanese fervor subsided enough to permit Dan or Henry Einaga to resume traveling, Frank used his old driver, Homer Baker, as a personal valet.

The arrival of World War II and the restricted travel placed on Japanese in the United States disturbed both Dan and Henry. Both men were fiercely patriotic, as loyal to America as they were to Frank and Jane and the Phillips family. Henry had even registered for the draft during World War I. Now they weren't allowed to go near the train station in Bartlesville or even travel as far as Tulsa without the consent of a district judge. Whenever people questioned Henry and asked where he was from, he always proudly said, "I'm an Okie." So was Uncle Frank—a proud and stubborn Okie. As of January 1, 1942, Frank and Jane Phillips became legal residents of the state of Oklahoma. Back in September, Frank had announced his intention to give up his residency in New York when he chartered an organization to handle his personal affairs called the Phillips Investment Company. Because of the higher income taxes imposed in New York, it would be an Oklahoma company. After January 1, Frank and Jane gave up their permanent suites in the Ambassador but continued to spend a great deal of time in New York and used the Park Avenue hotel as their eastern base of operations.

"Father dearly loved Oklahoma and always fought to keep the Phillips headquarters in Bartlesville," said John Phillips, Jr. "But the main reason he established their domicile back in Oklahoma was for tax purposes, pure and simple. It just was good business."

Frank was now caught up not only in a growing wartime business boom but with patriotic fervor. Within a few weeks after war was declared, Frank issued a statement to his employees:

"We are at war. No longer are we concerned only with preparations; we now must add the responsibility of doing anything and everything to strengthen our defenses and to achieve victory, complete and unqualified. Until that goal is reached, other interests must be secondary."

That didn't apply to everyone Frank cared about. On January 2, Frank met with Lewis Fisher, the young son of Tillie, the former Phillips family nanny, and Clayton, Frank's chauffeur. Lewis remained one of Frank's favored young people, and when the young man graduated from high school as a member of the National Honor Society, Frank saw to it that he was one of the first recipients of a university scholarship.

Lewis could recall the day—shortly before the big sixty-sixth-birthday celebration—when Frank and Don Emery, a Phillips vice president and general counsel, told him they were establishing the scholarship program. Frank told Lewis he wanted him to show everyone else how smart he was and to do the best job possible in his studies. Lewis did just that.

"I was very prejudiced when it came to Frank Phillips," said Lewis. "I thought the world of him and would do anything he asked."

Lewis treasured the gold pieces, the expensive cuff links, and other gifts he'd received from Frank over the years. He still had the fancy model boat made of teak wood that Jane had given him in the 1930s.

Now he was about to earn his degree at the University of Oklahoma, and during their meeting that bitterly cold January day, Frank implored Lewis to stick it out and wait until he graduated before enlisting in the military. As always, Lewis paid attention to Frank and went back to the university to complete his last semester.

Immediately after he graduated, Lewis joined the Army, attended officer candidate school, and received a commission as a second lieutenant. Before he was shipped overseas, Lewis came back to Bartlesville to bid his parents goodbye and introduce Uncle Frank to Gerda, the pretty young coed from Chickasha, Oklahoma, Lewis had courted and wed.

"Lewis was about to go to war and I had to meet Mr. Phillips," said Gerda Fisher. "He asked me straight out, 'Young lady, are you pregnant?' I told him I wasn't and he said that was good. He was so interested in Lewis and his welfare and Lewis really wanted to please him and pattern his life after Mr. Phillips and be just like him."

As Bartlesville's young sons marched off to combat in Europe and the Pacific, Frank made sure his company did its part to keep the engines of war fueled. Hycar Chemical Company, the Phillips-Goodrich synthetic-rubber joint venture, was enlarged to three times its initial capacity and the Phillips plant at Borger supplied butadiene and carbon black to make the rubber stronger and more resistant to wear. The Phillips research staff also announced several new developments in processes for producing special aviation gasoline.

There was also a dramatic change in the Phillips advertising program. The company's ads changed their focus to the critical issues of conservation and patriotism. "You Are a Soldier in the Battle of Transportation," read one of the new ad headlines. Another ad told customers to preserve their cars and reminded them to visit their Phillips 66 dealer at least once a week. "Care for Your Car—for Your Country," said the headline. Driving unnecessary miles aided the enemy. "Every bit of rubber and gasoline you save on the home front is a contribution to the combat needs of our fighting men on every battle front," said the ad. "You can hasten the day of victory by confining your driving to a patriotic minimum. Use your car only for going to and from work . . . for needed shopping . . . for wartime activities like vegetable gardening . . . for travel to and from places without transportation facilities." The closing slogan on every ad was the same: "For Victory . . . Buy U.S. War Bonds and Stamps."

As chairman of the Petroleum Industry Council for a fifteen-state region, Frank headed scrap-rubber drives and preached for sacrifice at home. He encouraged all Phillips workers to grow Victory gardens and he wrote letters for *PhilNews* urging everyone to buy bonds and participate in such activities as the Red Cross and Civil Defense. "No longer are front-line bravery and combat strategy the only essential factors which determine victory or defeat," wrote Frank. "Ten or twenty workers behind each soldier and sailor, together with as many others in civilian life, win or lose today's battles. It is they who either do or do not permit our firing-line defender to have the things he needs to protect himself and defeat our enemy."

On April 28, 1942, during the annual meeting of the Phillips stockholders and directors, Frank walked into Boots' office, where everyone was gathered, and spoke of the uncertain future and the need for the company to continue to back the war effort one hundred percent. "Winning the war is our immediate problem," said Frank, "and Phillips Petroleum has geared itself to war operations. I cannot speak freely about our contracts and negotiations with the government for the production of war munitions, such as aviation gasoline, synthetic-rubber components, explosive chemicals, and other petroleum products. We are bound by secrecy agreements . . . It is my opinion, however, that in proportion to its size, the company is engaged in essential war activity to a greater degree than most any other oil company in America."

On June 13, 1942, Phillips Petroleum Company turned twenty-five years old. The firm that was founded during one great world war now found itself enmeshed in another global conflict. The times did not justify a big celebration to commemorate the silver anniversary of Phillips Petroleum, or so Frank thought. He was more concerned about the 750 Phillips employees serving in the armed forces than he was about blowing out candles on big cakes or slurping down champagne. So instead of the traditional party, Frank sent each one of the workers on military leave from Phillips a birthday package.

Despite Frank's decision for the company not to officially mark the twenty-fifth anniversary in a conventional manner, the Bartlesville Chamber of Commerce believed the occasion had to be observed in some fashion. A couple of weeks before the big date, they hosted a banquet to honor the company but mostly to pay homage to Uncle Frank, the founder and the chairman of the board. Six hundred Bartians showed up at the Civic Center to eat baked ham, green beans, and strawberry shortcake. They listened to endless speeches about the man who wanted to become a Horatio Alger hero and succeeded. Chamber president John Cronin turned the tables on Frank and presented him with a silver dollar bearing the date 1873, the year he was born on the Nebraska frontier. Frank cracked some jokes with his cronies and puffed on cigars. The audience sang "America," "Let Me Call You Sweetheart," and "God Bless America." Bill Wedlin, a local violinist, and an accordionist named Happy Fenton, dressed in gypsy outfits, provided the music. Eight different radio stations broadcast the program.

Later that year Frank demonstrated how generous and sincere about the war effort he was at a big bond rally in Oklahoma City. Bette Davis, the two-time Oscar winner, came from Hollywood with a red and sniffly nose, and a determination to sell as many War Bonds as she could during a celebrity auction. Frank was all for it. After the plates were cleared at the Skirvin Hotel, Davis got down to business and stepped up to the microphone to start the show. The first item on the list was a prized Hereford.

"How do I do it? Do I just say how much?" asked Davis.

"Five thousand dollars!" roared a voice from the back of the room, and Davis broke into a smile.

"That's wonderful!" she said. The Hereford eventually sold for $100,000.

Next she held up a can of Phillips 66 motor oil and asked, "How much?"

Nobody believed she'd get as much for a can of oil as she did for a prize-winning hunk of beef. Even if it was Uncle Frank's oil. A bid came in for $10,000 and the crowd turned to see who made such a big offer. Another came in for $25,000 and everyone really got excited. Then when a bid of $75,000 was offered, Davis almost slammed down the gavel. But the bidding wasn't over.

"I bid a hundred thousand," said a familiar voice from near the front of the room. It was Uncle Frank. He topped them all and got the oil. Bette Davis autographed the War Bonds and the oil can and left Oklahoma City with more than one million dollars in bonds sold.

At the close of 1942, Frank was still very much in harness—leading Phillips in the company's contribution to the war as well as on the domestic scene. The Phillips research staff increased to one thousand employees, but the number of Phillips personnel in uniform jumped to more than two thousand. To help aid the war program, Phillips, like many other firms, began to hire more women for many of the laboratory, technical, and manual jobs which before the war had always been filled by men. This move was a significant change for a company involved in one of the most male-dominated industries in existence. Even Uncle Frank admitted that women accomplished surprising results when they were given the opportunity. "Our company has lagged behind industry in general on this important matter of utilizing 'womanpower,'" said Frank. "Further delay will not only seriously hamper our war operations but may make it impossible to obtain the kind of women employees our operations require."

Fern Butler had to smile when she learned about Phillips Petroleum's decision to increase its number of female employees. Although none of them were slated for top executive posts, she felt it was a step in the right direction. Fern was still restless since leaving the New York office and was looking for something constructive to do with her spare time. She corresponded with Frank and saw him occasionally, but it just wasn't the same. She felt disconnected and alone. Her situation wasn't improving. A stable fire destroyed her beloved horses—animals she had raised from colts—and Fern was on the verge of collapse. Her dogs and her horses had become her life. Now her horses were gone.

When Frank learned about the death of Fern's horses, he immediately reached out. He remembered a young colt that Fern had ridden and liked during one of her visits to the ranch. It had grown into a spirited stallion and was beautifully marked with black and white patches. Frank gave the horse to Fern and promised to ship him to her as soon as she was able to find a new stable. He kept the young stallion in the high-grass pastures and allowed only the best cowboy on the ranch—Huck Swift—to ride him. Huck was lithe and agile and weighed only one hundred and twenty pounds, Fern's exact weight.

Frank had his ranch manager, John Seward, send Fern a letter verifying that she owned the horse. "He has felt that there might be some hesitation on your part in claiming the horse should something happen to him," wrote Seward. Fern tucked the proof-of-ownership letter in her safety-deposit box and wrote Frank a formal thank-you note for his generous gift. "Someday I hope to be in a position to take him," wrote Fern, "and in the meantime it makes me very happy to know he is mine." The following year, the horse was ready to travel and Fern was prepared to give him a good home in Connecticut.

Joe Billam had replaced Seward as ranch manager and he personally loaded the horse in a boxcar and watched the train head east. But the stallion was wild and frightened. He had never been off the F.P. Ranch and he kicked the boxcar to pieces. Fortunately he wasn't harmed and the trip continued. When he was finally unloaded in Connecticut, Fern was there at the depot. She gentled the nervous stallion and calmed him. She also hired the best trainer she could find to work with the animal every single day for six months. When they finished, Fern could have ridden that horse English, western, or bareback.

Every time she called the horse's name, Fern thought of Frank and she saw in her mind the hills and the shimmering lakes on the ranch. Frank's gift horse bore the name that Fern had come up with herself years before. The name Frank had given to his Osage hideaway and to the daring Goebel's airplane. Fern's horse was named Woolaroc. She loved the big Osage pony and she rode him faster than the wind.

CHAPTER THIRTY

While the hideous war roared and rampaged around the earth, Frank found simple acts like giving a country boy, ironically named Frank Phillips, a pony or presenting Fern with a spunky Osage stallion gratifying and these gestures made everything seem sane and easy and honorable. They were the moments he most treasured during that terrible time.

When the tides of war seemed to turn in favor of the forces of good, Frank began following the war dispatches and studied both the European and Pacific theaters as though he were a Stateside general itching to get into the fray. He reveled when the last Nazi units surrounded at Stalingrad surrendered to the Russians, marking the start of Germany's long retreat from its Eastern Front. He got a satisfied glow when he learned that in the Pacific battle-weary U.S. marines recaptured Guadalcanal, a critical step both psychologically and strategically for the Allies.

As the American GIs pushed deeper into enemy territory, Frank and Boots—like a pair of field commanders—saw to it that Phillips Petroleum extended its exploration for oil and gas into the Rocky Mountains and the southeastern United States. The company filed hundreds of patent applications as a result of exhilarated wartime research and launched the construction of another new research laboratory at Phillips, Texas. They also sent off more employees—now more than 2,500—to the armed forces, and loaned a large number of key executives to government agencies.

In the spring, when his stomach began causing him some unusual discomfort, Frank checked into Barnes Hospital in St. Louis for a month-long observation. His physicians were not pleased with the results. They were concerned with Frank's continued gastric problems and insisted that he tender his resignation from the Petroleum Industry Council. Frank complied without a whimper.

Sore guts weren't the only source of pain for Frank. A couple of weeks after his stay in the hospital, he was experiencing more discomfort while back in New York, where he tried to stay fit with daily walks on Park Avenue and in the Wall Street district. This time the problem was Frank's feet. Fortunately, it was an ailment with an easy remedy. All it took was Paul Endacott and his new shoes.

Already the assistant to Boots and chairman of the operating committee, Endacott was made a vice president in April and within seven months he'd become a member of the executive committee and a company director.

"I was a new vice president and was in New York to meet with Mr. Phillips," said Endacott. "I was called over to the Ambassador on a Sunday and he was really out of sorts. His feet were killing him. He even had a doctor up there, but even he couldn't help." Shoe rationing had started that winter and new shoes were hard to come by. Endacott had just managed to come up with a new pair and he was wearing them for the first time that day.

"I looked at his feet and said, 'Mr. Phillips, it appears to me that those shoes you have on are too tight. I bet that's your trouble. I had the same problem until I got these new shoes and now my feet feel great.' Well, he looked my new shoes over and said, 'Your feet are bigger than mine. Let me try on those shoes of yours and see how they feel.' So I took my new shoes off and he put them right on and picked up his cane and walked out the door, took the elevator downstairs, and walked around the block. He came back, and son of gun if he didn't just keep those new shoes of mine! There I was with no shoes! Luckily I still had my old pair, so I walked down the hall in my stocking feet and took the elevator back to my room." Endacott may have lost his new shoes, but he saved the chairman's feet.

Later that same month, word reached Frank that young Army lieutenant Lewis Fisher—one of his surrogate sons—had been wounded in action during a tank battle somewhere in North Africa. Frank was assured that Fisher's wounds were not severe and he'd soon be returned to duty. Frank shared the news about Lewis Fisher and other choice morsels he'd been saving when he ducked out of the office to visit with Fern. Whenever Frank was in New York the old friends would occasionally meet for lunch and catch up with each other's lives. Mostly, they met only for a hamburger and a glass of milk, but their time together was important to both of them. Frank spoke of his family and the war and developments at the company. Fern told him about Connecticut and mutual friends and Woolaroc. She also talked about some of the good changes in her life.

Fern had decided to become involved in the war effort. Like Frank and Lewis Fisher and all the others, she wanted to do her part. She volunteered to drive an ambulance, and after undergoing intensive training as an emergency service driver and aide, she was assigned to help in the New York hospitals. She was issued a uniform and credentials from the U.S. Coast Guard. Three times each week, Fern commuted from her home at Owenoke Park to the city. It reminded her of the good old days. She really no longer needed her large Park Avenue apartment, so she leased it to Kitty Carlisle, who was engaged to Moss Hart, the Pulitzer Prize-winning playwright. Fern took a smaller apartment for the nights when she had to work late and couldn't get back to Connecticut.

She went on to work in the ambulance service throughout the war. Often the emergency calls took her into the meanest neighborhoods in the city. Sometimes a cop would have to go in with a medical team while Fern stayed locked inside her big ambulance. She didn't care. Fern knew she was as tough as the next one. She had battled Wall Street brokers, bankers, and oil men for thirty years. Some back-alley punk didn't give Fern Butler the slightest pause. She became caught up in the rhythm of the city and the time.

Back in Bartlesville the rhythm and pace were building; the drums of war were beating louder. During the first days of summer, John Phillips, Jr., barely eighteen years old and just weeks out of military school, stayed at his grandparents' town house and courted Shirlee Evans, the teenaged daughter of a man who worked in the Phillips transportation department. Johnny barely had time to unpack his bags when his orders arrived from the Army instructing him to report to officer candidate school.

While Johnny was sweating through his OCS program, his big sister, Betty, met and married Major Henry D. Irwin, an artillery officer and graduate of West Point. Betty was still the absolute apple of everyone's eye in the Phillips family, especially Frank and Jane. She and her young warrior, whom everyone called Hank, were married in the town house, and Frank gave the bride away. On New Year's Eve 1943, Johnny Phillips graduated from OCS at Fort Knox, Kentucky. His mother, Mildred, was there to pin on his gold second lieutenant's bars, and he was assigned to the 10th Armored Division at Camp Gordon, Georgia. When he wasn't training, Johnny spent most of his time writing or calling Shirlee, who had enrolled at the University of Arkansas at Fayetteville. Their relationship was getting more serious, and when John Jr. was given a two-week furlough in April 1944, he headed straight to Arkansas to propose to his sweetheart. As far as his family was concerned, the timing couldn't have been worse.

In Bartlesville, L. E. Phillips was dying.

For the past few years, L.E. had been slowly sinking. He managed to help Waite with a Red Cross War Fund Drive in 1943 and spent time at Philson Farms, where his Japanese chauffeur, Harry, had taken up residence until the smoke of war cleared. But during the scorching summer of that year L.E. suffered the first of two strokes which left him with only partial sight. The second stroke came a short time later and was more damaging. He became partially paralyzed and lost his speech and had to scribble out any communications.

At the fifty-third meeting of the Anchor Club, the organization L.E. and his best boyhood pals formed back in Iowa, he was pale and seemed in a daze much of the time. L.E. grew progressively worse. He became irritable and difficult to handle. Through it all, Node stuck by his side. She was with him, as well as their three children, including Phil, on leave from the Navy, when L.E. died quietly on Sunday morning, April 16, 1944, at his home on Cherokee. He was sixty-seven years old.

Condolences and telegrams poured in from across the country. In his wire Amon Carter told Node: "I was shocked this morning to hear of the death of sweet old L.E. Bless his heart, all of his friends will miss him. He was a sweet character with a great sense of humor and I have never known anyone I enjoyed being with more . . ."

John and Mary Kate flew in from Dallas for the funeral. "John not very sober," according to Jane's diary. He was trundled off to bed before dinner so he'd be in shape for the memorial service the next morning. Most of Bartlesville paused to honor L.E. and remember the man who helped Frank found Phillips Petroleum Company. The courthouse shut its doors at noon, and the Bartlesville banks closed out of respect for the old banker. So did the Phillips Petroleum offices. Work was stilled in the Phillips field areas during the funeral hour. A wide cross section of his friends and associates, ranging from top-flight executives, church and political figures to lease pumpers, bootblacks, and farmers showed up at the First Christian Church to hear the eulogy.

L.E.'s oldest sister, Jenny, had died the year before in Des Moines only about a week after her seventy-second birthday. But his two surviving sisters and all of his brothers—Frank, Waite, Ed, and Fred—were at the burial in Memorial Park to say so long.

While L.E. was being lowered into his grave, Johnny Phillips meanwhile was down in Arkansas mustering up the courage to face his grandparents. He had convinced Shirlee to marry him, but even though he was a commissioned officer in the Army, he and his bride-to-be were still underage and needed parental consent before they could be wed. They drove to Tulsa and approached Mildred. She passed the buck to Shirlee's parents in Bartlesville. Mildred also warned them that Frank and Jane Phillips would not be in favor of the marriage because of Johnny and Shirlee's ages.

Mildred was correct. Although the Evanses gave Johnny and their daughter their blessing, the young couple ran into a brick wall on Cherokee Avenue. John Sr. was against the marriage and so were Frank and Jane. They believed the couple was much too young. Disappointed by his grandparents' attitude but still in love, Johnny returned to Tulsa and married Shirlee. They had a church wedding attended by the Evans family, Mildred and her mother, Bobby Phillips, and Betty and Hank, who came over from Camp Chaffee, Arkansas. The newlyweds honeymooned in the Bliss Hotel and then returned to Georgia, where Johnny resumed his training.

Frank and Jane eventually reconciled themselves to John Jr.'s marriage. They liked his pretty young wife and they remembered their own wedding forty-seven years before and how some people, including John Gibson, had suggested that perhaps they were too young.

Frank was beginning to spend more time thinking about his age and about the men and women he had worked and played with over the years. Many of them were gone—Obie Wing, Pawnee Bill and May Lillie, Tom Mix, Will Rogers, Wiley Post, E. W. Marland, the Miller boys, John Gibson, H. V. Foster, and L.E. The faces that moved through Frank's life had changed.

The oil business was changed too. It was different. The old-school ways of doing things in the patch were long forgotten. Frank thought for the most part that was good. But many of the new policies and programs that companies were forced to comply with bothered him. So did some of the young bucks who now worked for the company. Boots Adams, the rising star of Phillips Petroleum, who had pulled himself up from warehouse clerk to company president, was getting on Frank's nerves. He confided in some members of his family that there was something about Boots that he couldn't quite put his finger on, that he didn't quite trust.

"Boots had many lucky breaks. He was a fantastic guy, but once he got into power it was heartbreaking," said Mary Low, Frank's oldest foster daughter. "The company really changed when Boots took over. In fact, he almost did the company in. It was his ambition and his greed for power and personal wealth. He's the only person I know of who ever managed to pull the wool over my mother's eyes. She stuck up for Boots with my father. But she wasn't alone, Boots managed to fool all of us."

Frank began to notice problems with his handpicked president back in 1943 when Boots put together the Panhandle Eastern Pipe Line deal. Phillips acquired 202,163 shares or 25 percent of the company's common stock and Boots had the transaction all ready for approval by the Phillips board, but something bothered Frank. It may have simply been that he thought Boots was getting too big for his britches.

Regardless of the reason, Frank let Boots know prior to the board meeting that the vote would not be as routine as Boots expected. Uncle Frank was still the chairman and chief executive officer. Frank opened the discussion about the pipeline vote by announcing: "To start with, gentlemen, I'm against it." Boots then described the merits of the acquisition in detail to the board, then Frank called for a vote of the board members. When it was Boots' turn to vote, Frank stopped him. "You can't vote, Boots," said Frank.

"You can't keep me from voting, Mr. Phillips," said Boots. "I'm a director too."

Frank let him go ahead and cast his vote. The results were predictable. Every director present except Boots, knowing how Frank felt, voted "nay." Boots sat stunned and deflated.

Frank leaned back in his chair and smiled. He had made his point with Boots and now Frank was ready for the deal to be approved. "Now, gentlemen, it seems to me that we have a problem here. We either ought to approve this thing or get ourselves a new president, and I'm not ready to do that. Let's take another vote." Panhandle Eastern passed.

"Mr. Phillips would always test the board and the executive committee," said Endacott. "He liked to see who his yes men were and who had the power."

Frank's distrust of Boots continued to heighten and hadn't diminished as time went on. "Frank Phillips lost confidence in Boots," said Endacott. "And Boots was also concerned because he was aware that Mr. Phillips hadn't made him an executor on his will or put him on the board of the Frank Phillips Foundation. That bothered Boots."

Boots wasn't the only one that was bothered. So were several of the company's top executives. But they weren't bothered by Frank Phillips—they were bothered with Boots Adams. When Frank learned that four of his top operating department heads were disenchanted with Boots and had lined up behind Don Emery, the vice president and general counsel, Frank seriously considered putting Emery in as the new president of Phillips Petroleum.

"Boots was a handsome guy with a lot of push," said Billy Parker. "But we had a vice president in charge of the legal department named Don Emery who aspired to be president."

When Boots, already troubled by Frank's change of attitude toward him, learned of the discontent in the executive ranks, he made a beeline for the chairman's office. So did Emery. The issue came to a head one Sunday when Frank held court at his Bartlesville town house. Boots came by on two occasions for private sessions with Frank. Three executive committee meetings, without Boots in attendance, were also held in the sunroom.

Frank was still trying to make a final decision about the presidency of his company. Should Boots remain as president or should the job be given to Don Emery?

"That was the real issue," said Endacott, who attended all three executive sessions and in between meetings drove out to Boots' country place to check on his condition. "Boots was out there at his farm waiting in absolute agony," said Endacott. "He was pacing around and watching them cut alfalfa just to keep his mind off things. He knew what was going on and he didn't imagine that he would survive. He felt Mr. Phillips was sympathetic with Don Emery."

After the series of meetings on Sunday, Frank continued his conferences on Monday and conferred with Boots at the town house later that evening.

Ultimately, after he was convinced by other executives that Boots' blind ambition and obsession for personal gain could be kept under control, Frank decided to not replace the man he had picked to be president. Boots would stay on. The notes from the executive committee meeting held that Monday offer additional insight into what some called the "Boots Crisis," but do not adequately reflect what actually took place over what seemed for Boots an eternal weekend.

"Frank Phillips presided at the meeting. Chairman stated that following an informal conference with certain members of the executive committee held on Sunday, July 23, and several succeeding conferences between himself, K. S. Adams, and Don Emery, as to the several serious differences which have arisen between Adams and Emery, he was happy to report that an agreement had been reached as to their future status with the company. Frank Phillips said both showed 'a broad-minded and generous attitude' during the conferences."

Boots breathed a sigh of relief.

Satisfied he'd made the correct decision, Frank took off the next afternoon and went to the ranch with Doc Hammond to fish the lakes and cleanse his mind. A couple hours of catching crappie and bass, or a ride on Buster or Pancho, was about the best way Frank knew of for keeping wise and clear-eyed. Out at Woolaroc, Frank could prowl the museum with Pat Patterson or sit on the front porch in the summertime and sip Dan's mint juleps.

In the evenings at Woolaroc, Frank and Jane visited with Patterson and talked about the latest development at the museum. If they weren't entertaining, they'd eat supper with their ranch foreman, Joe Billam. They'd go over bills and talk ranch business.

Joe's wife, Grace, was another favorite of the Phillipses. Grace Billam first started working for the family in 1936 when they needed help for a big dinner party. Eventually she became the Woolaroc hostess and managed the entire dining room and lodge. The Phillipses felt comfortable with Joe and Grace. Joe had worked around the ranch for many years before he became the manager. He was a first-class wrangler.

"I started out at the ranch as a hired hand for a dollar a day and room and board," said Joe. Grace, born the year Phillips Petroleum was founded, grew up on the ranch property and felt at home on the land. She was the daughter of Mark Barbee, an original Phillips hand who worked alongside his father, Ben Barbee, the old patriarch of Lot 185. Bill Noland, the man who pushed Frank out of harm's way when the big Lot 185 gusher came in and the drilling tools were flying through the air, was Grace's uncle.

"The Phillipses were good people and it was a pleasure to work for them," said Grace.

"Uncle Frank treated me like I was his son," said Joe. "I liked them both and I spent a lot of time killing frogs in Clyde Lake and shooting birds once I found out that Aunt Jane loved frog legs and that he liked to eat quail. But we had our disagreements. One day we got into a dispute over some ranch matter and he turned to me and said, 'You're fired until two o'clock.' He'd get into fights with Clayton Fisher—his driver—too.

They got in a big argument when he wanted Clayton to wear a uniform and cap when he drove. Clayton hated the idea. Well, he fired Clayton because of that. But ol' Clayton came right back. He wore the uniform too."

When John and Mary Kate visited from Dallas, Grace's orders were to not serve any alcoholic drinks, in deference to John's inclination to sip too much. "What would happen is that everybody would head back to the butler's pantry to get their own private drink," said Grace. "The family would be gathered out in the lodge and Jane would come back and have a drink, and then somebody else would come back for one, and before you knew it John would be back there to get himself a drink. Pretty soon everybody was drinking away and thinking they were the only ones."

During the war years the well-stocked butler's pantry at the lodge was a popular spot. The Phillipses continued to entertain in style, hosting the usual cadre of executives and celebrities as well as many key military and political figures, including Edward Wood, the Earl of Halifax, and former leader of the House of Lords and Foreign Secretary, who served as Britain's wartime ambassador to the United States.

Frank remained focused on the political scene—both abroad and on the home front. In the autumn of 1944, he watched, with more than passing interest, as the Republican nominee for President, Governor Thomas E. Dewey of New York—one of the youngest men ever proposed for the presidency—challenged Franklin Roosevelt, the firmly entrenched Democrat who sought an unprecedented fourth term of office.

Dewey launched his cross-country campaign tour in September. He made few platform stops and his speeches were aimed at radio listeners rather than the audiences which crammed into auditoriums to take a look at the former prosecutor with the distinctive black mustache.

Bill Skelly headed up the Republican Party in Oklahoma and he sought Frank's support for the GOP candidate. Frank knew that Dewey had married an Oklahoma native and that he and his wife enjoyed singing duets. Frank liked that. He also liked the fact that Dewey was a Mason, like himself, and a pretty fair poker player, again like himself, and that he appeared calm and collected no matter how torrid the weather, once again very much like Frank Phillips. The fact that Dewey was the choice of several Old Guard Republicans, including Frank's friends Herbert Hoover and Alf Landon, bode well for the candidate.

Skelly arranged for Frank to meet Dewey in Tulsa, the site of one of his campaign speeches. A meeting was scheduled for an afternoon in late

September at the Tulsa Hotel. The night before, in Oklahoma City, Dewey delivered a scathing speech filled with barbs aimed at F.D.R., the New Deal, and the Administration's war effort. The 15,000 delirious Dewey supporters, packed into Municipal Auditorium, screamed their approval with cries of "Pour it on!" Dewey lieutenants promised the best was yet to come and told the faithful to stay tuned in to Tulsa.

When the morning train arrived in Tulsa more than 20,000 people were crowded into the Union Depot to see Tom Dewey. Up in Bartlesville, Frank picked out his best bow tie and stuffed his coat pocket with cigars for the occasion.

Political insiders believed Dewey would now take off the gloves and deliver some serious blows to one of F.D.R.'s most vulnerable areas—his prior knowledge of the Japanese attack on Pearl Harbor in 1941. The Republicans smelled blood—and victory—in the Oklahoma air. The question had been raised long before by many of Roosevelt's critics concerning F.D.R.'s responsibility for allegedly goading the Japanese into war. More important, they also made allegations that Roosevelt and others knew about the Japanese fleet moving toward the Hawaiian Islands well in advance of the attack because the Japanese codes had been broken and decoded. Stories about the Japanese code and the Administration became a favorite topic of conversation at Washington cocktail parties, some of which were attended by Frank Phillips. The rumors intrigued Frank and piqued his interest in the 1944 campaign.

Accompanied by Boots and Endacott, Frank joined Skelly in Tulsa, and the foursome then proceeded to the hotel to meet Dewey and his campaign manager, Herbert Brownell. "Our meeting was set for two P.M.," said Endacott. "We got there early and a few minutes before two o'clock, Mr. Phillips started fidgeting. He was such a stickler for punctuality. By two o'clock, when there was no sign of anyone, he wanted to leave right then and there. We calmed him down and pretty soon Brownell stuck his head out and apologized and told us that a crisis had come up and to bear with them. It would be only a short wait. Mr. Phillips grumbled a little, but we stayed."

Finally, after close to a half-hour wait, the door to Dewey's suite opened and the Bartlesville delegation along with Skelly was ushered in to meet the candidate. "I'll never forget that scene," said Endacott. "We walked in and there sat Dewey. He was pale—very pale. He just sat there looking very much as though he was in a state of shock." Frank strode into the room and started the conversation with the question that was most on his mind. "Glad to meet you," said Frank, pumping Dewey's hand. "Now, let me ask you this—what about all this business of Roosevelt knowing about Pearl Harbor before it was attacked?"

Dewey continued to sit there dumbfounded, and then he and Brownell began to explain what had happened just moments before. It appeared that others besides the Republican insiders anticipated Dewey giving his version of the story behind "the day of infamy" when he spoke in Tulsa. Back in Washington, the U.S. military hierarchy, including General George Marshall, the U.S. Chief of Staff, was sensitive to the story about U.S. code breakers having knowledge of the movement of the Japanese fleet prior to December 7, 1941.

Citing national security, the military establishment had implored Dewey to make these allegations off-limits during his campaign. But Marshall learned that certain sources were feeding the Dewey organization pertinent data that would revive the Pearl Harbor question and make it a campaign issue. After he heard the reports about Dewey's fiery Oklahoma City speech, Marshall determined he had to act as quickly as possible to prevent the candidate from saying any more. Marshall, reportedly on his own and without any coaching from the White House, dispatched Colonel Carter C. Clarke, head of the War Department's cryptographic intelligence unit, to Tulsa. Clarke carried with him a three-page letter stamped "top secret." His orders were to deliver it in person to Tom Dewey.

"What I have to tell you is of such a highly secret nature," the letter read, "that I feel compelled to ask you either to accept it on the basis of your not communicating its contents to any other person and returning this letter or not reading any further and returning the letter to its bearer."

Dewey stopped reading and handed the letter back to Clarke. "Franklin Roosevelt knows about it too," said Dewey during his exchange with Clarke. "He knew what was happening before Pearl Harbor, and instead of being reelected he ought to be impeached." But Dewey realized not only that he was caught between a rock and a hard place but that he could not nor would not compromise himself by making what he called "blind commitments." Even though Clarke assured him that he had been sent by General Marshall, Dewey went on to grill him at length about his mission. He was certain the top-secret-letter ploy was a scheme cooked up by Roosevelt and was intended to silence Dewey.

Clarke, when he felt his mission had been accomplished, did an about-face and returned to Washington.

"That Army colonel got to Dewey's hotel room just ahead of us," said

Endacott. "We walked into his suite just after the colonel left. Dewey explained about the secret letter and said that he wouldn't read it. He said he wasn't sure what he'd do next. But it was clear to all of us that he was very shaken by the incident."

Dewey's speech in Tulsa didn't mention the Pearl Harbor theory, nor did it have the passion of his address the night before in Oklahoma City. Instead he spoke of sweeping out all the government bureaucrats who he said "have fattened themselves on your pocketbook and mine for twelve years." The next day he appeared in Springfield, Missouri, and continued to work his way back East to wait for the day of reckoning. The issue of the Japanese attack on Pearl Harbor and the Roosevelt's administration's handling or mishandling of the matter was always there like a hand grenade just waiting to explode. But it never did.

Frank predictably threw his support and a hefty contribution to the Dewey campaign. He had done the same for Hoover, Landon, and Willkie. As with those candidates, it did no good. Four straight dry holes. On November 7, 1944, Franklin Roosevelt won his fourth term with an electoral college majority of 432 to 99 and a popular margin of more than 3,500,000 votes. He carried thirty-six out of forty-eight states.

Frank was beginning to wonder if he'd ever see the end of the Roosevelt reign. He would. F.D.R. would serve only until mid-1945, when he would die of a cerebral hemorrhage and the Vice President, Harry Truman, a visitor at Woolaroc whom Frank respected, would take the office and would later go on to beat Tom Dewey again.

Frank had other matters to occupy his mind besides politics, though. Just a couple of weeks after the election, the loyal Phillips employees pulled a fast one on Uncle Frank and surprised him for his seventy-first birthday. The Frank Phillips Men's Club organized the event, billed as "Hookey Day."

At two o'clock sharp an emergency fire siren in the Phillips Building sounded and the employees, as prearranged, left their desks and gathered in front of the building. Led by the high school band and twenty-one parade marshals, they struck off for the Phillips town house on Cherokee. As they marched four abreast they were joined by more workers from the research lab, garage, airport, and other office buildings scattered around town. Soon a column of more than two thousand was headed toward Uncle Frank's house.

When the "hookey players" reached his house, Frank was just getting up from his usual after-lunch nap. He heard the commotion and, with

Aunt Jane at his side, stepped out on the front porch, only to find the huge
crowd on his lawn and in the streets singing "Hail, Hail, the Gang's All
Here."

The resolution for Frank, a document written on a fancy scroll, read
in part:

> We love, honor, and respect him for his great qualities of
> heart and mind; we follow and obey his commands because
> we know there is something, not seen with the ordinary
> eyes, that drives him on; because we have faith in his leader-
> ship; because we know that his heart and mind are devoted
> to the idea of getting the right thing done; because we know
> that he would never, under any circumstances, mislead us or
> let us down; because we know that he always has our inter-
> ests at heart; and because we feel intuitively that he has in
> him all the elements that qualify a man for leadership. Such
> men are few, and they are priceless and irreplaceable. Such a
> man is Frank Phillips, and there is not a man or woman in
> the entire Phillips Petroleum Co. organization who does not
> earnestly and sincerely hope and pray that he may live long
> to guide and direct the destinies of the great organization
> that was created by him.

Frank stood as erect as an Indian, and his eyes were filled with tears.
He was visibly moved by the crowd.

The big crowd sang "Happy Birthday." Afterward, as natural as the
wind, they broke out into "Let Me Call You Sweetheart." It never sounded
better. Everyone laughed and cheered some more. A few screamed Rebel
yells. Later several of the employees who came that cold November day
said they could actually feel the love between the crowd and the old man
on the porch. Frank stood there for a long time and tried to look at each of
their faces. He saw nothing but smiles. He wanted to memorize every one
of them and hold on to the moment and never let it go.

Before it was over he spoke: "God has been good to me." Then the
crowd left. The men pulled their overcoats up around their throats and the
women fixed their scarves. They marched away, down Cherokee Avenue,
with the drums keeping cadence.

Back on the porch, Frank blew his nose and noticed his cigar was
cold. He laid it in an ashtray and took one more look down the street

before going inside. The marchers were out of sight, but he could still faintly hear laughter and talk and, he thought, music. It didn't matter. He saw all of them anyway. They had stood on the lawn there, singing and cheering. Frank saw every last one. He had memorized their smiles.

CHAPTER THIRTY-ONE

Time had become precious to Frank Phillips and for good reason. It was running out. When he was a young man, life was a sweet concoction and he thought it would last forever. But as Frank stared at the calendar and approached his seventy-second year, he realized that time was not some commodity that could be bought, sold, or traded. It was more precious than all the gold, silver, and oil that any one person could possess. It was something to savor.

For someone who had gulped in great quantities of life, like a thirsty roughneck, it was difficult to slow down and take small sips, but he did. Frank had entered the age of prudence, and careful management was required. Discretion in living, something he had tried to master through the years, was now a must. He found that life was still a heady brew, even in measured spoonfuls, and he didn't want to spill a single drop. Like some fine old condor, Frank conserved his energy and planned his movements. He relied on instinct and reputation to keep him in the hunt.

Frank also hedged his bets. He took afternoon naps, started reading the Bible, and hired a young lady with the voice of an angel to come to the town house to sing hymns.

The first of these prudent tactics that Frank put into immediate practice was his insistence that the Phillips board of directors reduce his annual salary from $50,000 a year to $1.00. It was the lowest rate of pay he had ever received in his life. Even lower than when he was a farm boy and pulled cockleburs for fifteen cents a day. It was necessary, he explained, because of federal and state income taxes. When combined with his other income, Frank was receiving only $309.36 in actual salary as chairman of Phillips.

As far as Frank could see, the company's future appeared bright. Phillips' accomplishments in the war effort were very impressive. By the end of 1944, the company was producing a gallon of aviation fuel for every two gallons of automotive fuel. Phillips also conducted flight tests for the Army Air Corps. Using "flying laboratories," a variety of aircraft fuels were tested under all sorts of conditions from high above the desert Southwest to the frigid Alaskan skies. "There's no telling how many lives were saved by the information we got from our laboratories in the sky," said Billy Parker.

Phillips continued to be the nation's largest marketer of liquefied petroleum gases, mainly butane and propane, which were used by industry to turn out bombers, tanks, jeeps, and other important machines of war. Seventy-five percent of the company's Philgas output went directly to war production by 1945, and there was an increase in demand for household consumption. Even Roosevelt's Hyde Park residence was a Philgas account.

Near Borger, the Philtex plant produced lubricating oil additives, detergents, and insect repellents. In the field, Phillips workers kept oil and gas production flowing. The company drilled four times as many gas wells in 1944 as in any previous year. By 1945 the net crude oil production had leaped to 93,800 barrels a day, up considerably from the 65,300 level in 1940. During that same period profits also increased. In fact, they damn near doubled—from $11,590,300 in 1940 to $22,571,500 in 1945.

With the company's economic picture so bright and his own financial interests and investments in superb shape, Frank could well afford to cut his salary to a buck a year. He found he could also afford to spend a little more time with friends and family, especially his Betsie. Even more than Frank, Jane had become increasingly aware of her own mortality. Her heart was growing weaker and her general health was poor, but she quietly slipped medicine under her tongue and faced each day as if there was no tomorrow.

She cheered herself hoarse as the Phillips basketball team chalked up their championship seasons, and she kept up with the activities of the Jane Phillips Sorority chapters established throughout the nation. She worked jigsaw puzzles with the intensity of a surgeon and played marathon games of gin rummy with friends until all hours of the night. She dictated letters by the score to Rowena Snell, the young woman who became her secretary, and she gave dinner parties and hosted luncheons that would have made Perle Mesta green with envy. Jane also did her best to maintain communication with the various family members scattered across the country.

When Waite and Genevieve Phillips left Tulsa during the winter of 1945 and moved to Los Angeles, where Waite invested in real estate and continued his philanthropic work, Jane made sure that Frank and his younger brother continued to share their open exchange of ideas and opinions.

She also followed the movements and lives of her two foster daughters and their growing broods. Mary Low and her husband, Mark, still a Phillips employee, had moved from Chicago to Iowa and then to Bartlesville, where they were raising their three sons. Jane and Frank Begrisch, the parents of two daughters, maintained their home in Rye, New York.

Jane talked often with John and Mary Kate in Dallas or hosted them in Bartlesville. She kept up with Betty and Hank Irwin, who were on the move with the Army; Bobby, away at prep school, in the footsteps of his big brother; and Johnny and his wife, Shirlee, like the Irwins bouncing between Army posts.

On June 7, 1945, at Johns Hopkins Hospital in Baltimore, Shirlee gave birth to a six-pound-thirteen-ounce son. He was first child born to Shirlee and John Jr. and the Phillipses' first great-grandchild. They named him Frank Phillips III, after Frank and in deference to Bobby's twin, who had died. The news of Frank III pleased everyone in the family—especially Frank, in New York at the time, and Jane, back in Bartlesville, who felt like telling the world about the newest Phillips addition. She did the next-best thing and sat right down and wrote President Truman a letter with the news about her great-grandson. It was a happy time. The war in Europe had ended in May and throughout the summer American troops island-hopped closer to Japan. Soon all the fighting would be over. In August, Frank and Jane were in Atlantic City on holiday. Pat Patterson, Rowena Snell, and Jane's good friend Zoe Moore went along too. It was one of the Phillipses' best vacations. They stayed in bed all morning and read the papers, strolled on the Boardwalk in the afternoons or sat in the solarium, got dressed up for dinner, and played gin until the wee hours.

One morning Frank and Pat hired a tuna boat and a skipper and took to the sea for a day of fishing. "We got several miles out and ran smack into a run of albacore," said Patterson. "They were huge fish and we started catching them left and right. Each one felt like you were pulling in a horse! Well, Mr. Phillips was having a ball. That old son of a bitch was in his glory. He wouldn't quit. He'd say, 'Hell, they're still bitin'—let's fish some more!' When we finally stopped there were six hundred albacore on the decks of that boat. After we got back to the dock, Frank took one fish to give to the chef at the hotel. Then he turned to the captain and told him that he could have the rest. Good God, the man almost fell over! That was a month's salary!"

Jane celebrated her sixty-eighth birthday in August. Gifts began flowing in. There were enough flowers alone to fill the hotel. Bouquets arrived from Oral Wing and Boots and Blanche Adams; there were orchids from John, asters from Betty, yellow roses from Doc and George Hammond, and red roses from Marjorie Karch. Zoe Moore gave Jane flowers, and Pat brought her small red roses to wear in her hair.

Frank gave Jane a fox stole, but he also had another surprise. He had Pat ask Jane what she wanted most for her birthday. In an instant, Jane replied, "I want to see my friend Nan Du Page and a Marine band!" Nan Du Page was one of Jane's closest friends and Pat hoped that he'd be able to track her down. The Marine band would be a difficult order to fill, but he was willing to try. Frank told him to not spare any expense.

"I got right on the telephone and tried to find Nan Du Page," said Pat. "I called all over—Seattle, San Francisco, New York. I finally tracked her down in Paris and reached her there. When I explained about Aunt Jane wanting her to come to her birthday party in Atlantic City, she was delighted and promised to come at once. Which she did."

That took care of one of Jane's birthday wishes. There was still one more—Pat had to round up a Marine band. "I looked all over and I found this great orchestra named the Guardsmen. There were twenty-two of them and they had just played for the Miss America contest in Atlantic City. I paid them scale plus all the booze they could handle and I had a costume company in New York ship me some Marine uniforms. It was one helluva party. Frank had a few drinks and he was happy as hell and Jane got a bit tight too and she ended up leading a conga line through the hotel with that entire Marine band following her."

The next morning everyone was exhausted. They stayed in their robes and swigged lemonade. But while Jane and Frank, and most of the twenty-two members of the Guardsmen orchestra, nursed mild hangovers, a decisive blow to end the war was delivered thousands of miles across the globe. The city of Nagasaki, Japan, was partly destroyed by an atomic bomb. It was the second A-bomb dropped by the United States on Japan within days. On August 6, a lone B-29 bomber had dropped an atomic bomb on the city of Hiroshima, destroying almost three-fifths of that city. Jane's birthday celebration had been sandwiched between the pair of historical explosions.

That evening Jane, Frank, and their guests gathered in the Phillipses' rooms to listen to President Truman discuss the bombings he had authorized in an effort to bring the war in the Pacific to a conclusion. Now it was only a matter of waiting for the response of the Japanese government to this show of force. It was a short wait, for the very next day the Japanese sued for peace. By August 14, they accepted the Allied peace terms. In September, formal surrender documents were signed aboard the USS *Missouri* in Tokyo Bay.

World War II was over. Only "the shocking news that Boots and Blanche Adams had gotten a divorce," as Jane noted in her diary after getting the latest gossip from Mildred Rowson, received more attention at Bartlesville cocktail sessions that month. "It was clear to everyone that Boots got rid of his first wife, Blanche—who was a heavy drinker—by threatening to have her put away," said Pat Patterson. "He got her ancestral Indian lands away from her. I always knew he could be a mean and vicious son of a bitch, but when that happened, I really lost all respect for the man."

But even a juicy divorce scandal couldn't outlast all the talk about the war's end. Soon the boys would be marching home.

Lewis Fisher, the young man Frank always doted on, was one of the first. He came back as a captain and a hero with a chestful of medals, including the Distinguished Service Cross, Bronze Star, and Purple Heart. Lewis had fought across North Africa, Sicily, Italy, France, Belgium, and finally Germany. He landed on the French beaches on D Day, and the Distinguished Service Cross, which was pinned on his tunic for bravery under fire, was put there by none other than General George Patton. Frank swelled with pride when he saw Lewis in his uniform and heard the stories of valor.

After the young man had some time to adjust to civilian life, Frank saw to it that Lewis went back to school on a scholarship for a while, and by January 1947 the young war hero was working for Phillips Petroleum as a salesman in Des Moines. "I got up there and started to notice that I was getting raises and promotions," said Lewis. "I was going up the ranks very fast and I couldn't figure it out."

Later, after Lewis resigned from Phillips when he had an opportunity to buy into a business of his own, he learned that his Bartlesville connections may have had something to do with the royal treatment he had received. "I found out that Aunt Jane had called up there to Iowa about once a week to ask about me. She'd call the office to 'check on her boy' and see how I was getting along. Without my saying a word, I think my superiors quickly got the idea that I had friends in high places."

Others came home throughout the first two years of America's postwar transition. John Jr., after making the rounds of several Stateside Army bases, finally ended up attached to occupation forces in Germany, where he was seriously injured in a jeep accident near Stuttgart in March 1946. But his fractured skull, broken ribs, and punctured lung were repaired, and after convalescing at hospitals in New York and El Paso, Johnny returned to Bartlesville and went to work for his grandfather's company.

Phillips Petroleum was now a leader in not one but three growing industries—petroleum, natural gas, and chemicals. The company was on the brink of a period of unparalleled growth. Since the oil industry experienced less difficulty than most businesses in converting from war to peace, there were relatively few work interruptions. During the war, capital expenditures were concentrated largely in the producing and refining divisions. After the war, more emphasis was placed on the rehabilitation of marketing properties and the revamping and expansion of petroleum transportation facilities.

In 1944, the company had unveiled its foreign department and launched exploration activities in Venezuela, Colombia, Canada, and Mexico. By the following year Phillips crews began drilling in Venezuela, and within two years they struck oil, marking the first foreign discovery for the company. Over the course of the next two decades, Phillips not only would continue to prosper from its foreign exploration but would go on to establish and break several drilling records, including drilling the deepest wells ever drilled onshore and the farthest wells ever drilled offshore.

Phillips went on to acquire the outstanding 50 percent interest in Hycar Chemical in 1945, and the name of the wholly owned subsidiary was changed to Hydrocarbon Chemical Company. Phillips also purchased for investment approximately 27 percent of the common capital stock of the Shamrock Oil and Gas Corporation for almost $5 million, and sold all of the company's holdings of Panhandle Eastern Pipe Line Company common stock, bought two years before, at a profit of close to $5.5 million before income taxes.

Because of wartime shortages there was a great demand for Woolaroc brand freezers and other consumer products, including washing machines and radios, which were sold in Phillips 66 service stations after the war.

The appliances were made by local suppliers. Sales were brisk for a few years until Phillips decided not to compete with manufacturers which offered a greater product selection and repair service.

Of the 3,088 Phillips employees who ended up in the military service during the war, almost everyone who returned home safe and sound went right back to work for the company.

They returned to an America where moviegoers wept through the poignant *The Best Years of Our Lives*. They whistled along as Bing Crosby crooned "Blue Skies, Smilin' at Me." The baby boom was on and Dr. Benjamin Spock, a prominent New York pediatrician, wrote a book to help guide the thousands of new parents. It was a gentler time. A time for healing and romance. Barney Oldfield, America's pioneer automobile racer, died in 1946 just as cars and gasoline were making a comeback after a period when everything had been geared to national defense.

Bill Angel was one of the former Phillips employees who came back after the war. A Kansas native who went to business college and learned how to type and became a whiz at shorthand, Angel started to work in Phillips' inventory department in 1930. Over the years, he had worked in several departments and spent some time in Oklahoma City when Boots and the others were trying to open that oil field for the company drilling crews. Angel's wife, Sue, an early leader in the Jane Phillips Sorority, started working for Phillips in 1929.

Within days after Pearl Harbor, Bill entered the Army and spent three years overseas. When the war ended he was happy to come home and resume his career with the company.

"I got back to Bartlesville and someone asked me if I wanted to go to work as Mr. Phillips' secretary," said Angel. "I said no way. I had been gone for years and wanted to stay home for a change."

Angel had worked as a secretary for several top Phillips executives and knew the job had some perks but was never easy. "I just didn't want any part of it. I remembered that back in the early thirties I did some work for L.E. It was less than pleasant. For instance, one Christmas Eve, Phil Phillips called me up at home and told me to come over and take some dictation for his father. L.E. was a boss and you had to jump, so I went over and there was the whole family sitting at the table having a big dinner. I stood there while they ate and I took dictation. When he finally finished, L.E. said, 'Well, I guess it is Christmas,' and he pulled out two dollars and handed them to me. I was poor as a church mouse, and if it wasn't for the fact that I needed my job so bad, I would have told him what he could have done with those two lousy dollars."

Angel thought about how demanding the job of working for Frank Phillips would be, but when he was told that Marjorie Karch had retired and the fellow hired to act as Frank's traveling secretary couldn't make a business trip because of illness in his family, Angel reluctantly consented to go in his place. "Mr. Phillips was going out to spend some time in Phoenix, and I said that I'd go only until the other fellow could get out there. I brought just one extra shirt."

On the Phillips airplane that day, Doc Hammond came back to Angel and told him to go up front. "The boss wants to see ya," said Doc. Angel went forward and visited with Frank for a half hour. They talked about his experiences in the war and then he returned to his seat.

"A little bit later Doc Hammond came back again and said, 'Congratulations, the boss wants you to stay on with him!' I went back to Mr. Phillips and he told me that he decided I should be his new secretary and that once we got to Phoenix I was to wire the other fellow and tell him not to come. Well, I didn't have much choice. I needed a job and I sure didn't think it would be too smart to turn down an opportunity like that. I'm real glad I didn't. I got to like Mr. Phillips very much. He was different from anyone else I ever worked for. He was more like a father to me than a boss." Angel became another key member of Frank's personal entourage, along with Dan Mitani, Clayton Fisher, and Doc Hammond.

While Bill traveled around the country with Uncle Frank, Sue Angel was taken into Jane's personal circle and was able to accompany her on some out-of-town trips. "I went shopping with Aunt Jane in New York and I'll never forget her buying expensive shirts for Uncle Frank and his monogrammed handkerchiefs, which cost $120 for a dozen," said Sue Angel. "I thought to myself that every time Mr. Phillips blows his nose, that's a ten-dollar bill. But all the travel and the people we met gave us a real exposure that we wouldn't have ordinarily experienced. The Phillipses 'took us to school,' as it were. Mrs. Phillips exposed us to foods and dishes we'd never heard of and showed us so much that was new and different."

Even though he always kept a thousand dollars hidden on his person, Frank made sure Angel carried large amounts of money with him so he could dole out hefty tips to waiters, bell captains, porters, and orchestra leaders who would have their musicians break into "Let Me Call You Sweetheart" anytime the Phillipses appeared.

"We were coming back on a train from out West on one occasion and some of the railroad workers were out moving our baggage from one train to another," said Angel. "It was a bitterly cold day and Mr. Phillips got hold of me and said, 'Bill, get out there and find out how many fellows are on the crew and give each one of them five dollars so they can get something to warm up with when they finish with us.' That's just how Frank Phillips behaved."

In 1946, Boots announced his engagement to Dorothy Glynn Stephens, a Texas-born beauty. Shortly before they married, Frank and Jane hosted a party for the couple in New York. "We all went to the Stork Club," said Angel, "the Phillipses, Boots and Dorothy Glynn, Paul Endacott, Rowena Snell, and myself. Well, we ran up quite a bill. We had a big dinner and there were lots of drinks and song requests for the orchestra. I had $750 in my pocket and still had to borrow another $250 from Rowena and $250 from Boots to come close to covering the tab." Never one to spare expense when it came to celebrating an occasion, when Boots and Dorothy Glynn married, Frank sent the couple off to California in the company plane for a short honeymoon trip.

Back in Bartlesville, the honeymoon was long over between Boots and detractors who criticized Adams for helping set up his son Bud, just back from the Navy, in a venture in Houston, first called the Adams Oil Company. When Frank found out about Bud's new firm, he became concerned and told Boots not to allow the use of his last name since he felt it was a conflict of interest and that it could also cause some confusion. Boots acquiesced, and the name was changed from Adams to Ada, the first three letters of the family surname but apparently enough of a change to satisfy Uncle Frank.

"Boots helped Bud establish a company to handle the distribution of Phillips Petroleum products in that region," said John Phillips, Jr. "A short time later, Father learned from some of his loyal employees that Boots permitted some Phillips workers to go to Houston to assist in getting Ada started, and all the time they were employed by Phillips and drawing full compensation. There were apparently some other shenanigans being pulled too."

After learning about a few more details concerning the formation of Ada Oil, John Jr. and some other Phillips family members were in the town house and witnessed an emotional session between Frank and Boots. "Father actually fired Boots and told him he wanted him to get out of the company," said John Jr. "Father was sitting in a chair in the library and Boots ran in front of the chair, sobbing loudly and kneeling before Father.

He begged and pleaded with Father not to fire him and he promised to be loyal to the company forever. The next morning Father told me he was going to give Boots Adams another chance."

The following month, a smiling and confident Boots was amidst the scores of Phillips friends and relatives who gathered at Woolaroc lodge on February 18 to celebrate the fiftieth wedding anniversary of Frank and Jane. Boots presented the couple with a gold tray engraved with the signatures of one hundred and thirteen key executives, and after the buffet dinner of barbecued buffalo and roast turkey, the guests sang old-fashioned songs. There was a cake six feet high and four feet in diameter and another large cake for Jane and Frank Begrisch, who were celebrating their tenth wedding anniversary that same night.

Aunt Jane put on her lavender-gray lace wedding dress and Mary Low trimmed her mother's bonnet in blue ribbon. To make herself look even more old-fashioned, Jane carried her mother's wedding parasol. Uncle Frank wore formal dress, including a fine top hat. He carried a gold-headed cane.

Jane presented her husband with a bronze statue of himself created by Bryant Baker, and the highlight of the evening came when Chief Fred Lookout conducted a tribal ceremony and bestowed upon Jane an Osage name, Wa-Ko-Do-He, which translated to "Highly honored and generous lady."

It was a special day for the Phillipses, and Jane recorded her thoughts about the celebration in her diary: ". . . I have truly had 50 Golden years and am satisfied with my husband and children. They are all pure gold."

Guests at the fiftieth anniversary bash were able to peek inside the Woolaroc Museum. The big stone building had been closed for many months because of extensive additions and improvements being carried out. Frank was anxious to show off the new building additions.

"We made the Golden Wedding celebration, but just barely," said Pat Patterson. "Everything was finished except for the paint on the walls of the new dome room, but who could tell if it was painted or it was unpainted white plaster?"

Frank drove Jane up to the front of the museum in a horse and buggy, just like the kind they used on their wedding day in Creston. The museum was filled with thousands of roses and great baskets of gladiolas.

"As he escorted his lady so proudly through the museum followed by the guests that had come to the party, I realized for the first time that Frank Phillips loved the museum as much as I did," said Patterson. "He couldn't hide the excitement he felt as he pointed out the many, many changes we had made."

Impressed by others who had created great public attractions through their philanthropic foundations, in 1944 Frank had donated Woolaroc to the Foundation.

"Those of us who have been more fortunate have a debt to society which I believe can best be paid by training and educating the youth of the nation," wrote Frank at the time of the dedication. "I dedicate this museum to the boys and girls of today—the fathers and mothers of tomorrow. May they profit by a knowledge of man's past and be enabled to plan and live a happier future."

Frank had sent the Foundation trustees a letter spelling out his thoughts about the management of the Foundation's affairs after his death. He made it very clear that he wished Paul Endacott to succeed him as chairman of the Foundation and for John Jr. to be elected a member and trustee to fill the vacancy created. "During my lifetime, I derived a great deal of pleasure in building Woolaroc Ranch and Museum," wrote Frank. "Through its medium I tried to preserve and perpetuate a part of the country I knew as a young man. In this Museum are relics of many milestones in the progress of this country and of the Company which bears my name. I hope the Ranch and Museum can be preserved and maintained as a monument to the West as I knew it in the days before Phillips Petroleum Company was organized."

At the annual stockholders meeting of Phillips Petroleum on April 28, 1947, the musuem was for ready for a formal reopening, and by May 4 public tours were started.

In the late afternoons, one of Frank's favorite pastimes was to sit on the front porch of the lodge and greet visitors who walked down to look at the big log building after their tour of the museum collections. He and Jane also enjoyed showing Woolaroc off to friends and guests, and they frequently checked with the staff to get an accurate count of how many people passed through the museum doors on a particular day. "The Phillipses always loved people," said Mary Kate. "And they never put on airs and got all stuck-up like some people do when they have so much money."

To the end of their lives, they still found great satisfaction in using their money to please each other. For her seventieth birthday Frank had a greenhouse erected outside the town house so Jane could always have fresh flowers, and for Christmas 1947 Jane's gift to Frank was an elevator installed in the town house. Now the aging oil man wouldn't have to climb the mahogany staircase.

"Father appreciated that elevator," said John Jr., "but one evening he was going up all alone and the door became stuck. It just wouldn't open. Father started bellowing like a wild bull, and everybody was scurrying around. Poor Dan and Henry were beside themselves and, of course, Granny was worried and said we had better call for some help. But finally Dan worked on it with a screwdriver, and after some prying and pulling, all of a sudden the door opened! We all rushed over to see what had happened to Father. There he was—calm and cool. He was leaning on his cane and had his cigar poised just waiting for someone to light it."

During 1947 and 1948, Frank and Jane also found pleasure in watching the Phillips basketball players earn a reputation as the greatest team the company ever put on a court. Bob Kurland, a former Oklahoma State All-American with fiery red hair, was a team leader along with the scrappy Gerald Tucker.

At Madison Square Garden the 66ers won their first Olympic playoff by defeating a great University of Kentucky squad, 59–49. Kurland was brilliant as he completely dominated Alex Groza, the Wildcats' star center, before the largest crowd to that time ever to see a basketball game in the Garden.

Aunt Jane fussed over the basketball players and their wives and girlfriends that month when the Phillipses hosted the team for a party at the ranch. After the dinner was over and Frank and Boots made their speeches, several guests walked over to the museum to see the fancy mosaic tile façade being installed around the front entrance. The life-size images of Indians in full ceremonial and hunting regalia and the other inlaid symbols were designed by Winold Reiss and were created by Italian craftsman L. V. Fascotti, who spent more than a year preparing the glass mosaic for placement. Behind the massive doors in the dome room was the imposing bronze of Frank by Bryant Baker.

Throughout the spring and early summer of 1948, Jane kept up her busy schedule, but those closest to her realized something was wrong. She didn't look well and was spending more time in bed. The phrase "I stayed home all day" frequently appeared in her diaries. Her spirits were lifted when John Jr. and Shirlee's second son arrived. To Jane's delight, they named the baby John Gibson Phillips III. She was also happy that Betty and Hank Irwin had moved to Tulsa and she was once again close to her granddaughter. Jane's family had always kept her going, but even they couldn't help as the summer continued and her health declined.

"Granny didn't tell anyone about her heart problems," said John Jr. "It had troubled her for at least two years. Pat Patterson probably knew more than some, but even he didn't really know how serious it was."

Jane interviewed a new male nurse, whose job was to look after Frank. She liked the man and seemed greatly relieved that there was someone else to help Dan, Clayton, and all the others tend to her husband.

By late July, Jane was very ill and a physician was summoned to the town house. He diagnosed her illness as a kidney infection. "Spent the day in bed and was really ill," Jane scrawled in her diary. "Frank went to the office and ranch for p.m. and dinner . . ." It was Jane's last diary entry.

On July 29, Jane suffered a heart attack. Rowena was on vacation, but there were a few servants at the town house and her daughter Mary Low was visiting.

"Luckily I was there the afternoon she had the attack," said Mary. "We got her heart medicine and called the doctor. That was the longest twenty minutes of my life, waiting for him to come. Mark rushed over too and we all got her in a chair and into the elevator and brought her back to her room. The doctor left me in charge and I stayed with her most of the night. They all wanted to put her in a hospital, but she wouldn't have anything to do with it. She wanted to stay at the town house."

When he found out about his wife's attack, Frank, in his usual blunt manner, charged into Jane's room and blurted out: "Well, Betsie old girl, I hear you nearly died!" Mary and the others quickly ushered him out.

"Mother was at the point where she wasn't sure if Father was kidding or not," said Mary. "They barred him from the room and we got him out to the ranch. I was the one who had to keep him out and I could see the expression on his face. I saw that look he had. It nearly killed me."

The next day Jane rallied. She seemed to be her old self again. Mildred Rowson and a nurse helped stand watch and the doctor came by to check on Jane's condition. Fresh flowers from the greenhouse out back brought a smile to her face. "I was going to fly my mother and Shirlee out to Colorado in my plane when we heard Granny was ill," said John Jr. "We decided to postpone the trip. She wouldn't hear of it. She told me that she'd be all right and she wanted us to go and enjoy ourselves. So we left."

On Sunday, August 31, 1948, at 1:30 A.M., Jane Phillips died. She was a week away from her seventy-first birthday.

The exact moment she died, a nurse was with Jane. She had no last words. "I was staying the night," said Mary Low. "Near the end Mildred Rowson came and got me, but by the time I reached Mother's side, she was already gone."

Within a short time, people began arriving at the town house. It was decided to call up Dan Mitani out at the ranch so he could wake Frank and tell him about the death of his wife of a half-century. But Dan refused. "Morning is soon enough," he told the others. Dan's words of wisdom prevailed, and he then fetched Pat Patterson up at his museum apartment and they waited for dawn.

"I broke the news to him that she was gone," said Patterson. "I waited until first light and went upstairs in the lodge and woke him. We helped him sit up and then I put it to him straight. I just couldn't see to do it any other way than to tell him straight out what happened. Mr. Phillips, your wife has passed away," Patterson said.

"When?" asked Frank.

"Just a little bit ago. Now we have arrangements to make."

Frank just sat there on the edge of his cowboy bunk in the dim Sunday-morning light.

"It was as though somebody came along and hit that old man with a sledgehammer right between the eyes," said Patterson. "It floored him. He was stunned. He couldn't move."

Finally Frank found some words. "Help me up," he said to Patterson. "Don't leave me."

Out in Denver the telephone rang in John Jr.'s room at the Brown Palace. It was Henry Einaga calling from the town house. "I was shocked and horrified," said John Jr. "I burst into tears." Too upset to fly his own plane, John Jr. asked Billy Parker to come out to Denver and bring them home.

"I'm kind of superstitious, but when Jane Phillips died I was up in Kansas City having some surgery performed in a hospital," said Sue Angel. "One afternoon Mrs. Phillips sent me a dozen American Beauty roses. They were gorgeous. The very next morning I opened my eyes and looked over and those roses were completely wilted. It wasn't five minutes later that Bill called me and said that Aunt Jane had passed away."

All of Bartlesville mourned her death. Floral displays filled every room of the town house. President Truman sent a message of sympathy to Frank from the summer White House at Independence. "You have the deep sympathy of Mrs. Truman and myself in your great bereavement. I remember Mrs. Phillips well. She was a grand woman."

Jane's body was taken to Neekamp Funeral Home, where she laid in state for two days. When her old friends and the members of the Jane Phillips Sorority came to pay their respects, they found Aunt Jane at rest upon the chaise longue taken from her bedroom. For the funeral service Jane was placed in a casket and returned to the town house. On her funeral day, the county courthouse was locked up tight and the Phillips offices, and the business houses and banks in town closed for the service. Woolaroc Museum was ordered closed during the entire month of August in memory of Aunt Jane.

John and Mary Kate were in the Orient and could not return in time for Jane's funeral. Node and Phil Phillips were in the Hawaiian Islands and were also unable to attend. Jane was the last of her immediate family. Both of her brothers were dead. More than two hundred family members and close friends gathered for her memorial service. Frank was among them.

"Mr. Phillips really didn't comprehend his wife's death for some time," said Paul Endacott. "The morning of the service, Frank came downstairs and walked into the library and saw the casket set up over near the fireplace. He looked at it and then looked around the room and he had a sort of blank expression. He looked at me and said, 'What's going on here today?' I didn't know how to answer."

Rev. James E. Spivey of the First Presbyterian Church conducted the service. His message was simple and brief. He spoke of the high esteem in which Jane was held by her friends and family and of her place in the community. He told of her love for flowers. In the crowd servants and loved ones sobbed. Some thought to themselves how pleased she would be to see her fine town house brimming with so many blossoms.

At intervals during the service there were songs. Frances Yates, a friend of Jane's for many years and an avid jazz musician who composed the "Twelfth Street Rag" when she was a girl, served as the accompanist. Jimmy Walker sang Jane's favorite hymn, "Nearer My God to Thee," and a quartet, composed of Ruth Welty Stewart, Edith Fountain Cook, Harold Chapman, and Jimmy Miller sang "Lead Kindly Light" and "Abide with Me."

Spread over her casket were peacock feathers, saved by Grace Billam and others out at the ranch. They had been quickly sewn into a blanket as Jane had requested. "I remember looking up and seeing those peacock feathers on her coffin as they were carrying it out the front door," said

Grace Billam. "I thought of how beautiful Mrs. Phillips had always been. I could still see that snow-white hair and those snappy brown eyes."

The funeral procession slowly wound itself through the city streets until it reached White Rose Cemetery and the mausoleum where Jane's remains would stay—surrounded by many she knew and loved—until a permanent resting place was prepared at the ranch in the Osage.

The minister recited another prayer and then the relatives and friends returned to their automobiles. The first lady of the Phillips empire was at peace.

Back at the town house, the odor of gladiolas and roses and orchids and gardenias sweetened the air. While others bustled about fixing plates of food for the living, Frank sat and thought about the dead. It was a Saturday night long ago at the town house and everyone was young and happy.

He saw his brothers L.E. and Waite. John was just a boy. He heard music and someone was playing the piano and there was singing:

> Let me call you sweetheart, I'm in love with you.
> Let me hear you whisper that you love me too.
> Keep the love light glowing in your eyes so true.
> Let me call you sweetheart, I'm in love with you.

He saw his Lady Jane. She was smiling.

CHAPTER THIRTY-TWO

Frank Phillips, the oil man with barbed-wire nerves and the courage of a wolf, didn't realize his own capacity to love until Jane was gone. He no longer heard her smoky laughter in the town house. He missed her hand on his shoulder. Some mornings he awoke with a shudder as though his heart had been ripped out. At night, in his old man's bed with only ghosts and the cloudy dreams that come with age to keep him company, Frank's soul ached.

"He simply did not realize just how much he adored my grandmother," said John Jr. "Right after she died, he'd come over to my house and talk about her. He believed he had taken her for granted. He loved her so very much."

Jane's eleven-page last will and testament was admitted to probate with Frank appointed executor in accordance with her request. It was a predictable document, naming as heirs of her estate Frank; her son John; her two foster daughters; her grandchildren and several other close relatives and friends, including the children of Obie and Oral Wing; Mildred Rowson; Jane Thomas Beecher, the daughter of childhood friend Minnie Hall Thomas; and her son's wives—past and present—Mary Kate and Mildred.

According to the terms of the will, ownership of the town house—which had always been in Jane's name—was to go to Betty with the clear understanding that Frank was able to remain there for the rest of his life. Just before her grandfather's seventy-fifth birthday, Betty and Hank Irwin disposed of their home in Tulsa and moved into the town house on Cherokee Avenue to help care for Frank.

John Jr. and Shirlee and their two young sons moved from the old

John Gibson home on Delaware, where they'd been living, to the residence still in John Jr.'s name across the street from the town house.

Those who were aware of their relationship, believed that Frank would turn to Fern Butler and perhaps even marry his longtime lover. But it was much too late for Fern and Frank. Their time together had come and gone. Fern was not quite sixty and, like Frank, she was set in her ways. She remained at her home in Connecticut with her dogs and Woolaroc and tried her best to stay in contact with her old friend.

After Jane's death, those around him soon discovered that Frank really had but one desire—to build a mausoleum out at the ranch to serve as a final resting place for Jane and himself.

"He had picked out the spot with Mrs. Phillips sometime before she died," said Paul Endacott. "One evening I was out at the ranch for dinner with the Phillipses, and as we were driving back to town, Mr. Phillips had Clayton Fisher stop the car and he pointed to the side of a hill and said, 'Betsie, that's where we're going to be buried.' "

Inspired by the memorial built for his friend Will Rogers at Claremore, the mausoleum soon became the primary focus of Frank's life.

"Mr. Phillips and I made many trips to Claremore to inspect the Rogers memorial," said Pat Patterson. "He was impressed with the permanence of the structure and he told me he wanted the same thing but on a less obtrusive scale."

A month after Jane's death, a resolution was passed by the trustees of the Frank Phillips Foundation to execute a warranty deed for twenty-three acres for a mausoleum to be held in perpetuity for Frank for the sum of one dollar.

"He decided that the mausoleum should be built just exactly where it was built," said Patterson. "I refused to start working on it while both of them were alive. I didn't think that was healthy, but once we started construction, he came out there every single day just to see how the work was progressing."

Frank wanted the mausoleum to be within easy walking distance of the Woolaroc lodge and museum yet far enough away to stay out of the public eye. He also felt it was important for the design to blend with the natural surroundings. Workmen cut into the hillside, which faced south and overlooked Elk Lake. A fit home for dragonflies and cattails, the lake was also one of Frank's fishing spots and was spanned by a gently curving sandstone bridge.

Built of native stone with no cut edges, piled layer upon layer, the tomb appeared to have sprung from the side of the hill as if it was part of the terrain. There was no formal landscaping, except for some flower boxes holding red geraniums on the flagstone terrace in front of the entrance. Blackjack oaks, typical of the Osage, grew on both sides, and the brush was thick with pokeweed and Virginia creeper. Black snakes sunned on the rock ledges.

Workmen, oblivious to Frank watching from his limousine parked on the bridge, blasted through eighteen feet of solid rock to form the burial chamber. To achieve even more permanence than was provided by the natural rock, the twenty-four-square-foot room was lined with a twelve-inch, steel-reinforced concrete wall. Special waterproofing material was used in construction in order to eliminate moisture, and the chamber was air-conditioned and a telephone was installed.

It was as though Frank intended to keep doing business after he was dead and buried. He even joked with some of the ranch hands that after he was gone they shouldn't be too surprised if he stepped out and joined them sometime when they were out trying to catch a stringer of perch in Elk Lake.

Above the huge bronze doors which opened into the chamber "Phillips" was chiseled in the stone. Inside was a circular rotunda—outlined by eight columns of St. Cecilia marble imported from Italy—which rose ten feet to a dome with indirect lighting to reflect the eggshell hue of the ceiling. The walls were covered with thousands of mosaic tiles ranging from light red to maroon, with designs created by other mosaic pieces colored gold, green, and blue. Several types of marble, including Carthage, were used in the chamber, and in the center of the room was an eight-pointed star formed by the use of different shades of marble.

Construction took about a year. As soon as it was finished, Frank had Jane's casket brought from White Rose and a brief memorial service was held at the new Woolaroc mausoleum.

"Once the work was started," said Patterson, "Uncle Frank seemed to be filled with a great urgency to see it completed quickly. When the last stone was in place and Mrs. Phillips was entombed out at the ranch, he appeared to be happier than at any time in recent years."

Accompanied almost everywhere he went by Dan Mitani, Bill Angel, Clayton Fisher, and a series of male nurses, Frank tried to live up to the title of chairman he was so proud to carry. At the ranch, he enjoyed stopping by at the employee picnics to share a joke or a plate of beef and to hand out cigars.

"Clayton would drive us down to the picnic grounds and everyone would spot Mr. Phillips and wave and want him to eat with them or have a drink or just stop and visit," said Angel. "Some of them would be fishing or in boats and some gambling. Uncle Frank would lay down a twenty-dollar bill and shoot the dice with them. He'd keep shooting until he lost the money and then he'd move to another table and shoot the dice until he lost there."

In New York, Frank kept up with his old contacts on Wall Street and enjoyed lunches at one of the clubs, or cocktails with Cardinal Spellman at the Ambassador.

"Mr. Phillips always hated restaurants where the lights were turned down low," said Angel. "He'd say to the maître d', 'What's the matter, can't you pay your light bill?' Well, one night I took him to a fancy place I heard about that was real dark, but I understood the food was so good that it was worth taking a chance. I explained all that to Mr. Phillips and he agreed to keep his mind open about the lights. We went, checked our coats and hats, and walked in. It was very dark. Mr. Phillips took one look and turned to the headwaiter and let him have it with his standard line about not being able to pay the light bill. I tried to reason with him, but it was no use. I gave the waiters and hat-check people a bunch of money and we left."

Outside the restaurant, Angel found Frank was in a sour mood. He didn't want to go to the Stork Club or any of the other spots he normally preferred and it was too late to try to get a reservation anywhere else. "Finally, I came up with a solution. I told him I knew of a place where we could walk right in and get served immediately and the lights were bright. He was all for going and off we went. I took him to nothing but a glorified Automat. All we could get were hot dogs and orange juice and we had to stand, but the lights were shining like the noonday sun. He ate his hot dog and said, 'This is great.' Course, I knew he was disappointed but that he'd never admit it.

"When we got back to the hotel that evening, he went straight into his room and I quickly called down to room service and ordered up some oyster stew, melon, and a glass of milk. I didn't mention it to Mr. Phillips when he came out and sat down to read the evening newspapers. Pretty soon he started fidgeting and he said, 'Bill, are you hungry?' I said, 'Why, no, Mr. Phillips, not after that fine dinner we had tonight in that nice, well-lighted place.' He kind of grumbled and went back to his paper. In less than a minute he spoke up again and asked if I'd like a snack. I repeated my same answer. This time he shook the newspaper and frowned. Finally, a couple of minutes later he shouted, 'Damn it all, Bill, I'm hungry!' The timing was perfect. No sooner had those words left his mouth than there was a rap at the door and in came room service with the big tray of food. Mr. Phillips was so tickled with what I had pulled on him. I can still see him sitting there with the napkin tucked around his neck and he'd scoop up a big spoonful of stew and get to laughing and spit it back. He laughed till he cried."

Although he had days when he felt strong and clear, in truth Frank wasn't the man who once controlled one of the world's largest independent oil firms. Enfeebled by arteriosclerosis—hardening of the arteries—he withdrew more and more from the company business and let the reins of Phillips Petroleum slide into the hands of Boots Adams and the others who emerged from the ranks of corporate leadership.

With each passing day Frank's body weakened and his mind sometimes took off on its own, with no particular place to go. People whispered that he was getting senile. "He liked to reminisce," said Endacott. "He could remember clear as a bell something that happened out in the oil field fifty or sixty years ago, but he'd have trouble recalling what was discussed just a few minutes before."

Frank gave up active management in the company which he founded in 1917 on March 16, 1949. He resigned as chairman of the board of Phillips Petroleum Company. In his letter of resignation sent to Boots Adams, Frank congratulated the management and employees for their current successful operations and expressed his great confidence in the company's promising future.

"The time has come when I feel that I must be relieved of active responsibility as regards management of the Company," wrote Frank. "As I can't keep up with you young fellows anymore I feel that I should not be a candidate for active membership on the Board of Directors and that my name should not be included as one of the proxies on whose behalf solicitation will be made to the stockholders. I am sure you will understand my position clearly and concur fully with my wishes in this matter.

"You and the entire organization are doing a wonderful job, and the future of the Company is indeed promising. Since I expect to be around for a long while, I assure you of my continued interest, my full support, and my great pride in the Company's every achievement."

In response to Frank's decision, Boots issued a statement: "Naturally, those of us in the Company who have been privileged to work side by side with Mr. Phillips for these many years regret that his personal circumstances and reasons are such as to cause him to request withdrawal from further responsibilities of active management. He already has done much more than his share. Certainly he has earned the right to take things a little easier . . ."

At the next regular meeting of the board of directors, Frank was elected Honorary Chairman and Honorary Director without voting or management responsibilities. Even after he severed his last vestige of power, Frank found it difficult to stay away.

Sometimes, in the middle of the night at the town house, Frank would stir from his dreams and summon Dan to his room to help him dress and prepare to go downtown to his old office. Dan loved the old man and never wanted to disturb him, so he'd telephone Clayton and ask him to come by with the car and they'd play out their charade. When Clayton pulled up with the limousine in the circular driveway, Dan would help Frank, carrying his trusty cane, out the door of the town house and into the back seat. Clayton usually headed toward the office but would then turn off and start driving around the neighborhood. He'd drive a few blocks west, then turn down another street and go east. After several minutes he'd glance in the rearview mirror and see if Frank had nodded off. Then Clayton would start driving slowly back to Cherokee Avenue. As he turned into the driveway, Clayton would say, "Well, Mr. Phillips, we put in another hard day at the office. We went to work in the dark and now we're coming home in the dark." Dan would be waiting at the curb to open the door and take Frank back upstairs to return to his bed and his dreams.

"Father's memory was going. He had trouble the last couple years of his life," said Jane Begrisch. "He'd say to me things like 'Where's Betsie? Have you seen Betsie?' "

But Frank was in good hands as long as he was with Dan or the Angels or the Billams or Clayton Fisher.

Toward the end of his life Frank would go by the office and look at his mail and chat with some of the executives and then he'd go home for lunch. After Frank took a short nap, Dan would call down to the office for Bill Angel and say, "On our way," and Bill would rush downstairs to the front door of the Phillips Building. Pretty soon, Clayton would pull up with Dan and Frank and they'd all head out to Woolaroc for the rest of the afternoon and evening.

In good weather they'd drive down to one of the lakes and fish. Sometimes when the sun was bearing down, Dan would hold an umbrella over

Frank's head while he waited for a fish to take his bait. "We'd usually go down on the dock at Clyde Lake and fish," said Angel. "It was me and Clayton and Mr. Phillips and Danny. We'd catch crappie and everything else that came along. Mr. Phillips would get upset if we were all catching fish and he wasn't, so sometimes one of us would talk to him and keep him occupied while someone else slipped a fish on his line and tossed it out. Then we'd yell that he had a bite and he'd get all excited. After we fished, we'd go back up to the lodge and then Sue and some of the others would come out and we'd sit around and play the player piano and have something to drink and visit before dinner."

Frank liked fishing with his friends, but best of all he loved to sit on the front porch of the lodge. He would sit there for hours at a time, smoking his cigars and waving at visitors from the museum. Throughout the spring of 1950 Frank spent as much time as possible sitting on the porch at Woolaroc watching life come back to the trees and grass. There was hope in the warm air and Frank felt good about the prospects for his upcoming two-month summer holiday in Atlantic City.

As much as he enjoyed the Osage, Frank also liked to escape the blistering heat for the cool breezes of the Jersey shore. Atlantic City was the dowager resort—a city of hotels, restaurants, pavilions, saltwater taffy, and gaming tables. It was the home of the great Boardwalk with its miles of sun-bleached herringbone pine planks. It was beauty queens and high rollers. It was at once naughty and nice. It was a place that prided itself on its ability to deliver pleasure.

Frank checked into the elegant Traymore Hotel on the Boardwalk. From his window he could watch the changing of the tides and see the first blink of dawn over the ocean. The first several weeks at the seashore were tranquil and soothing. It was a much more leisurely holiday than some Frank had spent in Atlantic City. This time he didn't go fishing for albacore or hire an orchestra to dress like marines. This time there was no Lady Jane.

Waite came all the way from his home in Beverly Hills to spend some time with Frank. But he didn't like the way his older brother looked. Frank seemed weak and not himself. Waite was worried and he told Bill Angel to keep him advised. Bill was also concerned about Frank and watched him closely. But the vacation was almost over. In only a few days they would be packing up and returning to Oklahoma. The thought of going home cheered Angel and he hoped Frank would be in shape to make the trip.

By mid-August, Frank was feeling very low, complaining of acute abdominal pains. Angel immediately contacted Dr. James Mason, a local surgeon, and he also called Dr. Keith Davis, Frank's personal physician in Bartlesville. The discomfort had not let up, and on Sunday, August 20, Frank was admitted to Atlantic City Hospital with what was diagnosed as a severe gallbladder attack.

A Phillips Petroleum airplane was dispatched and brought Dr. Davis to Atlantic City, with a stop in St. Louis to pick up Dr. Roland Keifer, another surgeon. Angel met their plane and took the two physicians directly to the hospital to see Frank and confer with Dr. Mason.

On Monday evening, when Frank showed no improvement, an emergency gallbladder operation was performed. Frank survived the surgery, but his condition was poor. First reports over the wire services on Tuesday said that "Phillips is slightly improved" but that after the gallbladder surgery "a heart weakness caused complications." Later in the day updated wires changed his condition to "critical." Then: "gravely ill." Finally: "near death."

Out in Colorado Springs, John and Mary Kate had just checked into the Broadmoor after a trip to Hawaii when word came concerning Frank. The message was to come to Atlantic City as quickly as possible.

"John immediately got in touch with Amon Carter, and he sent one of his airplanes for us and took us to Atlantic City," said Mary Kate. "We arrived there in the morning, and John went on to the hospital while I went to the hotel. In only a short time John was at the hotel to get me. He said, 'My father is dying.' "

At the hospital, Bill Angel, weary and frightened, stayed by Frank's side. He had assurances from all the doctors that everything possible was being done. He never gave up hope that Uncle Frank's condition would improve and that they'd be able to finish packing and go home. "He looked pretty bad, but I sat there with him and we talked for a while. Then I told him he'd better get some sleep. I said I'd be back in a little bit and I went over to the hotel."

When John and Mary Kate arrived at the hospital, they hurried to Frank's room. "He didn't say anything," said Mary Kate. "He had an oxygen mask over his face for breathing. He looked so peaceful."

John wasn't sure his father was still conscious, but then their eyes met and the two men looked straight into each other's eyes. At that instant all the years of misunderstandings, disappointments, and failed dreams didn't mean a thing anymore. All that mattered was that Frank knew his son was with him and he would not die alone.

On August 23, 1950, at 1:20 P.M.—the very afternoon he was scheduled to return to Oklahoma—Frank Phillips took his final breath and died. He was seventy-six years old.

Frank had risen from being a farm boy, with a Horatio Alger dream, to become an oil tycoon with a $625 million corporation and more than 18,000 employees. Somehow it was fitting that such a man—an old risk taker—should die in a city which made its living by luring gamblers with dreams of finding fortunes.

"I was back at the hotel," said Bill Angel. "I was just sitting there and here comes Mary Kate crying. She told me Mr. Phillips died just minutes after I left." Stunned, but trained not to betray the chain of command Uncle Frank had carefully built over the course of more than three decades, Angel instinctively went to the telephone and called Boots.

Uncle Frank was going home. A pair of Phillips Petroleum planes left Atlantic City on Thursday, August 24. One carried John and Mary Kate and the other bore Bill Angel and Frank. The DC-3s touched down in Bartlesville at 2 P.M.

"Betty and I met the plane when it arrived," said Mary Low. "I'll never forget seeing that body covered with a white sheet on the stretcher. It was so hard for me to believe that Father was really gone. We never knew he was that ill when he left for Atlantic City."

Besides loved ones, a delegation of local officials met the plane and formed a procession of cars behind the hearse as it slowly drove to Neekamp Funeral Home. When the hearse passed the Phillips Building, a policeman stood at attention on the sidewalk with his cap over his heart. Throughout the downtown area echoed the musical chimes from the First Presbyterian Church. Every flag in the city was at half staff.

Frank was eulogized in Congress, and editorials and obituaries praising his legacy appeared in newspapers ranging in size from the New York *Times* to little dailies and weeklies throughout the nation. Many of the headlines mourning Frank's death were in huge front-page type. There were scores of speeches, and rivers of tears.

A Kansas City *Star* editorial said that Frank "was a tycoon, daring, generous on the big scale and delighted with his own prejudices. Everything he touched succeeded from the first barbershop to an oil empire . . . He lived hard and thrived on the life. He plunged into backstage politics. He was excited by the rise of aviation and liked to back spectacular flights. He could act on careful calculation or impulse as he chose."

An editorial writer at the San Angelo *Standard* in Texas stated: "The death of Frank Phillips in Atlantic City removes from the American scene another of that rapidly thinning group of individualists who saw an opportunity and dared to make the most of it."

And Paul Hedrick, oil editor of the Tulsa *World,* wrote: "Very soon Frank and Jane Phillips will be reunited at the impressive mausoleum. Oil men come and go but the industry will miss Frank Phillips—oil man, public citizen and friend."

On Friday, August 25, thousands of friends, neighbors, and employees paid their final respects to Uncle Frank. They moved in a steady stream through the funeral-home chapel past Frank's body lying in a solid seamless copper casket with its hand-tufted beige silk velvet interior. The floral arrangements which banked the casket included a basket of flowers that were picked that morning out at Woolaroc. Members of the Frank Phillips Men's Club served as attendants at the funeral home.

Included in the multitude who passed by Frank's casket were whole families and individuals, retired roughnecks, and workers from rival companies. Among the thousands of floral offerings and telegrams were such signatures as Baruch, Skelly, and Du Pont, Truman, Spellman, Vallee, and Landon. Politicians, state officials, Indian chiefs, corporate presidents, religious leaders, ambassadors, and historians paid homage to Uncle Frank.

"The passing of our great friend Frank Phillips is a source of deep sorrow to all of us who have been privileged to know him personally and to be associated with him through the development of Phillips Petroleum Company," said Boots Adams in his statement on behalf of the company. "Mr. Phillips was a man of great vision and public spirit but he always kept the common touch with employees and his limitless circle of friends everywhere. His many contributions to the company, the oil industry and to humanitarian causes will always serve as an appropriate monument to him."

In response to the tremendous outpouring of love, the Phillips family announced that Frank's friends could make contributions to the Jane G. Phillips Memorial Hospital, which was soon to be built in Bartlesville in tribute to Aunt Jane.

After the immediate family had visited the funeral home, Frank was taken to the town house for the final memorial service. An honor guard of two thousand Boy Scouts lined the route between the funeral home and the residence on Cherokee. As the hearse passed, each Scout saluted.

Phillips Petroleum installations and offices throughout the operating area closed for twenty-four hours to honor Frank. Every business in Bartlesville locked its doors for his funeral. Buses stopped running. Even doctor's offices closed except for emergencies. Funeral wreaths were hung on shops and homes throughout the city. Some businesses posted handmade signs: "Closed for Uncle Frank." All of Bartlesville was in mourning. During the service musical selections were played for the first time from the electric carillon atop Phillips Petroleum's new twelve-story Adams Building, named for Boots. "Rock of Ages" was the first song and the carillon tones carried across the city and beyond.

Frank's casket was brought into the library and placed exactly where Jane Phillips' coffin had been only two years before. Again, the town house was filled with flowers. Among the five hundred floral displays the Frank Phillips Men's Club sent a huge 66 shield made up of more than three thousand red and yellow roses.

About two hundred close friends and relatives were in the town house for the private service. Bill Skelly, the rough old teamster who wildcatted in the Osage alongside Frank, was there. So were the old Phillips directors who served on the board when Frank was running the company. Art Goebel, the pilot who caught Frank's fancy and stirred the country's imagination, flew in from Los Angeles.

Fern Butler didn't attend. She stayed in Connecticut, and when the service was going on and the minister was reading his prayers, she was on Woolaroc and the big Osage horse was galloping into the wind. Fern went into mourning. She put on weight, didn't tend to her hair, and was as mean as a hornet's nest. After a whole year passed, she finally came to Oklahoma. Relatives could see the pain was gone from her eyes. She was her old self. She went to the ranch, and Joe Billam took her up to the mausoleum. She brought flowers for Frank and Jane. With tears rolling down her face, Fern then said her goodbyes.

At the town house the day of his funeral, fifty-five of Frank's relatives showed up to say their goodbyes. The children, grandchildren, and great-grandchildren were all present and so were Frank's brothers and sisters.

Outside a silent crowd of more than three thousand surrounded the residence. They came to hear the service as it was broadcast from loudspeakers mounted on the roof. On the east side of the property was a sea of khaki, blue, and green formed by the Scout uniforms of the boys and girls who came to say so long to the man who supported them. The crowd had started gathering early in the morning. Many of them had worked for Uncle Frank and some had stood on the lawn several years before when all the workers played "hookey" and marched down Cherokee Avenue singing their hearts out for Frank Phillips. Now the scene was different. There was no laughter, no marching band. From the beginning of the chimes to the benediction, only muffled crying and the rustling of the trees could be heard.

Rev. James E. Spivey, pastor of the First Presbyterian Church and the man who preached Jane's eulogy, conducted the service. " 'Hearts that long have ceased to beat remain to throb in hearts that are or are to be,' so said Longfellow. As we gather in this service of memorial to our beloved Uncle Frank Phillips, the meaning of those words has significance and through the years will have increasing significance," said Spivey. "It was because his heart long beat in affection and interest for his family, his associates in business, his community, for youth, that we have this great sorrow in his going . . .

"He was part of the pulsating life of the thundering oil boom days, but his was the genius who transformed that semi-chaos into a throbbing, humming, rhythmic business enterprise, who was able to catch the ear and the heart of youth and of children, of his friends the Indians, until all began to feel themselves in harmony with him . . ."

Between the minister's Scripture readings and prayers, there were songs. The same quartet which sang for Jane rose and stood near the coffin.

Shall we gather at the river where bright angel feet have trod
With its crystal tide forever, flowing by the throne of God.
Yes, we'll gather at the river, the beautiful, the beautiful river,
Gather with the saints at the river that flows by the throne of God.

Gerald Tucker—the All-American basketball player, Olympic athlete, and Phillips 66er whom Frank loved to watch—sang "The Lord's Prayer."

It was a brief service, lasting only thirty-five minutes. Frank would have been pleased. When the benediction was spoken and the choral amen was sung, the pallbearers moved to the coffin. Boots Adams, Paul Endacott, Don Emery, Stanley Learned, Henry Koopman, R. C. Jopling, Taylor Gay, Clarence Clark, Bill Angel, and John Cronin grasped the handles, lifted the casket, and shuffled out the front door.

It was a sunny August afternoon, but outside the house Waite felt a chill as he watched the casket slide into the rear of the hearse. Choked with tears, he had to turn away. "Frank may have had his faults but his averages were high," said Waite. "There goes my best friend—I loved him so much."

The crowd's silence was their final tribute to Uncle Frank. A small girl plucked a bloom from a floral spray—a remembrance to be pressed in the family Bible. An elderly priest whose welfare benefits had been financed by Frank since 1934 recited quiet prayers for Frank's soul.

As the family moved toward the shiny automobiles parked along Cherokee to start the funeral procession to Woolaroc, Pat Patterson spied Dan Mitani in the shadows of the house. The man Dan had cared for and worried over for so many years was gone. Dan's body shook as he stood there crying. "If anyone in the world ought to be in this procession, it's you," said Pat as he put his arm around Dan and pulled him toward the street. "You're riding in the lead car with me."

The hearse and the limos pulled away from the curb, and an older man, who looked like a roughneck, waved at the long line of cars. "That's the last of Uncle Frank," he said. The procession moved slowly past landmarks—past Phillips 66 stations and the big office buildings, past the homes of faithful employees and executives, past churches whose debts were paid by the dead oil man, past the field where Frank had the circuses set up their big top for the children.

Outside of town the procession headed toward Woolaroc and the mausoleum. It was Frank's favorite drive in all the world—up the steep hills and past the land once worked by Foster and Getty. Out in the Osage where Phillips Petroleum was born.

Just before the procession reached the ranch gate, there was some movement in the brush near the road. It was a cowboy, gnarled and tanned by a lifetime of prairie living. He broke out of the scrub oak on his big stallion and dismounted. He took a pair of old boots from his saddle and stuck them backwards in each of the stirrups, the symbol of a horse with no rider.

The man stood next to the horse and turned to face the road. He stood at attention, and as the hearse passed him the cowboy swept off his hat and held it over his heart. Then he put on his hat, mounted up, and with one snap of the reins, disappeared back in the brush.

The procession turned in at the gate. New clouds—white as bleached bones—rolled in the sky. Change was stirring in the Osage. Thousands of yellow bitterweed blossoms carpeted the pastures and Charlotte spiders spun great webs in the high reaches of the lodge's front porch.

Frank Phillips was home to stay.

EPILOGUE

Frank gave away most of his personal wealth. That was clear within a few weeks after his death when his last will and testament was opened in a Bartlesville courtroom and it was revealed that the oil man and philanthropist had given away more than three-fourths of his tremendous personal wealth.

Only eight million dollars remained to be distributed. Half was left to the Frank Phillips Foundation, and the rest was divided among twenty-nine relatives, friends, and employees. In 1934—during the height of the Depression—Frank had already established substantial trusts for his foster daughters, his son John and John's three children, as well as Fern Butler.

"Father didn't want any of us to starve, but it was also quite clear he wanted all his heirs to make it on their own and not spend their time living high on the hog," said John, Jr.

Woolaroc still contained the original 3,500 acres which were enclosed with miles of high, steel-mesh fence. Two adjoining cattle ranches on the north and south sides, which had increased the size of the spread to 14,265 acres by the time of Frank's death, were disposed of in the settlement of his estate.

Frank's death devastated John Phillips, Sr. It had such a profound effect on him that he even sobered up for a while. It didn't last. Neither did John. His grief over the passing of his parents, along with the years of heavy drinking, took its toll. On January 17, 1951—less than five months after Frank's death—John Gibson Phillips, Sr., only fifty-two years old, died of a heart attack aboard the *Queen Mary* en route from Spain to New York. "He and Mary Kate wanted to spend Christmas in strange lands where he wouldn't be reminded of Father's absence," said John, Jr. "He missed my grandparents so very much that I think he had no more reason to live. He tried hard to mend his ways and I truly believe he was becoming more responsible but his body just played out." John Phillips, Sr., was entombed in the family mausoleum at Woolaroc.

The Frank Phillips Legacy

• 1951. Boots Adams was elected chairman and chief executive officer of Phillips Petroleum Company; Paul Endacott elected president; Stanley Learned elected executive vice president and assistant to the president; Bill Keeler elected as director, vice president, and assistant to the executive vice president.

• 1951. Two Phillips research chemists (J. Paul Hogan and Robert L. Banks) developed a strong and easily molded plastic (crystalline polypropylene) that became the foundation of a multimillion-dollar business for the company.

• 1952. Phillips became the first United States oil company approved by the U.S. Department of the Interior to drill in Alaska. This initial foothold in Alaska led to the establishment of a $200 million liquid natural gas project and participation in the development of the Prudhoe Bay oil field.

• 1953. Phillips launched a major marketing expansion that would culminate in 1967 with a Phillips 66 service station in every one of the fifty states (and a total of 23,400 throughout the nation).

• 1954. Phillips was involved in a landmark U.S. Supreme Court decision *(Phillips vs. Wisconsin)* that established the authority of the Federal Power Commission to regulate the price of natural gas in interstate commerce.

• 1956. The company's gross income exceeded one billion dollars for the first time, with capital expenditures at an all-time high of $257 million.

• 1962. Paul Endacott became vice chairman of the company, a position he held until he retired in 1967 after forty-four years of service. Stanley Learned succeeded Endacott as president of Phillips and, two years later, when Boots Adams retired, he became chief executive officer.

• 1964. On January 19, Waite Phillips died in Los Angeles. He was eighty-one.

• 1967. Bill Keeler, a Phillips employee since before the Depression and Principal Chief of the Cherokee Nation since his appointment by President

Harry Truman in 1949, was elected president and chief executive officer, and a year later he became chairman. John Houchin, a Phillips vice president since 1958, succeeded Keeler as the company's president.

• 1968. Sydney Fern Butler died in Muskogee, Oklahoma, where she moved the year before in order to live with her niece, Nancy Marshall. Fern was a little more than a month shy of her seventy-fifth birthday when she passed away.

• 1969. Phillips discovered the Ekofisk field in the Norwegian North Sea, the first giant oil field in Western Europe and a pioneer effort in the hostile environment of the North Sea. When completed, the Greater Ekofisk Development was the largest offshore complex in the world—a $6.7 billion project.

• 1973. Betty and Hank Irwin deeded the Phillips town house to the Oklahoma Historical Society.

Bill Martin, the former Phillips 66ers basketball star who rose through the company's financial ranks and became president in 1971, took over as chief executive officer, a title he held until 1980. Martin became chairman in 1974. Bill Douce, who joined Phillips at the Borger refinery in 1942, became president in 1974 and in 1980 succeeded Bill Martin as chief executive officer. Douce was named chairman in 1982.

• 1975. Boots Adams died on March 30 at the age of seventy-five.

• 1979. Betty Phillips Irwin made national headlines when she was ordered to pay $1,600 a month in alimony to Hank Irwin, the first man in New York State to win a major alimony award since the U.S. Supreme Court struck down all state laws prohibiting alimony payments to men. Betty died as the result of a brain tumor in 1985.

• 1981. A Phillips partnership discovered the Point Arguello field offshore California, considered one of the largest U.S. oil fields since Alaska's Prudhoe Bay.

• 1982. C. J. "Pete" Silas, executive vice president since 1980, became president and chief operating officer. At forty-nine he was the youngest man since Boots Adams to hold the job. In 1985 he became chairman and chief executive officer and Glenn Cox became president of Phillips Petroleum.

• 1983. Phillips acquired the General American Oil Company of Texas for
$1.14 billion.

• 1984. Phillips acquired Aminoil Inc., the energy subsidiary of R. J.
Reynolds, for $1.6 billion.

• 1984–1985. Phillips became the target of two investor groups attempting
hostile takeovers of the company. One was led by T. Boone Pickens, a
former Phillips employee; the other by corporate raider Carl Icahn. Both
efforts failed and Phillips carried out a financial plan involving in part an
exchange of debt securities for approximately half the company's stock.
The transaction gave shareholders both an immediate return and a continuing equity interest in the company. Old-timers, who remembered the
days of Henry Wells and the Shelton Gang, said Pickens and Icahn
wouldn't have stood a chance against Uncle Frank. Though anxious and
scarred, Bartlesville and Phillips Petroleum survived and picked up the
pieces. Frank Phillips would have been proud.

INDEX